SAUNDERS COMPLETE PACKAGE FOR TEACHING ORGANIC CHEMISTRY

Ternay: **Contemporary Organic Chemistry**

Francis: **Student Guide and Solutions Manual to Ternay's Contemporary Organic Chemistry**

Moore and Dalrymple: **Experimental Methods in Organic Chemistry** — *Second Edition*

Pavia, Lampman and Kriz: **Introduction to Organic Laboratory Techniques: A Contemporary Approach**

Banks: **Naming Organic Compounds: A Programed Introduction to Organic Chemistry** — *Second Edition*

Weeks: **Electron Movement: A Guide for Students of Organic Chemistry**

NAMING
ORGANIC
COMPOUNDS

SECOND EDITION

A Programed Introduction
to Organic Chemistry

JAMES E. BANKS, Ph.D.
Montana State University

 *SAUNDERS GOLDEN SUNBURST SERIES*

W. B. SAUNDERS COMPANY
Philadelphia, London, Toronto

W. B. Saunders Company: West Washington Square
Philadelphia, Pa. 19105

1 St. Anne's Road
Eastbourne, East Sussex BN21 3UN, England

833 Oxford Street
Toronto, M8Z 5T9, Canada

Naming Organic Compounds: ISBN 0-7216-1536-8
A Programed Introduction to Organic Chemistry

Last digit is the print number: 9   8   7   6   5   4   3   2

PREFACE

This self-instructional program is designed to help you understand the rules used by chemists to name and represent structurally the compounds dealt with in organic chemistry. My goal is to help you to learn to derive the names of organic compounds from their formulas, and to write structural formulas which correspond to a given name. As you go through the program, you will have frequent opportunities to test your understanding. In the early chapters the exercises are mostly multiple-choice questions. You will probably benefit from reading the discussion that follows the incorrect responses, as well as the correct one. Questions are more widely spaced in the later chapters so you can progress at a faster pace.

It is a truism that trivial names identify compounds while systematic names represent structures. For its Ninth Collective Period (1972–1976), *Chemical Abstracts* has abandoned most trivial names and adopted systematic nomenclature. Because the IUPAC rules still permit the use of many trivial names, there may be two acceptable names for a given substance. Consequently, this second edition is virtually new. Chapters 1, 2, and 3 have been revised and expanded. Chapters 4 and 5 have been rewritten completely.

Although it is not an integral part of the program, the appendix on Wiswesser Line-Formula Notation is strongly recommended. All of today's students of organic chemistry will undoubtedly use it during their professional lifetimes.

Success in the study of organic chemistry demands an understanding of the rules that govern organic nomenclature. If you read the program diligently and practice the principles of nomenclature in your daily work, you will gain the satisfaction of becoming increasingly familiar with the language of organic chemistry.

The author wishes to express appreciation to those who have given their time and thought to the preparation of this second edition. As he did for the first edition, Dr. A. J. Barnard, Jr., read the entire manuscript and suggested changes that improved the program greatly. Professors Frank K. Cartledge, John S. Swenton, and Andrew L. Ternay also reviewed and commented upon the entire program. Dr. Kenneth Lipkowitz proofread the final manuscript, and Anne Banks assisted with the galley proofs. All errors, ambiguities, and incongruities that appear herein are, of course, the responsibility of the author. Finally, sincere thanks to Joan Garbutt of W. B. Saunders Company for her patience and good humor.

JAMES E. BANKS

Bozeman, Montana

CONTENTS

LIST OF FIGURES

LIST OF TABLES

Chapter One

CLASSIFICATION OF ORGANIC COMPOUNDS

The function of a chemical name is similar to the function of a personal name. It provides the chemist with a word or set of words that is unique to the substance. The systematic name of a compound conveys at least its empirical formula and also, if possible, its main structural features.

1.1 HYDROCARBONS

Compounds that contain only carbon and hydrogen are known as *hydrocarbons.* There are literally millions of hydrocarbons and their derivatives. Although most organic compounds contain elements besides carbon and hydrogen, these derivatives are all related to the hydrocarbons.

You are familiar with empirical formulas that specify the ratios of the atoms of the elements in a compound. For most inorganic compounds the empirical formula is sufficient identification. Several organic compounds may, however, share the same empirical formula. Molecular formulas are needed to show not only the ratios but also the actual numbers of atoms in the molecule. These molecular formulas, for example, share the same empirical formula, CH_2O: $C_2H_4O_2$, $C_3H_6O_3$, and $C_4H_8O_4$.

Sometimes a given molecular formula may represent more than one compound. Such compounds are called *isomers* [ĭ′ sŏ mĕrz]. Although they share the same molecular formula, isomers are distinctly different compounds. They have specific chemical and physical properties that distinguish them from one another. For example, there are 35 isomers with the molecular formula C_9H_{20}. One chemist, who is also a mathematician, calculates that there could be 4,111,846,763 isomers with the formula $C_{30}H_{62}$.

As we will see in the next few pages, isomers sometimes differ in the way the atoms are bound to one another. By assigning a unique name to each compound, the chemist can avoid having to specify the properties of the particular compound he has in mind.

In the formulas below, which pair represents two hydrocarbon isomers? (Check your answer on the next page.)

A ———————————————————————

C_2H_6O $\qquad\qquad$ C_2H_5OH

B _____

C_4H_{10} C_4H_8

C _____

C_5H_{12} C_5H_{12}

• •

A _____

You are wrong. Your answer is that C_2H_6O and C_2H_5OH represent two hydrocarbons that are isomers. They are indeed isomers, because they do have the same molecular formula. However, they do not fulfill the other condition: they are not hydrocarbons. Hydrocarbons are organic compounds that contain *only* carbon and hydrogen. Go back and choose another answer.

B _____

You are incorrect. Your answer is that C_4H_{10} and C_4H_8 represent two hydrocarbons that are isomers. Since each formula contains only carbon and hydrogen, they are both hydrocarbons. They are not isomers, however, because isomers are compounds that have the same molecular formula. For instance, C_2H_6O and C_2H_5OH are isomers. Go back and select another pair of formulas.

C _____

You are correct. Since they contain only hydrogen and carbon, and since they have the same molecular formula, C_5H_{12}, two hydrocarbons that are isomers are represented. Go on to the next section.

Molecular formulas obviously do not provide enough information to distinguish between isomers. Structural formulas do provide this information. They indicate the order in which the atoms are bound to one another. A dash is used to represent a covalent bond, that is, a shared pair of electrons. For example, here is a structural formula for one isomer with the formula C_3H_8:

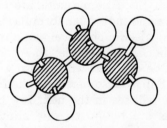

The drawing beside the formula shows one kind of model of the same molecule.

You may find it difficult to see that the model and the structural formula represent the same molecule. The trouble is caused by the fact that the structural formula depicts a three-dimensional object projected on a two-dimensional piece of paper. There are bound to be some distortions. Even the simplest hydrocarbon, methane, CH_4, cannot be accurately shown because the four covalent bonds on the carbon atom define a tetrahedral shape rather than lie in a plane. Below are four different models, all of which represent CH_4.

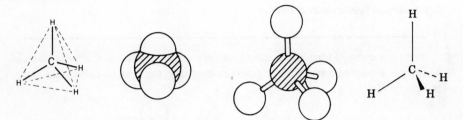

The model on the far right illustrates a convention used when it is desirable to emphasize bond orientation. A wedge indicates a bond projecting out of the page. The base of the wedge is considered to project toward the reader. A dotted line indicates a bond projecting behind the page away from the reader. This model is similar to a Fischer projection (see p. 89).

You can make a model of a tetrahedron by following the directions given in Figure 1.1, on page 5. The vertices of the tetrahedron coincide with the bonds on a carbon atom. For convenience, chemists usually write structural formulas with the bonds at right angles. The structural formula for methane is written

$$H - \overset{\displaystyle H}{\underset{\displaystyle H}{\overset{|}{\underset{|}{C}}}} - H.$$

Consider this structural formula:

$$H - \overset{\displaystyle H}{\underset{\displaystyle H}{\overset{|}{\underset{|}{C}}}} - \overset{\displaystyle H}{\underset{\displaystyle H}{\overset{|}{\underset{|}{C}}}} - \overset{\displaystyle H}{\underset{\displaystyle H}{\overset{|}{\underset{|}{C}}}} - H$$

Are all the atoms in the same plane? That is, does it represent a "flat" molecule?

A _____

Yes

B _____

No

• •

A _____

You are not correct. Always bear in mind that real molecules are three-dimensional. Moreover, the four covalent bonds that can be formed by a carbon atom are directed toward the vertices of a tetrahedron, as shown:

In order to write formulas on paper, bonds are usually drawn as if they were at right angles to each other. Remember that the important feature of structural formulas is that they show the *order* in which atoms are bound to one another. Continue and read answer *B* below.

B ─────────────────────────────────

You are correct. Writing its structural formula on a flat piece of paper does not make a molecule flat. If you compare carefully the structural formula and the model for C_3H_8

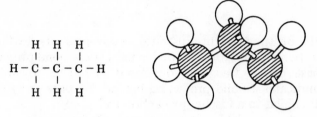

you will see that carbon and hydrogen atoms are bound in the same order in both. That is all a structural formula shows; it does not depict spatial relationships accurately. Go on to the next section.

The compound with the molecular formula CH_3Cl, represented by this model

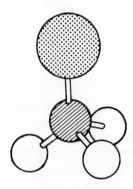

may have its structural formula written in many ways. Here are a few of them:

$$\underset{\underset{H}{|}}{\overset{\overset{Cl}{|}}{H-C-H}} \quad \underset{\underset{H}{|}}{\overset{\overset{H}{|}}{Cl-C-H}} \quad \underset{\underset{H}{|}}{\overset{\overset{H}{|}}{H-C-Cl}} \quad \underset{\underset{Cl}{|}}{\overset{\overset{H}{|}}{H-C-H}} \quad \underset{\underset{H}{\backslash}}{\overset{\overset{H}{\backslash}}{Cl-C-H}} \quad H-\underset{\underset{H}{/}}{\overset{\overset{Cl}{/}}{C}}-H$$

Structural formulas cannot give information about the orientation of the molecules in space. Any structural formula that shows three hydrogen atoms and one chlorine atom bound to a carbon atom is a satisfactory structural formula for this compound.

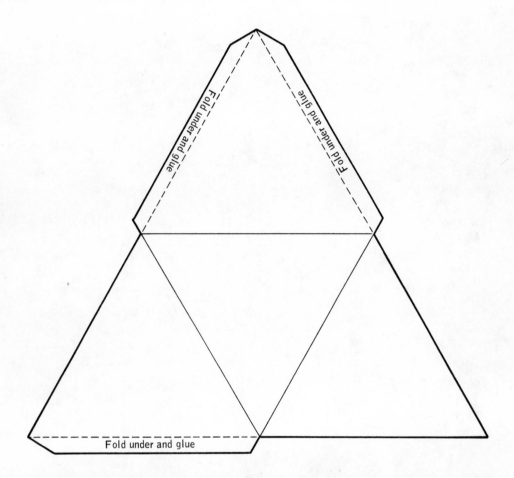

Figure 1.1 Model of a tetrahedron. If the center of the tetrahedron is taken to represent the nucleus of a carbon atom, the vertices will show the orientation of the four covalent bonds of the atom.

To assemble the model, cut out along the solid lines around the edges. A sturdier tetrahedron results if the cutout is cemented to light cardboard. If desired, additional tetrahedra can be cut out and will be useful. Fold the other solid lines over a ruler. Glue the faces together with the aid of the three glue flaps.

Which of these pairs of structural formulas represents the same molecule?

A _____

```
     H   H   H              H   Cl  H
     |   |   |              |   |   |
 H - C - C - C - H      H - C - C - C - H
     |   |   |              |   |   |
     H   H   Cl             H   H   H
```

B _____

```
     H   H                  H   H
     |   |                  |   |
 H - C - C - H          H - C - C - Cl
     |   |                  |   |
     H   Cl                 H   H
```

C _____

```
     H   H   Cl             H   H   Br
     |   |   |              |   |   |
 H - C - C - C - H      H - C - C - C - H
     |   |   |              |   |   |
     H   H   H              H   H   H
```

● ●

A _____

You are wrong. The structural formulas

```
     H   H   H              H   Cl  H
     |   |   |              |   |   |
 H - C - C - C - H      H - C - C - C - H
     |   |   |              |   |   |
     H   H   Cl             H   H   H
```

do not represent the same molecule. If you look at them closely, you can see that there is a chain of three carbon atoms in each formula that could be written as $- C - C - C -$. In the first formula the chlorine atom is bonded to one of the terminal carbon atoms. The chlorine atom in the second formula is bonded to the carbon in the middle of the chain. Since not all the atoms are bonded in exactly the same order, the formulas represent two different substances, or molecules. You should recognize, however, that they are isomers. Go back and choose another answer.

B _____

Correct. These two formulas represent exactly the same molecule. Since all the covalent bonds formed by the carbon atom are chemically identical, it makes no difference how they are oriented when the structural formula is written. Go on to the next section.

C _____

You are incorrect. Look at the formulas carefully again. You will find that they do not even represent isomers. The molecular formula of one is $C_2 H_5 Cl$ and the other is $C_2 H_5 Br$. Choose another answer.

Atoms joined by a single pair of electrons in a covalent bond are commonly considered as being able to rotate about the axis of the bond — that is, about the line joining the two atoms. Although there is some difference in energy associated with the different orientations, it is small and is not a barrier to the rotation at ordinary temperatures.

Structural formulas do not attempt to show the spatial orientation of the molecule or the atoms within the molecule. The following three drawings, for example, all show the carbon chain of the same molecule having the molecular formula C_5H_{12}.

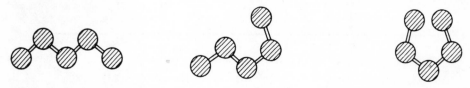

The structure of molecules does not change as they move about or change their shape by rotation about bond axes. It is customary to write structural formulas as straight as possible. Bends and crooks in them have no more meaning than bends and crooks in the molecule insofar as structure is concerned. These three structural formulas all represent the same molecule:

$$
\begin{array}{lll}
\underset{\underset{H}{|}}{\overset{\overset{H}{|}}{H-C}-\underset{\underset{H}{|}}{\overset{\overset{H}{|}}{C}}-\underset{\underset{H}{|}}{\overset{\overset{H}{|}}{C}}-\underset{\underset{H}{|}}{\overset{\overset{H}{|}}{C}}-\underset{\underset{H}{|}}{\overset{\overset{H}{|}}{C}}-H
& &
\end{array}
$$

The atoms in all three are connected to one another in the same order.

Choose the pair of structural formulas that represents two different compounds rather than the same molecule in different orientations.

A _____

B _____

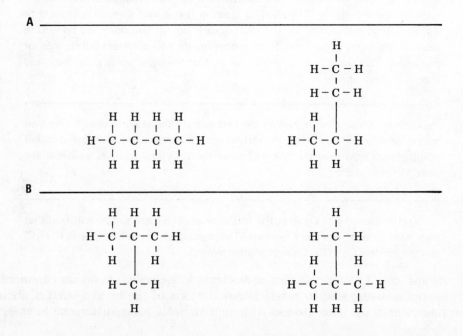

C _____

```
      H   H   H                          H   H   H
      |   |   |                          |   |   |
  H – C – C – C – H                  H – C – C – C – H
      |   |   |                          |   |   |
      H   H   |                          H   |   H
              |                              |
          H – C – H                      H – C – H
              |                              |
              H                              H
```

• •

A _____

You are incorrect. The two structural formulas you chose do represent the same compound. The only difference between them is that the chain of carbon atoms in the second is bent near one end. Remember that molecules can change their shapes without changes in structure. Since the structural formula specifies structure but not shape, it can be written in different orientations without a change in structure. Go back and choose another answer.

B _____

Wrong. These two structural formulas are identical. The only difference between them is that the second has been rotated 180° around the axis of the chain of carbon atoms. This is just a difference in orientation, not a difference in structure. Remember, structural formulas do not give information on spatial orientation. They represent structure only. Go back and choose another answer.

C _____

You are right. These two structural formulas represent different compounds. In the first, four carbon atoms are attached to one another in a continuous chain. In the second, the continuous chain has only three carbon atoms with the fourth attached to the middle carbon atom in the chain. This is a structural difference, not just a difference in orientation. Note also that the dashes, which indicate covalent bonds, may be shown with whatever length is needed to fit all the atomic symbols.

Molecules with the same structure that differ in spatial orientation as a result of rotation around a carbon-carbon bond are said to have different *conformations* and are called *conformational isomers* or *conformers*. Although differences in conformation are not reflected in the names of substances, you need to be aware of the phenomenon.

In 1954 Melvin S. Newman proposed a special kind of formula to delineate different conformations. Because of its utility, it was adopted quickly and is termed a Newman projection.

To prepare a Newman projection of a molecule from its structural formula, one first chooses the carbon-carbon bond about the axis of which the remainder of the molecule will be thought to rotate. This axis is considered as being perpendicular to the page. The two carbon atoms are represented as a circle. Lines from outside the circle toward the center represent bonds to other atoms. The lines for bonds to the nearer carbon atom pass through the circle to its center. Lines for bonds to the farther carbon atom stop at the circumference of the circle. This convention is illustrated by the following two examples. The projections on the left are perspective, or "saw-horse" representations.

Perspective   *Newman Projection*   *Perspective*   *Newman Projection*

In the classical interpretation, rotation about single bonds only was considered to lead to different conformers. Today some chemists extend the consideration to include bonds of higher order, including double bonds. Most chemists view rotation around the axis of a double bond as leading to isomers, not conformers (see p. 72).

Types of conformers are designated by the adjectives *eclipsed, staggered,* and *skewed.* Look at these three Newman projections for

$$H - \underset{\underset{H}{|}}{\overset{\overset{H}{|}}{C}} - \underset{\underset{H}{|}}{\overset{\overset{H}{|}}{C}} - H.$$

eclipsed   *staggered*   *skewed*

Notice that when bonds to both of the carbons would be coincident in the eclipsed conformer, they are drawn at a small angle to one another. The staggered conformer has the maximum angular separation of the hydrogen atoms from one another. The skewed conformer is intermediate between the other two types. Only one eclipsed and one staggered conformer of this molecule exist, but there is an infinite number of skewed conformers. For this molecule, the staggered conformation has the lowest energy and is the most thermodynamically stable form.

Newman projections are valuable in the study of cyclic compounds, including complex natural products such as steroids.

Which of the following is a correct Newman projection for the staggered conformer of

$$H - \underset{\underset{H}{|}}{\overset{\overset{H}{|}}{C}} - \underset{\underset{H}{|}}{\overset{\overset{H}{|}}{C}} - \underset{\underset{H}{|}}{\overset{\overset{H}{|}}{C}} - H?$$

A

B

C

•••

A _____

You correctly perceived that $H-\underset{\underset{H}{|}}{\overset{\overset{H}{|}}{C}}-$ can be attached to one of the carbon

atoms represented by the circle in a Newman projection. The bonds are separated by 60° in the staggered conformation. Go on to the next section.

B _____

No. The bonds are properly staggered, but count the number of atoms. Do you find three carbon atoms and eight hydrogen atoms? If not, go back and choose another answer.

C _____

You are incorrect. The projection represents a staggered conformation, but count the number of atoms. Remember that the circle stands for two carbon atoms. Do you find the proper total of three carbon and eight hydrogen atoms? If not, choose another answer.

Summary

1. Structural formulas are used to show the order in which the atoms of an organic compound are bonded to each other.

2. Structural formulas do not show the actual three-dimensional shape of organic molecules because such formulas are two-dimensional representations of three-dimensional objects. They do show the bonding sequence of the atoms in the molecule.

3. Molecules are in constant motion. Ordinary structural formulas, therefore, do not give information about the spatial orientation of the molecules they represent.

4. Rotation of atoms around the axis of the covalent bond that joins them allows molecules to assume different shapes without changing their structure. Structural formulas, too, may take different shapes without reflecting different structures.

5. Different spatial orientations resulting from rotation around the axis of a covalent carbon-carbon bond are called conformers. Newman projections can be used to display them.

staggered   *skewed*   *eclipsed*

Careful attention to these points will enable you to compare structural formulas or Newman projections and thus recognize true structural differences as opposed to different ways of writing the same formula.

Following are several pairs of formulas. Decide whether they represent the same or different molecules. The correct answers are just below the formulas. Cover the printed answers until you have decided on your own response.

Do the pairs of structural formulas given below represent the same or different molecules? (Answers are on the next page.)

1.

```
     H   H   H                    H   H   H
     |   |   |                    |   |   |
 H - C - C - C - Cl          H - C - C - C - H
     |   |   |                    |   |   |
     H   H   H                    H   H   Cl
```

2.

```
     H   H   H   H                H   H   H   H
     |   |   |   |                |   |   |   |
 H - C - C - C - C - Cl      H - C - C - C - C - H
     |   |   |   |                |   |   |   |
     H   H   H   H                H   Cl  H   H
```

3.

```
     H   H   H                    H   H   H   H
     |   |   |                    |   |   |   |
 H - C - C - C - H           H - C - C - C - C - H
     |   |   |                    |   |   |   |
     H   H   |                    H   H   H   H
             |
         H - C - H
             |
             H
```

4.

```
          H                              Cl
          |                              |
    H     |     H                H      |      H
      \   |   /                    \    |    /
        \ | /                        \  |  /
   H ----(   )---- H            H ----(   )---- H
        / | \                        /  |  \
      /   |   \                    /    |    \
   Cl     |     H                H      |      H
          |                             |
          H                             H
```

5.

```
     H   H   H                    H   H   H
     |   |   |                    |   |   |
 Cl- C - C - C - H           H - C - C - C - Cl
     |   |   |                    |   |   |
     H   H   H                    H   H   H
```

6.

```
     H   H   H   H                H   H   H
     |   |   |   |                |   |   |
 H - C - C - C - C - H       H - C - C - C - H
     |   |   |   |                |   |   |
     H   H   H   H                H   |   H
                                      |
                                  H - C - H
                                      |
                                      H
```

••

1. Same. All four covalent bonds on a carbon atom are equivalent.

2. Different. The chlorine atoms are bonded to different carbon atoms in the chain. (Note: the two compounds are isomers.)

3. Same. Bends and crooks do not indicate different structures.

$$\begin{array}{ccc} & H & H \\ & | & | \\ Cl - & C - & C - H. \\ & | & | \\ & H & H \end{array}$$

4. Same. Two staggered conformations of $Cl - C - C - H$.

5. Same. Molecules are in constant motion through space. These two formulas can be superimposed if one is rotated $180°$ on a vertical line through the middle carbon atom.

6. Different. The first formulas shows a continuous chain of four carbon atoms while the second has a continuous chain of only three with the fourth carbon atom attached to the middle one.

Review any question that you missed to be sure you understand the correct answer. Then go on to the next section.

In the early days of organic chemistry, compounds were given names related to their natural source. For instance, acetic acid derives its name from the Latin word for vinegar, *acetum*. As soon as compounds began to be synthesized from materials other than natural products, the problem of naming them became acute. Chemists soon recognized that some sort of system was needed. One system was adopted at an international congress of chemists held in Geneva in 1892. Although it has been modified and extended since then, it is still in use under the name "Geneva system." Today, rules for naming organic compounds are established by the International Union of Pure and Applied Chemistry (IUPAC). These rules are followed throughout the world. In this program the IUPAC rules will be emphasized. Where the custom of American chemists differs from these rules, the usage of *Chemical Abstracts* will be indicated. Many of the non-systematic, or trivial, names accepted by the IUPAC rules will be mentioned. Systematic chemical nomenclature is not fixed for all time — changes and improvements do occur occasionally. The practice of *Chemical Abstracts* underwent a pronounced change in 1972 and few trivial names are now retained for indexing purposes. Trivial names are still widely used in the conversations and writings of chemists, as well as in merchandising chemicals, and thus they are of importance to students of chemistry.

The first step in classifying organic compounds is to separate them into two main divisions, open-chain or aliphatic [ăl′ ĭ făt″ ĭk] compounds and cyclic [sĭ klĭk] compounds. The word aliphatic is derived from the Greek word for fat; fatty acids are aliphatic compounds (see section 5.3). The division is shown by this diagram:

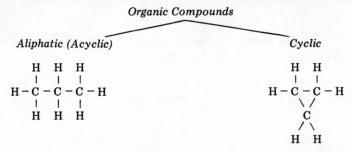

Notice that there is a closed ring, or cycle, of carbon atoms joined to one another in the cyclic compound, whereas the aliphatic compound has none. Aliphatic compounds are sometimes termed acyclic compounds.

Which of these structural formulas represents an aliphatic hydrocarbon?

A _____

$$
\begin{array}{ccccc}
 & H & O & H & \\
 & | & \| & | & \\
H- & C & - C - & C & -H \\
 & | & & | & \\
 & H & & H &
\end{array}
$$

B _____

$$
\begin{array}{ccc}
 & H & H \\
 & | & | \\
H- & C - & C -H \\
 & | & | \\
 & H & H
\end{array}
$$

C _____

●●

A _____

You are not correct. The compound represented by the structural formula

$$
\begin{array}{ccccc}
 & H & O & H & \\
 & | & \| & | & \\
H- & C & - C - & C & -H \\
 & | & & | & \\
 & H & & H &
\end{array}
$$

is indeed aliphatic since it does not contain any ring structures. It is not a hydrocarbon, however, since hydrocarbons contain only hydrogen and carbon. This compound contains oxygen. Go back and choose another answer.

B _____

You are correct. The formula

$$
\begin{array}{ccc}
 & H & H \\
 & | & | \\
H - & C - C & - H \\
 & | & | \\
 & H & H
\end{array}
$$

contains only hydrogen and carbon. There is no ring structure; consequently, it represents an aliphatic hydrocarbon. Go on to the next section.

C _____

Incorrect. The formula

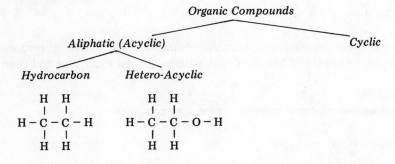

represents a hydrocarbon since it contains only carbon and hydrogen. However, it contains a ring formed by five carbon atoms connected to each other. Therefore it is a cyclic compound, not aliphatic. Aliphatic compounds do not contain a ring structure. Go back and select another answer.

Aliphatic compounds can be classed as hydrocarbons or hetero-acyclic compounds, depending upon whether elements other than carbon and hydrogen are present. Our classification scheme can now be extended as shown here:

Organic Compounds

Aliphatic (Acyclic)   *Cyclic*

Hydrocarbon   *Hetero-Acyclic*

$$
\begin{array}{ccc}
H & H \\
| & | \\
H - C - C - H \\
| & | \\
H & H
\end{array}
\qquad
\begin{array}{ccc}
H & H \\
| & | \\
H - C - C - O - H \\
| & | \\
H & H
\end{array}
$$

As we shall see in later chapters, the cyclic compounds may be further classified. The remainder of this chapter will be devoted to the names for hydrocarbons.

1.2 ALKANES

There are several families, or types, of aliphatic hydrocarbons. The simplest is called the *alkane* [ăl′ kāne] family. Alkanes are saturated aliphatic hydrocarbons. The word *saturated* indicates that they have no double or triple covalent bonds between carbon atoms. An alkane containing three carbon atoms can be represented by the structural formula

$$
\begin{array}{ccccc}
& H & & H & & H \\
& | & & | & & | \\
H - & C & - & C & - & C & - H \\
& | & & | & & | \\
& H & & H & & H
\end{array}
$$

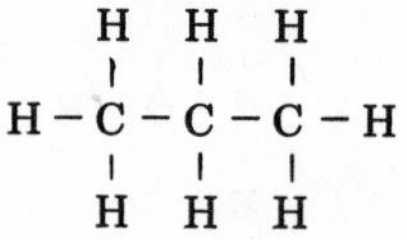

Detailed structural formulas of this sort use a great deal of space and are often not necessary. For this reason chemists often write condensed structural formulas in which all atoms attached to a given carbon atom are written after the "C" and on the same line. For example, the formula for the alkane shown above can be condensed to $CH_3 CH_2 CH_3$. Such a formula is also termed a *line* formula.

If other elements are present in the compound, they can be indicated in the same way. When two different kinds of atoms are attached to one carbon atom, hydrogen is usually written first. These three examples should make the point clear.

Structural formula	*Condensed structural formula*
$$\begin{array}{ccccc} & H & & H & & H \\ & | & & | & & | \\ Cl - & C & - & C & - & C & -H \\ & | & & | & & | \\ & H & & H & & H \end{array}$$	$CH_2 ClCH_2 CH_3$
$$\begin{array}{ccccc} & H & & H \\ & | & & | \\ Cl - & C & - & C & - F \\ & | & & | \\ & Cl & & F \end{array}$$	$CHCl_2 CHF_2$
$$\begin{array}{ccccc} & Cl & & H \\ & | & & | \\ Cl - & C & - & C & - H \\ & | & & | \\ & Cl & & H \end{array}$$	$CCl_3 CH_3$

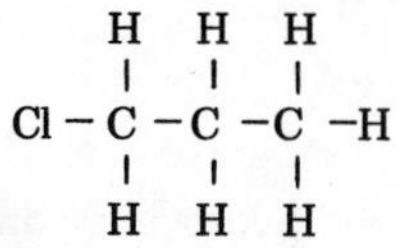

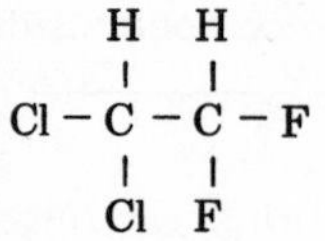

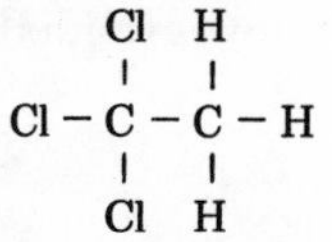

Following are three questions concerning condensed structural formulas. The answers are given together below the questions. Check each one and then go to the next.

1. Are these two structural formulas for the same compound?

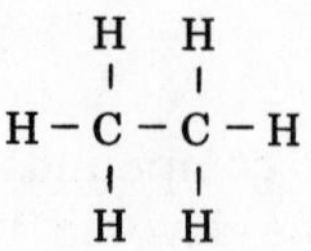

$$
\begin{array}{ccc}
& H & & H \\
& | & & | \\
H - & C & - & C & - H \\
& | & & | \\
& H & & H
\end{array}
\qquad CH_3 CH_3
$$

2. The structural formula for a compound is

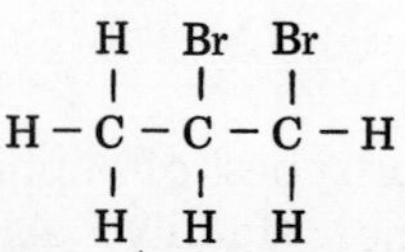

$$
\begin{array}{ccccc}
& H & & Br & & Br \\
& | & & | & & | \\
H - & C & - & C & - & C & - H \\
& | & & | & & | \\
& H & & H & & H
\end{array}
$$

Is this the correct condensed structural formula?

$$CH_3 CHBr_2 CH_2$$

   3. The condensed structural formula of a compound is

$$CH_3 CCl_2 CHClCH_2 Cl$$

Write its structural formula.

● ●

   1. Yes, both represent a compound with the molecular formula C_2H_6.

   2. No. All atoms attached to a given carbon atom are written after that carbon atom. The correct answer is $CH_3 CHBrCH_2 Br$.

   3.

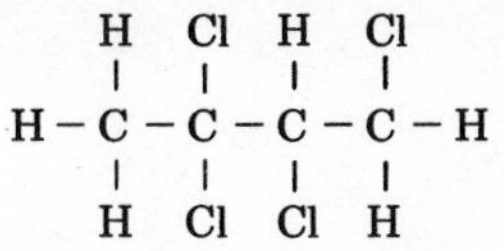

```
        H   Cl  H   Cl
        |   |   |   |
    H – C – C – C – C – H
        |   |   |   |
        H   Cl  Cl  H
```

The names of the alkanes are based on the number of carbon atoms they contain and on the way in which the carbon atoms are attached to one another. Here are the structural formulas for two alkanes. Each contains four carbon atoms and ten hydrogen atoms, yet they are not the same compound. Since they both have the molecular formula C_4H_{10}, they must be isomers. Since they differ in the way the atoms are attached to one another, they are *constitutional isomers.*

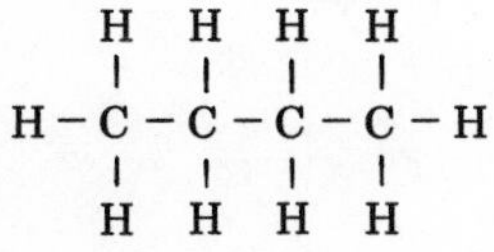

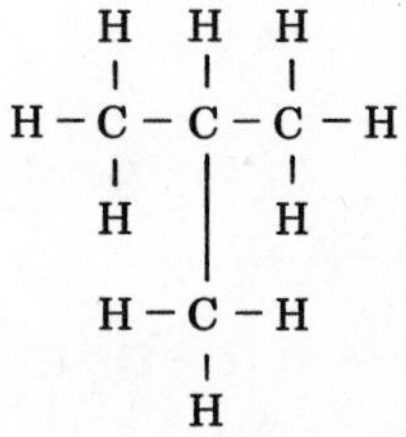

```
    H   H   H   H                        H   H   H
    |   |   |   |                        |   |   |
H – C – C – C – C – H                H – C – C – C – H
    |   |   |   |                        |   |   |
    H   H   H   H                        H   |   H
                                             |
                                         H – C – H
                                             |
                                             H
```

You can see that the four carbon atoms in the compound on the left are attached in a continuous chain of four atoms. The carbons on the right are in the form of a "T." Skeleton formulas are often used to make such differences more apparent. A *skeleton formula* shows all of the atoms in a compound except hydrogen. It is assumed that all of the remaining valences of each carbon are filled by hydrogen. To use skeleton formulas you must know that a carbon atom is capable of forming four covalent bonds and that a hydrogen atom forms only one. The skeleton formulas of the two alkanes shown above are

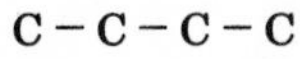

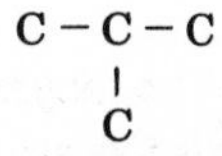

```
    C – C – C – C                        C – C – C
                                                 |
                                                 C
```

It is much easier to recognize the difference, isn't it?

Remember that a single covalent bond between two atoms does not greatly restrict the rotation of the atoms in most instances. Structural formulas can be written around corners and jogs without changing their structure. Likewise, formulas can be reversed from left to right and top to bottom without affecting the structural representation.

Following are some skeletons for you to compare. The answers are given together below the questions.

Indicate by yes or no whether each of these skeletons represents the same compound.

1.

$$C-C-C-C$$

$$\begin{array}{c} C \\ | \\ C-C-C \end{array}$$

2.

$$\begin{array}{c} Cl \\ | \\ C-C-C-C \end{array}$$

$$\begin{array}{c} C-C-C-C \\ \quad\quad | \\ \quad\quad Cl \end{array}$$

3.

$$\begin{array}{c} C \\ | \\ C-C-C-C \\ \quad | \\ \quad C \end{array}$$

$$\begin{array}{c} C \\ | \\ C-C-C-C \\ \quad | \\ \quad C \end{array}$$

4.

$$\begin{array}{c} C-C-C \quad C-C \\ \quad\quad | \quad | \\ \quad\quad C-C \end{array}$$

$$\begin{array}{c} C-C-C-C-C \\ \quad\quad\quad\quad | \\ \quad\quad\quad\quad C \\ \quad\quad\quad\quad | \\ \quad\quad\quad\quad C \end{array}$$

5.

$$C-C-O-C-C$$

$$\begin{array}{c} O \\ \| \\ C-C-C-C \end{array}$$

6.

$$\begin{array}{c} Cl \\ | \\ Cl-C-C-C-C \\ \quad\quad | \\ \quad\quad Cl \end{array}$$

$$\begin{array}{c} Cl \\ | \\ C-C-C-C-Cl \\ \quad | \\ \quad Cl \end{array}$$

7.

$$\begin{array}{c} C-C \quad C-C \\ \quad | \quad\quad | \\ \quad C-C \end{array}$$

$$\begin{array}{c} C-C-C-C \\ \quad | \quad\quad | \\ \quad C-C \end{array}$$

8.

$$\begin{array}{c} C \\ \backslash \\ \quad C-C-C \\ / \\ C \end{array}$$

$$\begin{array}{c} C \\ | \\ C-C-C-C \end{array}$$

● ●

1. Yes. Both have a continuous chain of four carbon atoms.

2. Yes. They can be superimposed by a 180° rotation left to right and top to bottom.

3. No. The longest continuous chain on the left is four atoms. The longest on the right is five atoms.

4. Yes. There are seven carbon atoms in each continuous chain.

5. No. Oxygen is between two carbon atoms on the left, but joined to only one carbon on the right.

6. No. Chlorine atoms are attached to adjacent carbon atoms on the left, but not on the right.

7. No. The compound on left is aliphatic (acyclic); the compound on right is cyclic.

8. Yes. Each has a chain of four carbons, with a fifth carbon attached to the carbon next to the end of the chain.

Review any questions you missed to be sure you understand your mistakes. Then go on to the next section.

Remember, an alkane is a saturated aliphatic hydrocarbon. It contains only carbon and hydrogen atoms and there are no double or triple covalent bonds.

The alkanes in which all of the carbon atoms are in a continuous ("straight") chain are known as *normal alkanes*. You should be able to recognize normal alkanes readily at this point. Alkanes that do not have all of the carbon atoms in a continuous chain are called branched alkanes. You can see that, of the two alkanes containing four carbon atoms, one is a normal alkane and the other is a branched alkane.

$$C-C-C-C \qquad\qquad \begin{array}{c} C \\ | \\ C-C-C \end{array}$$

Normal   *Branched*

Which of these skeletons does *not* represent a normal alkane?

A _____

$$\begin{array}{c} C-C-C \\ | \\ C \end{array}$$

B _____

$$\begin{array}{c} C \\ | \\ C-C-C \\ | \\ C \end{array}$$

C _____

$$\begin{array}{c} C-C-C-C \quad C \\ | \quad\ \ | \\ C-C \end{array}$$

• •

A _____

Incorrect. The skeleton formula $C-C-C$ *does* represent a normal alkane.
$$\begin{array}{c} \quad\quad\ | \\ \quad\quad\ C \end{array}$$

You should recall that carbon atoms joined by a single covalent bond can rotate freely around the axis of the bond. The structural formula can be written with any sort of bend or crook. The formula above might just as well be

$$
\begin{array}{ccc}
 & \text{C} \quad \text{C} & \text{C} \\
 & | \quad\; | & | \\
\text{C} - \text{C} - \text{C} - \text{C} \quad \text{or} \quad \text{C} - \text{C} & \text{or} & \text{C} - \text{C} \\
 & & | \\
 & & \text{C}
\end{array}
$$

They all represent the same compound. Go back and choose another answer.

B _____

You are right. No matter how you rearrange it, the skeleton formula

$$
\begin{array}{c}
\text{C} \\
| \\
\text{C} - \text{C} - \text{C} \\
| \\
\text{C}
\end{array}
$$

cannot be written as a continuous chain without breaking some carbon-carbon bonds. It represents a branched alkane. Continue on to the next section.

C _____

You are incorrect. Remember that any formula written on paper is a two-dimensional attempt to show a three-dimensional object. Since a single carbon-carbon covalent bond usually permits the atoms to rotate freely around the axis of the bond, the skeleton can be written with bends and crooks. The important factor is the order in which the carbon atoms are joined together. Therefore, the formula you have chosen

$$
\begin{array}{cc}
\text{C} - \text{C} - \text{C} - \text{C} & \text{C} \\
| & | \\
\text{C} - \text{C} &
\end{array}
$$

is the same as $\text{C} - \text{C} - \text{C} - \text{C} - \text{C} - \text{C} - \text{C}$. Here is another skeleton which contains the same number of carbon atoms:

$$
\begin{array}{c}
\text{C} \\
| \\
\text{C} - \text{C} - \text{C} - \text{C} - \text{C} \\
| \\
\text{C}
\end{array}
$$

It cannot be written as a single continuous chain without rearranging the order of attachment. Do you see the difference? When you do, go back and select another answer.

You are already familiar with the name of the simplest of the normal alkanes. It is methane [měth′ ăn], CH_4. The next in the series is ethane [ěth′ ăn], C_2H_6. The normal isomers of the alkane family up to ten carbon atoms are shown in Table 1.1.

TABLE 1.1 Names and Formulas of Normal Alkanes

Number of carbon atoms	Name	Structural formula	Molecular formula
1	Methane	$\begin{array}{c} H \\ | \\ H-C-H \\ | \\ H \end{array}$	CH_4
2	Ethane	$\begin{array}{cc} H & H \\ | & | \\ H-C-C-H \\ | & | \\ H & H \end{array}$	C_2H_6
3	Propane (prō′ păn)	$\begin{array}{ccc} H & H & H \\ | & | & | \\ H-C-C-C-H \\ | & | & | \\ H & H & H \end{array}$	C_3H_8
4	Butane (bŭ′ tăn)	$\begin{array}{cccc} H & H & H & H \\ | & | & | & | \\ H-C-C-C-C-H \\ | & | & | & | \\ H & H & H & H \end{array}$	C_4H_{10}
5	Pentane (pĕn′ tăn)	$\begin{array}{ccccc} H & H & H & H & H \\ | & | & | & | & | \\ H-C-C-C-C-C-H \\ | & | & | & | & | \\ H & H & H & H & H \end{array}$	C_5H_{12}
6	Hexane (hĕx′ ăn)	$\begin{array}{cccccc} H & H & H & H & H & H \\ | & | & | & | & | & | \\ H-C-C-C-C-C-C-H \\ | & | & | & | & | & | \\ H & H & H & H & H & H \end{array}$	C_6H_{14}
7	Heptane (hĕp′ tăn)	$\begin{array}{ccccccc} H & H & H & H & H & H & H \\ | & | & | & | & | & | & | \\ H-C-C-C-C-C-C-C-H \\ | & | & | & | & | & | & | \\ H & H & H & H & H & H & H \end{array}$	C_7H_{16}
8	Octane (ŏc′ tăn)	$\begin{array}{cccccccc} H & H & H & H & H & H & H & H \\ | & | & | & | & | & | & | & | \\ H-C-C-C-C-C-C-C-C-H \\ | & | & | & | & | & | & | & | \\ H & H & H & H & H & H & H & H \end{array}$	C_8H_{18}
9	Nonane (nō′ nàn)	$\begin{array}{ccccccccc} H & H & H & H & H & H & H & H & H \\ | & | & | & | & | & | & | & | & | \\ H-C-C-C-C-C-C-C-C-C-H \\ | & | & | & | & | & | & | & | & | \\ H & H & H & H & H & H & H & H & H \end{array}$	C_9H_{20}
10	Decane (dĕ căn)	$\begin{array}{cccccccccc} H & H & H & H & H & H & H & H & H & H \\ | & | & | & | & | & | & | & | & | & | \\ H-C-C-C-C-C-C-C-C-C-C-H \\ | & | & | & | & | & | & | & | & | & | \\ H & H & H & H & H & H & H & H & H & H \end{array}$	$C_{10}H_{22}$

Although you may not immediately see any regularity to the names, the prefixes from *pent* onward are derived from either the Greek or Latin words for the respective numbers. Memorize the names in this table. They are the basis for the rest of this program.

The names of the higher alkanes use the Greek or Latin prefixes to specify the number of carbon atoms.

C_{10}	decanes	C_{22}	docosanes
C_{11}	undecanes	C_{23}	tricosanes
C_{12}	dodecanes	C_{24-29}	tetracosanes, etc.
C_{13}	tridecanes	C_{30}	triacontanes
C_{14}	tetradecanes	C_{31}	hentriacontanes
C_{15}	pentadecanes	C_{32-39}	dotriacontanes, etc.
C_{16-19}	hexadecanes, etc.	C_{40}	tetracontanes
C_{20}	eicosanes	C_{41}	hentetracontanes, etc.
C_{21}	heneicosanes		

The family of normal alkanes is known as an homologous [hŏ mŏl′ ō gŭs] series. The compounds in an homologous series differ from one another by a certain specific structural unit.

What is the structural unit that is added to one normal alkane in order to form the next member of the homologous series?

A _____

One carbon atom

B _____

One carbon atom and two hydrogen atoms

C _____

One carbon atom and three hydrogen atoms

D _____

I don't know what you're talking about

● ●

A _____

You are partly right. Each member of the series has one more carbon atom than its predecessor. However, it has more hydrogen atoms, too. Compare the molecular formulas to see if you can find how many hydrogen atoms are added along with each carbon atom. Then choose another answer.

B _____

You are right. As the series grows from one member to the next, the

$$\text{structural unit} -\overset{\displaystyle H}{\underset{\displaystyle H}{\overset{|}{\underset{|}{C}}}}- \text{ is added each time. Go on to the next section.}$$

C _____

You are on the right track, but not entirely correct. Each member of the homologous series of alkanes has one more carbon than the one before it. Does it also have three more hydrogen atoms? Compare the molecular formulas of propane and butane. Then compare butane with pentane. Then choose another answer.

D _____

Well, let's see if we can help you to understand. If you compare the structural formulas of propane and butane, what is the difference?

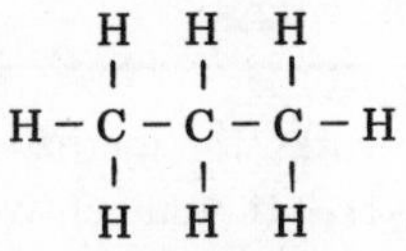

$$H-C-C-C-H$$

propane

$$H-C-C-C-C-H$$

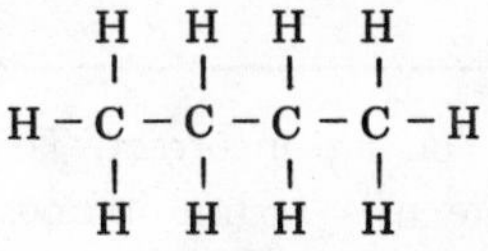

butane

You can see that butane is merely propane with an added

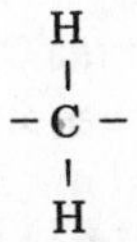

$$-C-$$

Now make a similar comparison between butane and the next member of the homologous series, pentane. When you have done this, go back and choose another answer.

Consider another homologous series for a moment. It is a series of alcohols as represented by these condensed structural formulas:

$$CH_3OH, \quad CH_3CH_2OH, \quad CH_3CH_2CH_2OH, \quad CH_3CH_2CH_2CH_2OH, \quad etc.$$

The unit of structural difference in this series is the same as in the normal alkane series. It is CH_2. Each member of the series differs from its predecessor by this unit.

Every member of an homologous series can be represented by a general formula in which a letter substitutes for the number of carbon atoms, and a combination of the letter and a number substitutes for the atoms of the other elements. If the letter "n" is used to represent the number of carbon atoms in one of the alcohols, the number of hydrogen atoms is always "2n + 2." Each alcohol has only one oxygen atom. Therefore, the general formula of the alcohols is $C_nH_{2n+2}O$. Verify this general formula by checking it against the molecular formulas for the alcohols that are shown above. You might say that the general formula is a sort of algebraic equivalent for any one of the series.

The molecular formulas of the first four normal alkanes are CH_4, C_2H_6, C_3H_8, and C_4H_{10}. What is the general formula for the series of normal alkanes?

A _____

$$C_nH_{4n}$$

B _____

$$C_nH_{3n+1}$$

C _____

$$C_nH_{4n-2}$$

D _____

$$C_nH_{2n+2}$$

• •

A _____

You are incorrect. The general formula C_nH_{4n} works for methane, CH_4, where n = 1, but it does not work for ethane, C_2H_6. If your answer were correct, ethane would have to be C_2H_8. A correct general formula must specify the molecular formula for every member of the homologous series if the proper substitutions are made. Go back and work out another answer.

B _____

You are wrong. The general formula C_nH_{3n+1} works for methane, CH_4, but it does not work for any other member of the series. Suppose you were to use it to predict the formula of propane. If n = 3, then $C_nH_{3n+1} = C_3H_{10}$. This is not the correct molecular formula for propane. A correct general formula must enable you to determine the molecular formula of any member of the series. Go back and work out the correct answer.

C _____

Your answer is that C_nH_{4n-2} is the general formula for the alkanes. Come on now. This general formula doesn't even work for the first member, methane. Since methane has only one carbon atom, n = 1, then $C_nH_{4n-2} = CH_2$. You should know that the correct molecular formula is CH_4, not CH_2. A general formula must allow you to predict the formula for any member of the alkane family. Go back and work out the right answer.

D _____

You are correct. The general formula for the alkanes is C_nH_{2n+2}. Go on to the next section.

So far you have learned to write six kinds of formulas: molecular, structural, condensed structural (line), skeleton, Newman projection, and general. Each has advantages and disadvantages. Molecular formulas give composition but no information about structure. A structural formula is tedious to write because all the $C-H$ bonds must be shown. Skeleton structural formulas show the important structural features, but they do not represent the entire substance. For this reason skeleton formulas are rarely used except to emphasize a particular structural feature. Newman projections are used to emphasize the spatial orientation of atoms or to illustrate conformers. Most textbooks use condensed structural (line) formulas to specify various compounds because they save space as well as give complete information about the structure of a compound. General formulas are used to represent all of the members of a homologous series.

Whenever a particular structural unit is repeated several times, parentheses can be used to condense the formula even more. A few examples will make this clear to you.

	Molecular formula	Condensed formula
Methane	CH_4	CH_4
Ethane	C_2H_6	CH_3CH_3
Propane	C_3H_8	$CH_3CH_2CH_3$
Butane	C_4H_{10}	$CH_3(CH_2)_2CH_3$
Pentane	C_5H_{12}	$CH_3(CH_2)_3CH_3$

Notice that the parentheses have exactly the same meaning as in inorganic compounds. For example, $Ca(OH)_2$ represents 1 Ca^{++} ion and 2 OH^- ions. The $(CH_2)_2$ in the butane formula represents $- CH_2CH_2 -$.

Which of the following is a condensed structural formula for octane?

A _____

$CH_3(CH_2)_8CH_3$

B _____

$(CH_2)_8$

C _____

$CH_3(CH_2)_6CH_3$

D _____

I need help.

• •

A _____

You are incorrect. Your answer is that $CH_3(CH_2)_8CH_3$ is a condensed structural formula for octane. Look at the formula carefully. How many carbon atoms are there? There are ten. The prefix *oct-* in the name should tell you that there are eight carbon atoms in octane. You have chosen the condensed formula of decane. Go back and find the formula for octane.

B _____

You are wrong. You have chosen $(CH_2)_8$ as the condensed structural formula of octane. This formula has the correct number of carbon atoms, all right, but what about the hydrogen? If you multiply to remove the parentheses, $(CH_2)_8$ becomes C_8H_{16}. If you substitute $n = 8$ into the general formula, $C_nH_{2n+2} = C_8H_{18}$. Your answer lacks two hydrogen atoms. If this is not clear to you, try drawing a complete structural formula for $(CH_2)_8$. Compare it with the structural formula for octane on page 21, then choose another answer.

C _____

Right. The condensed formula $CH_3(CH_2)_6CH_3$ represents octane, C_8H_{18}. You may wish to compare it with the complete structural formula on page 21. Continue on to the next section.

D _____

OK, here is some help. The use of parentheses in condensed structural formulas is nothing more than an attempt to save space by not writing repeating structures separately. In inorganic chemistry you write $Mg(NO_3)_2$ instead of

$MgNO_3 NO_3$. Similarly, in organic chemistry you can write $CH_3(CH_2)_3CH_3$ instead of $CH_3CH_2CH_2CH_2CH_3$. Read page 25 again and choose another answer.

To begin our discussion of branched alkanes, we'll look at their carbon skeletons. You know that a normal alkane is one in which there is a continuous chain containing all the carbon atoms. Although the skeleton may be written with bends and corners, the entire chain can be traced from one end to the other without lifting your pencil or retracing. This is not true for branched alkanes. Compare these two skeletons by tracing the carbon chains with your pencil.

$$
\begin{array}{ccc}
\text{C} - \text{C} - \text{C} - \text{C} & \qquad\qquad & \text{C} - \text{C} - \text{C} - \overset{\displaystyle \text{C}}{\underset{\displaystyle \text{C}}{\text{C}}} \\
\qquad\quad | & & \\
\qquad\quad \text{C} & &
\end{array}
$$

Normal alkane   *Branched alkane*

The skeleton on the right cannot be rearranged to include all of the carbon atoms in a continuous chain without breaking a carbon-carbon bond.

Here is a fairly difficult question for you. What is the minimum number of carbon atoms in a branched alkane?

A _____

3

B _____

4

C _____

5

• •

A _____

You are wrong. No matter how you write the skeleton, three carbon atoms must be in a continuous chain. Here are several ways of depicting a three-carbon chain:

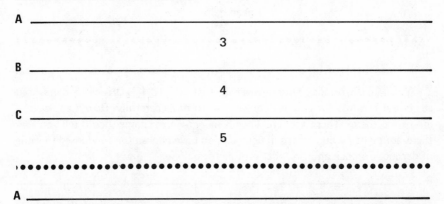

Since the carbon atoms can rotate freely around the covalent bonds, every one of these skeletons represents ordinary propane. Try another answer.

B _____

Correct. It is possible to write a branched alkane containing only four carbon atoms, but impossible to write one with two or three carbons. Go on to the next section.

C _____

You are incorrect. Suppose that we consider this skeleton for a branched alkane containing five carbon atoms:

$$\begin{array}{c} \text{C} \\ | \\ \text{C} - \text{C} - \text{C} \\ | \\ \text{C} \end{array}$$

Try removing any one of the outside carbon atoms. Isn't the remaining skeleton still a branched alkane? Sure it is. Try another answer.

Neglecting conformers and cyclic structures, there are only two different ways that a carbon skeleton of four atoms can be assembled. They are

$$\text{C} - \text{C} - \text{C} - \text{C} \qquad \text{and} \qquad \begin{array}{c} \text{C} - \text{C} - \text{C} \\ | \\ \text{C} \end{array}$$

The complete structural formulas of the two compounds are

$$\begin{array}{c} \quad\;\text{H}\;\;\;\text{H}\;\;\;\text{H}\;\;\;\text{H} \\ \quad\;\;|\;\;\;\;\;|\;\;\;\;\;|\;\;\;\;\;| \\ \text{H} - \text{C} - \text{C} - \text{C} - \text{C} - \text{H} \\ \quad\;\;|\;\;\;\;\;|\;\;\;\;\;|\;\;\;\;\;| \\ \quad\;\text{H}\;\;\;\text{H}\;\;\;\text{H}\;\;\;\text{H} \end{array} \quad \text{and} \quad \begin{array}{c} \quad\;\text{H}\;\;\;\text{H}\;\;\;\text{H} \\ \quad\;\;|\;\;\;\;\;|\;\;\;\;\;| \\ \text{H} - \text{C} - \text{C} - \text{C} - \text{H} \\ \quad\;\;|\;\;\;\;\;|\;\;\;\;\;| \\ \quad\;\text{H}\;\;\;\;\;\;|\;\;\;\;\;\text{H} \\ \quad\;\;\;\;\;\text{H} - \text{C} - \text{H} \\ \quad\;\;\;\;\;\;\;\;\;| \\ \quad\;\;\;\;\;\;\;\;\;\text{H} \end{array}$$

Since both compounds have the same molecular formula, C_4H_{10}, they are isomers. To distinguish between them, the first was named butane (sometimes *n*-butane for normal butane, the trivial name) and the second was called isobutane [i′ sō̇ bū′ tȧn].

As the number of carbon atoms increases, so does the number of possible constitutional isomers. How many constitutional isomers has the formula C_5H_{12}?

A _____

2

B _____

3

C _____

4

D _____

5

●●●

A _____

You are incorrect. There are more than two isomeric alkanes that contain five carbon atoms. Here are the skeletons of two of them:

$$C-C-C-C-C \qquad\qquad \begin{array}{c} C \\ | \\ C-C-C-C \end{array}$$

Try to find at least one more that is different from these two. Then choose another answer.

B _____

You are right. There are three isomeric alkanes that have the formula C_5H_{12}. Their skeletons and names are on this page. Go on to next section.

C _____

You are wrong. There are not four isomers that contain five carbon atoms. Perhaps you have written one skeleton in two different orientations. For instance

$$\begin{array}{c} C \\ | \\ C-C-C-C \end{array} \qquad \begin{array}{c} C-C-C-C \\ | \\ C \end{array} \qquad \begin{array}{c} C \\ | \\ C-C-C-C \end{array}$$

All three of these skeletons represent the same compound. They can be superimposed by rotation. Check the skeletons you have written to be sure that they are all different. Then go back and choose another answer.

D _____

You are incorrect. There are not five isomers that contain five carbon atoms. Perhaps you have written the same skeleton in two or more different orientations. For example, all of the skeletons written below have the same structure and represent the same compound. They can be superimposed by rotation.

$$\begin{array}{c} C \\ | \\ C-C-C-C \end{array} \quad \begin{array}{c} C-C-C-C \\ | \\ C \end{array} \quad \begin{array}{c} C \\ | \\ C-C-C-C \end{array} \quad \begin{array}{c} C-C-C-C \\ | \\ C \end{array}$$

Check the skeletons you have written to see which ones are alike. Then go back and choose another answer.

The three isomers containing five carbon atoms have these carbon skeletons:

$$C-C-C-C-C \qquad \begin{array}{c} C \\ | \\ C-C-C-C \end{array} \qquad \begin{array}{c} C \\ | \\ C-C-C \\ | \\ C \end{array}$$

pentane   *isopentane*   *neopentane*

The name pentane is both the trivial and the systematic name for the first compound. Isopentane and neopentane are the trivial names for the latter two, and are allowed by the IUPAC rules.

The number of possible isomers expands quite rapidly for the higher alkanes (i.e., those with more carbon atoms). New prefixes, if coined indefinitely, would soon become impossible to remember. For this reason, the only one in widespread use is *iso-* [ī′ sō]. This prefix is retained in the trivial names for all compounds having a *single* carbon branch on the carbon atom next to the end of a continuous chain. Thus the skeletons of isohexane and isoheptane are:

```
              C                                    C
              |                                    |
  C - C - C - C - C                    C - C - C - C - C - C
       isohexane                              isoheptane
```

Which of these skeletons represents a compound whose common name begins with the prefix *iso-*?

A _____

```
        C - C - C - C
        |           |
        C           C
```

B _____

```
        C - C - C - C
        |           |
        C           C
```

C _____

```
      C - C - C - C - C
              |
              C
```

D _____

```
                C
                |
      C - C - C - C - C
                |
                C
```

• •

A _____

You are correct. If you straighten out the kink, the skeleton you have chosen can be written

```
    C - C - C - C - C
            |
            C
```

It has a single carbon branch on the carbon atom next to the end of a continuous chain. Therefore, it is an iso- compound, specifically, isohexane. Go on to the next section.

B _____

You are incorrect. Perhaps you have forgotten that a single covalent bond between carbon atoms does not restrict the rotation of the atoms. In practical terms, this means that carbon skeleton formulas can be written with crooks that may be straightened out. The formula you have chosen

$$C - C - C - C$$
$$\quad | \qquad\quad |$$
$$\quad C \qquad\quad C$$

can also be written $C - C - C - C - C - C$ and is hexane. With this in mind, go back and find a skeleton which represents an iso- compound.

C _____

You are not correct. An iso- compound has a single carbon branch on the carbon next to the end of a continuous chain. The skeleton you have chosen has either (a) a two-carbon branch on the carbon next to the end of the chain, or (b) a single-carbon branch in the middle. Whether it is (a) or (b) depends on your viewpoint.

Continuous chain in capitals; branch in small letters.

$$C - C - C - C - C \qquad C - C - C - C \qquad C - C - C - C - C$$
$$\qquad\quad | \qquad\qquad\qquad\qquad | \qquad\qquad\qquad\qquad |$$
$$\qquad\quad C \qquad\qquad\qquad\qquad c \qquad\qquad\qquad\qquad c$$
$$\qquad\qquad\qquad\qquad\qquad\qquad\quad |$$
$$\qquad\qquad\qquad\qquad\qquad\qquad\quad c$$

Your Answer   *(a)*   *(b)*

D _____

Wrong. Your answer is that the skeleton

$$\qquad\qquad C$$
$$\qquad\qquad |$$
$$C - C - C - C - C$$
$$\qquad\qquad |$$
$$\qquad\qquad C$$

represents an iso- compound. By definition, an iso- compound has a single carbon branch on the carbon next to the end of a continuous chain. No matter how you choose the continuous chain in this skeleton, there are *two* carbon atoms that cannot be included. Therefore, it does not represent an iso-compound. With this definition of an iso- compound in mind, go back and select another answer.

The systematic names of all the alkanes are based on the number of carbon atoms in the longest continuous chain of carbon atoms. If the longest chain contains four carbon atoms, the compound is named as a butane. If it has five, it is named as a pentane, and so forth. For an illustration, look at the skeleton of isopentane again.

$$C - C - C - C$$
$$\qquad\quad |$$
$$\qquad\quad C$$

The longest continuous chain has four carbon atoms. Therefore, its systematic name will be based on butane.

The compound represented by this skeleton is named as a propane.

```
      C
      |
  C - C - C
      |
      C
```

There are four ways to count a chain of three carbon atoms, but no way to make a chain of four or five.

```
      C             C 1           C            C 3
  1   | 2 3         | 2       3   | 2 1        | 2
  C - C - C     C - C - C    C - C - C     C - C - C
      |             |            |             |
      C             C 3          C             C 1
```

You have seen this skeleton before. Even though its systematic name is based on propane, another acceptable name is neopentane.

Following are some carbon skeletons. Count the carbon chains in each skeleton, find the longest continuous one, and determine the name of the parent hydrocarbon. (The answers are given together below the questions.)

1.
```
          C
          |
  C - C - C - C
          |
          C
```

2.
```
  C - C - C - C
  |   |   |   |
  C   C   C   C
```

3.
```
  C   C - C - C
  |   |   |
  C - C   C - C
```

4.
```
  C           C
  |           |
  C - C - C
  |           |
  C           C
```

5.
```
      C           C
      |           |
  C - C - C - C - C
      |
      C
```

•••

1. Four carbon atoms in chain: butane

2. Six carbon atoms in chain: hexane

3. Seven carbon atoms in chain: heptane

4. Five carbon atoms in chain: pentane

5. Five carbon atoms in chain: pentane

Now let's look at two partially condensed structural formulas:

$$CH_3-CH_2-\overset{*}{C}H_2-CH_3 \qquad CH_3-CH_2-\underset{\underset{CH_3}{|}}{CH}-CH_3$$

Both have four carbon atoms in the longest chain and are named as butanes. How do they differ? Can you see that the difference is the replacement of one of the H's on the *-marked carbon atom by CH_3-? Replacement of this sort is called substitution, and CH_3- is here the substituent group. So, the second compound above is a substituted butane. Now we need a way to designate the CH_3-.

Groups of atoms that have an unshared electron are called radicals in the IUPAC rules. When a substituent group is in question, the unshared electron is represented by $-$ as in CH_3-. For free radicals it is represented by a $\cdot$, as in $\cdot CH_3$. To avoid confusion between the two, most chemists prefer to call the former *groups* and the latter *free radicals* (see p. 125).

CH_3- is a group that might be formed by removing a hydrogen atom from methane, CH_4. Likewise the group CH_3CH_2- could be formed from the normal alkane, ethane. A whole series of these groups can be formed from the normal alkanes. Collectively, they are known as *alkyl* [ăl′ kĭl] *groups*. They are named by changing the *-ane* suffix to *-yl*. For example

CH_3-	CH_3CH_2- or C_2H_5-	$CH_3CH_2CH_2-$ or C_3H_7-
methyl	*ethyl*	*propyl*

No hyphens are used in fully condensed, or line, formulas. Hence the line formulas for the two compounds shown above are $CH_3(CH_2)_2CH_3$ and $CH_3CH_2CH(CH_3)_2$.

What is the name of the group represented by all three of these formulas?

$$CH_3CH_2CH_2CH_2- \qquad CH_3(CH_2)_2CH_2- \qquad C_4H_9-$$

A _____

butyl

B _____

pentyl

C _____

diethyl

• •

A _____

You are right. All three of the formulas represent the group that remains when one hydrogen atom is removed from butane. Compare each formula before you go on to the next section.

Butane	$CH_3CH_2CH_2CH_3$	$CH_3(CH_2)_2CH_3$	C_4H_{10}
Butyl group	$CH_3CH_2CH_2CH_2-$	$CH_3(CH_2)_2CH_2-$	C_4H_9-

B _____

You are incorrect. Perhaps you need to review the names of the alkanes. Here is a list of their names, together with the number of carbon atoms in each:

1	methane	7	heptane
2	ethane	8	octane
3	propane	9	nonane
4	butane	10	decane
5	pentane	11	undecane
6	hexane	12	dodecane

With these clearly in mind, turn back and read page 32 carefully before you select another answer.

C _____

Where did you pick up the term *diethyl*? Certainly not in this program. Go back and choose an answer that uses something you have seen before in the program.

The groups derived from isoalkanes are named in the same way as those derived from normal alkanes. The -*ane* suffix of the name is changed to -*yl*. For example

$$CH_3 - CH - CH_3$$
$$\vert$$
$$CH_3$$

isobutane

$$CH_3 - CH - CH_2 -$$
$$\vert$$
$$CH_3$$

isobutyl group

$$CH_3 - CH - CH_2 - CH_3$$
$$\vert$$
$$CH_3$$

isopentane

$$CH_3 - CH - CH_2 - CH_2 -$$
$$\vert$$
$$CH_3$$

isopentyl group

Note carefully that the free valence of the groups is at the opposite end of the chain from the branch and that the branch, as before, is a single carbon atom attached to the carbon next to the end of the longest continuous chain.

There is one iso- group that does not have a corresponding isoalkane. Can you guess what it is? It's the isopropyl group. The formula of the isopropyl group is

$$CH_3 - CH -$$
$$\vert$$
$$CH_3$$

isopropyl group

You can see that the corresponding hydrocarbon is propane.

Which of the following formulas correctly represents the isohexyl group?

A _____

$$- CH_2 - CH - CH_2 - CH_2 - CH_2 -$$
$$\vert$$
$$CH_3$$

B _____

$$CH_3 - CH - CH_2 - CH_2 - CH_2 -$$
$$| $$
$$CH_3$$

C _____

$$CH_3 - CH - CH_2 - CH - CH_3$$
$$| \qquad\qquad |$$
$$CH_3$$

● ●

A _____

You made a mistake. Here is the formula you have chosen for the isohexyl group

$$- CH_2 - CH - CH_2 - CH_2 - CH_2 -$$
$$| $$
$$CH_3$$

It has the right skeleton and a free valence is shown on the proper carbon atom, but it is still not right. There are two hydrogens missing instead of one. Do you see that? When you do, go back and find the right answer.

B _____

You are right. Isohexane contains a total of six carbon atoms and the isohexyl group is formed by removing one hydrogen atom from the carbon atom at the end opposite the branch. Go on to the next section.

C _____

You are wrong. The formula you chose for the isohexyl group is

$$CH_3 - CH - CH_2 - CH - CH_3 \cdot$$
$$| \qquad\qquad |$$
$$CH_3$$

It has the right skeleton, but the free valence is in the wrong place. The free valence on iso- groups is located at the terminal (end) carbon atom opposite from the branch. Go back and select another answer.

The formulas for alkyl groups can always be recognized by the dash indicating the free valence. Again notice the difference between propane and a propyl group:

$CH_3 CH_2 CH_3$ $\qquad\qquad\qquad\qquad\qquad$ $CH_3 CH_2 CH_2 -$ or $C_3 H_7 -$

propane $\qquad\qquad\qquad\qquad\qquad\qquad\qquad\qquad$ *propyl group*

When groups of this sort are attached to a carbon atom in a carbon chain, they are written after the carbon atom in the structural formula and enclosed in parentheses if needed for clarity.

$$\qquad\qquad\qquad\qquad\qquad\qquad\qquad\qquad\qquad CH_3 \qquad\qquad CH_3$$
$$\qquad\qquad\qquad\qquad\qquad\qquad\qquad\qquad\qquad | \qquad\qquad\quad |$$
$$CH_3 CH(CH_3)CH_2 CH(CH_3)_2 \text{ represents } CH_3 - CH - CH_2 - CH - CH_3$$

In checking to make sure that you have interpreted a formula correctly, remember that each carbon atom forms four covalent bonds.

You may sometimes see a carbon atom that is bonded to one other carbon atom called a primary carbon atom, one bonded to two others a secondary carbon atom, one bonded to three a tertiary carbon atom, and one to four a quaternary carbon atom. The group $-CH_2-$ is named methylene [mĕth′ ĭ·lēn] or methylidene. The correct name for the $-CH{\textstyle <}$ group is methylidyne, although you may see the *incorrect* name methine [mĕth′ ĭn]. This set of formulas illustrates the terminology:

$$\overset{\displaystyle H}{\underset{\displaystyle H}{C-\overset{|}{\underset{|}{C^{*}}}-H}}$$ C* is a primary carbon atom CH_3- is methyl group

$$\overset{\displaystyle C}{\underset{\displaystyle H}{C-\overset{|}{\underset{|}{C^{*}}}-H}}$$ C* is a secondary carbon atom $-CH_2-$ is methylene group

$$\overset{\displaystyle C}{\underset{\displaystyle C}{C-\overset{|}{\underset{|}{C^{*}}}-H}}$$ C* is a tertiary carbon atom $-CH{\textstyle <}$ is methylidyne group

$$\overset{\displaystyle C}{\underset{\displaystyle C}{C-\overset{|}{\underset{|}{C^{*}}}-C}}$$ C* is a quaternary carbon atom

Find the structural formula of the compound represented by this line formula: $CH_3C(CH_3)_2CH_3$

A _____

$$\overset{\displaystyle CH_3}{\underset{\displaystyle CH_3}{CH_3-\overset{|}{\underset{|}{C}}-CH_3}}$$

B _____

$$\overset{\displaystyle CH_3}{CH_3-\overset{|}{CH}-CH_2-CH_3}$$

C _____

$$\underset{\displaystyle CH_3}{CH_3-\overset{|}{C}-CH_3-CH_3}$$

, • (

A _____

You are right. Here are the condensed and expanded structural formulas shown side-by-side:

$$CH_3 C(CH_3)_2 CH_3 \qquad\qquad CH_3 - \overset{\overset{\displaystyle CH_3}{|}}{\underset{\underset{\displaystyle CH_3}{|}}{C}} - CH_3$$

The trivial name for the compound, accepted by the IUPAC rules, is neopentane. Go on to the next section.

B _____

You are incorrect. The expanded formula which you chose does not represent the same compound as $CH_3 C(CH_3)_2 CH_3$. Let's see how your answer would be condensed. Your answer is

$$CH_3 - \underset{\underset{\displaystyle CH_3}{|}}{CH} - CH_2 - CH_3$$

Branches are indicated in condensed formulas by enclosing them in parentheses following the carbon atom(s) to which they are attached. In your answer the branch is $-CH_3$. It is attached to the second carbon atom. Therefore, the condensed formula is $CH_3 CH(CH_3)CH_2 CH_3$. You might omit the parentheses since it should be clear that the $-CH_3$ must be a branch. If it were in the chain, the carbon atom would have five covalent bonds: three to hydrogen and two to other carbon atoms. Go back and select another answer.

C _____

You are wrong. Your answer is

$$CH_3 - \overset{\#}{\underset{\underset{\displaystyle CH_3}{|}}{C}} - \overset{*}{CH_3} - CH_3$$

You seem to have forgotten that each carbon atom in an alkane forms four covalent bonds. If you will check your answer, you will see that the carbon atom indicated by # has only three, while the one marked with * has five. Go back and choose another answer.

Let's get back to isopentane and learn its systematic name. Its structural formula is

$$CH_3 - CH_2 - \underset{\underset{\displaystyle CH_3}{|}}{CH} - CH_3$$

Have you already guessed that its systematic name is methylbutane? After all, it has a methyl group, CH_3-, substituted on the butane skeleton, $CH_3 CH_2 CH_2 CH_3$. Note that the methyl group is substituted for one of the hydrogens in the butane molecule. Does the name methylbutane say everything that needs to be said? Is there another compound that could have the same name? What about this one?

$$CH_3 - \underset{\underset{CH_3}{|}}{CH} - CH_2 - CH_3$$

Although it might appear at first glance to be a different compound, you can see that this formula is the same as the first if you reverse it from left to right. Hence, methylbutane is a unique and satisfactory systematic name for the compound.

What happens when one more carbon atom is added to the end of the methylbutane chain? The new structural formula is

$$CH_3 - \underset{\underset{CH_3}{|}}{CH} - CH_2 - CH_2 - CH_3$$

The longest continuous carbon chain now contains five carbon atoms. Since a methyl group is substituted for one of the hydrogen atoms, the compound must be methylpentane. Is methylpentane a unique and sufficient name for this compound?

A _____

Yes

B _____

No

• •

A _____

You are incorrect. Your answer is that methylpentane is a unique and sufficient name for

$$CH_3 - \underset{\underset{CH_3}{|}}{CH} - CH_2 - CH_2 - CH_3$$

In order for this name to be a sufficient one, there must be no other compound that is a methylpentane. Consider this formula:

$$CH_3 - CH_2 - \underset{\underset{CH_3}{|}}{CH} - CH_2 - CH_3$$

It's methylpentane, too, isn't it? Can you superimpose it on the other formula? No, you cannot. Consequently, methylpentane alone is not a sufficient name for either one of them. Continue to the next section to learn how each is named.

B _____

Correct. Methylpentane is not a unique and sufficient name for the compound represented by the formula

$$CH_3 - CH - CH_2 - CH_2 - CH_3$$
$$|$$
$$CH_3$$

There is a different compound that could also be called methylpentane. Its formula is

$$CH_3 - CH_2 - CH - CH_2 - CH_3$$
$$|$$
$$CH_3$$

Continue to the next section to learn how to find a unique name for each of these two compounds.

The skeletons of the two methylpentanes are

$$C - C - C - C - C \quad \text{and} \quad C - C - C - C - C$$
$$| \qquad\qquad\qquad\qquad |$$
$$C \qquad\qquad\qquad\qquad\quad C$$

Clearly something more is needed to give each a unique name.

The distinction is made by the use of numbers. The carbon atoms in the chain are numbered consecutively, starting with 1 at the end of the chain nearer the substituent group. Study these examples:

1 2 3 4 5		1 2 3 4 5
or		but not
5 4 3 2 1		5 4 3 2 1

$$C - C - C - C - C \qquad\qquad\qquad C - C - C - C - C$$
$$| \qquad\qquad\qquad\qquad\qquad\qquad\qquad |$$
$$C \qquad\qquad\qquad\qquad\qquad\qquad\qquad C$$

3-methylpentane   *2-methylpentane*
                                          but not
                                   4-methylpentane

The numerals are used to locate the position of the methyl group on the pentane chain. Such numerals are termed locants. Numbering begins at the end nearest the branch so that the lowest number(s) will be used.

What is the systematic name for this compound?

$$C - C - C - C - C - C$$
$$|$$
$$C$$

A _____

3-methylhexane

B _____

4-methylhexane

C _____

isoheptane

• •

A _____

You are right. The systematic name for the compound with the skeleton

$$C - C - C - C - C - C$$
$$| $$
$$C$$

is 3-methylhexane. The carbon chain is numbered from the end nearer the branch. Go on to the next section.

B _____

You are incorrect. Your answer is that the compound having the skeleton

$$C - C - C - C - C - C$$
$$|$$
$$C$$

is 4-methylhexane. You have numbered the carbon chain from the wrong end. Always start at the end nearer the branch. For example

$$C - C - C - C - C - C - C$$
$$|$$
$$C$$

represents 3-methylheptane, *not* 5-methylheptane. Go back and select another answer.

C _____

You are wrong. Isoheptane is not the name of the compound represented by the skeleton

$$C - C - C - C - C - C$$
$$|$$
$$C$$

Let's see why. You should recall that iso- compounds have a branch at the carbon atom next to the end of the carbon chain. The skeleton of isoheptane is

$$C - C - C - C - C - C$$
$$|$$
$$C$$

How would you name isoheptane systematically? If the carbon chain is numbered consecutively from the end nearer the branch, the branch is seen to be attached to the number 2 carbon atom in the chain. Isoheptane is 2-methyl-hexane. Go back and pick another answer.

You may think that you have spent a long time learning to count carbon chains and number them. Nevertheless, this ability is vital to success in learning the nomenclature for other families of organic compounds. Your progress should accelerate from this point onward.

Now you need some practice in naming substituted or branched alkanes. Remember the three steps you have learned:

1. Find the longest continuous carbon chain.

2. Establish locants for substituent groups by numbering the carbon atoms consecutively from the end nearer to the branch.

3. Give the locant of the substituent group and name it according to the number of carbon atoms it contains.

Following are shown five carbon skeletons. The correct names are listed together following the carbon skeletons. If you miss more than one, you should go back and review.

1.
```
C - C   C - C   C
|   |   |   |
C - C   C - C
```

2.
```
C - C - C - C
        |
        C
        |
        C
```

3.
```
C - C - C - C - C - C
            |
            C
            |
            C
```

4.
```
C - C - C
        |
        C - C - C
        |       |
        C       C
        |
        C - C
```

5.
```
C - C - C - C - C - C - C - C - C
                |
                C
                |
                C
                |
                C
                |
                C
```

• •

1. nonane

2. 3-methylpentane

3. 3-ethylhexane

4. 4-propylheptane (Note: There are six different ways to number the chain. Each leads to the same answer.)

5. 5-butylnonane (Note: There are six ways to number the chain. All six lead to the same name.)

Suppose you encounter an alkane with two branches, like this one, for instance:

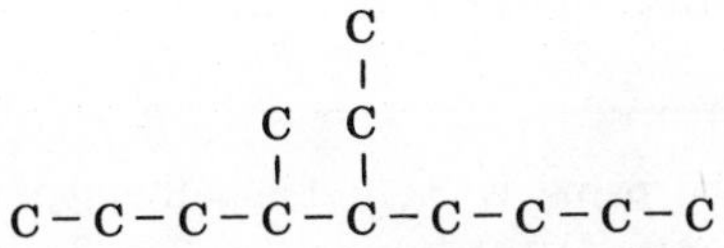

```
              C
              |
        C     C
        |     |
  C - C - C - C - C - C - C - C - C
```

Begin the same way. Find the longest carbon chain and assign locants from the end nearer to any branch. Thus

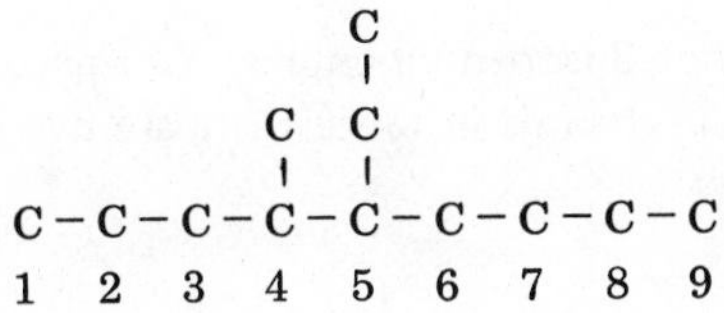

```
              C
              |
        C     C
        |     |
  C - C - C - C - C - C - C - C - C
  1   2   3   4   5   6   7   8   9
```

This compound is a substituted nonane. Now name and locate each of the substituent groups. This is 5-ethyl-4-methylnonane. Why isn't it 4-methyl-5-ethylnonane? Because the IUPAC rules prefer that substituents be recited in alphabetical order. Occasionally you will find them recited in order of the increasing size of the substituent groups. Of course, either way yields a unique name. The second way does not lend itself to satisfactory indexing, though.

How would you name this compound?

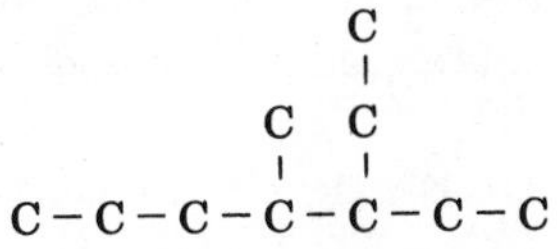

```
              C
              |
        C     C
        |     |
  C - C - C - C - C - C - C
```

A _____

4-methyl-5-ethylheptane

B _____

3-ethyl-4-methylheptane

C _____

3-isopentylpentane

D _____

isodecane

• •

A _____

You are wrong. Your name for the skeleton

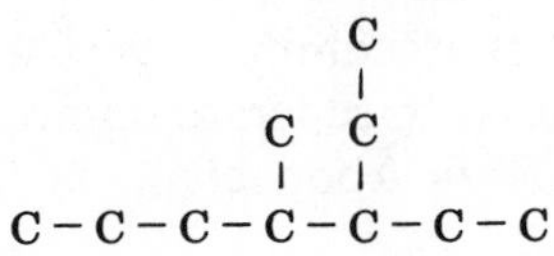

```
              C
              |
        C     C
        |     |
  C - C - C - C - C - C - C
```

is 4-methyl-5-ethylheptane. Most of your answer is right. It is a heptane and the branches are methyl and ethyl. Your first error was to number the heptane chain from the wrong end. Your second was to list the branches in order of size rather than alphabetically. Correct these two mistakes and try again.

B _____

Right you are. The name is 3-ethyl-4-methylheptane. The longest chain contains seven carbon atoms; the branches are an ethyl and a methyl group. They have the locants 3 and 4. Ethyl precedes methyl alphabetically and is named first. Go on to the next section.

C _____

No. The name is not 3-isopentylpentane. You need some more practice in finding the longest chain. In this instance there are two ways to number a chain of seven carbon atoms:

```
            C                              C 1
            |                              |
        C   C                          C   C 2
        |   |                          |   |
    C - C - C - C - C - C - C      C - C - C - C - C - C - C
    7   6   5   4   3   2   1      7   6   5   4   3
```

Go back and read page 41 carefully before you choose another answer.

D _____

You are wrong. First of all, an iso- compound is a singly branched compound with a methyl group substituted on the carbon atom next to the end of the chain. Isodecane has a total of ten carbon atoms with this skeleton:

```
    C - C - C - C - C - C - C - C - C
                                |
                                C
```

Read page 41 carefully again before choosing another answer.

Learn a few fine points and you will be able to name alkanes with the experts. Whenever there are two or more of the same groups substituted on the longest chain, the fact is indicated by the appropriate prefix: *di-* for two of them, *tri-* for three of them, *tetra-* for four, *penta-* for five, and so on. To illustrate, consider the compound commonly, but incorrectly, termed "isooctane." It is used as an antiknock additive in gasoline. Its carbon skeleton is

```
            C       C
            |       |
        C - C - C - C - C
            |
            C
```

According to our system, it is 2,2,4-trimethylpentane. The "tri-" indicates three methyl groups and the "2,2,4-" defines where they are — that is, it is a list of locants. Notice, too, that commas are used to separate numbers from numbers while hyphens are used to separate numbers from words. You can find several ways to number a five-carbon chain in "isooctane." Once you have made a choice, you must stick with it. Don't use one set to locate one branch and another set for a second branch. Here is another skeleton for you to name.

```
              C
              |
      C   C   C
      |   |   |
  C - C - C - C - C - C - C
```

It is 5-ethyl-2,4-dimethylheptane. Right? Note that the prefixes such as *di-* and *tri-* are *not* considered to be part of the group names when determining the proper alphabetical order.

What name would you give to this alkane?

```
          C           C
          |           |
      C   C   C       C
      |   |   |       |
  C - C - C - C - C - C - C - C - C - C
      |           |
      C           C
                  |
                  C
                  |
                  C
```

A _____

4,7-ethyl-2,2,3-methyl-5-propyldecane

B _____

4,7-diethyl-2,2,3-trimethyl-5-propyldecane

C _____

4,7-diethyl-8,9,9-trimethyl-6-propyldecane

• •

A _____

Close but no cigar. You missed the important point on the previous page. If there are two or more identical substituent groups, you must signal the fact with a prefix. For example

```
      C   C
      |   |
  C - C - C - C - C - C
```

This is 3,4-dimethylhexane, *not* just 3,4-methylhexane. The *di-* indicates that there are two methyl groups. Go back and choose the correct answer.

B _____

You are correct. You can have confidence now that you can name just about any alkane. Go to the next section.

C _____

Come on now. Once you have located the longest carbon chain, number it from the end *nearest any branch.* You did not do this.

```
        C           C                        C           C
        |           |                        |           |
    C   C   C       C                    C   C   C       C
    |   |   |       |                    |   |   |       |
C - C - C - C - C - C - C - C - C - C    C - C - C - C - C - C - C - C - C - C
    |           |                            |           |
    C           C                            C           C
                |                                        |
                C                                        C
                |                                        |
                C                                        C

10  9  8  7  6  5  4  3  2  1        1  2  3  4  5  6  7  8  9  10
        Your numbers                         Correct numbers
```

Can you see your mistake? The two methyl groups are closer to the end than the ethyl group. Go back and select another answer.

There is an easy way to check names to see if they are complete. Just add the number of carbon atoms specified by the name and see if it is the same as the total number in the formula. Let's return to a compound you saw a couple of pages back. Its skeleton was

```
                    C
                    |
        C       C   C
        |       |   |
    C - C - C - C - C - C - C
```

You named it 5-ethyl-2,4-dimethylheptane. Here is the way to check it:

Fragment of name	Number of carbon atoms
5-ethyl	2
2,4-dimethyl	2 (1 for each methyl)
heptane	7
	11

There are 11 carbon atoms in the skeleton, too.

A final note about writing names. All numerals are set off from group names by hyphens. Consecutive numerals are separated by commas. The last-named group and the base name are written without a space between them.

So far we have concentrated on figuring out the names from given formulas. It ought to be easy for you to reverse the process now. Which of these skeletons represents 5-ethyl-2,4-dimethyloctane?

A _____

```
                    C
                    |
        C       C   C
        |       |   |
    C - C - C - C - C - C - C
```

B _____

```
                  C
                  |
      C - C - C - C - C - C
              |       |
              C       C - C
              |       |
              C       C
```

C _____

```
                  C
                  |
      C - C - C   C - C - C
          |   |   |
      C - C - C - C   C
```

••

A _____

You are incorrect. From the name 5-ethyl-2,4-dimethyloctane you know that the longest carbon chain must have eight atoms. What is the longest chain in your answer?

```
                  C
                  |
      C       C   C
      |       |   |
      C - C - C - C - C - C - C
```

Only seven. This is the skeleton of 5-ethyl-2,4-dimethylheptane. Do you agree? When you do, go back and work out another answer.

B _____

You are right. The skeleton has a chain of eight carbon atoms and the branches are of the proper size and location to represent 5-ethyl-2,4-dimethyloctane. Note that the consecutive numbers are separated by commas, but that numbers are separated from group names by hyphens. The final group and the base name are combined into dimethyloctane. Go on to the next section.

C _____

You are wrong. The skeleton you have chosen does have a chain of eight carbon atoms and does represent an octane.

```
      8   7   6   3   C 2 1
                      |
      C - C - C   C - C - C
              |   |   |
      C - C - C - C   C
          5   4
```

However, it does not represent 5-ethyl-2,4-dimethyloctane. The name specifies that the methyl groups are in the 2 and 4 positions. In this skeleton they are both on number 2. Go back and find the completely correct answer.

In all the examples so far cited it's been easy to choose the number 1 carbon atom. What happens when there is a substituent group the same distance from each end of the chain, as in

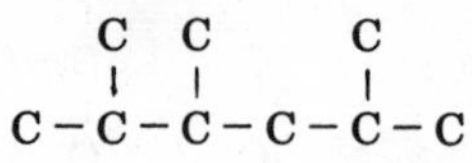

```
          C   C       C
          |   |       |
      C - C - C - C - C - C
```

According to the IUPAC rules, sets of numerical locants are compared term by term. The set that contains the lowest number at the point of the *first difference* is considered "lowest" and the one to be used for the name. This principle is applied irrespective of the nature of the substituent groups. The compound shown above is named 2,3,5-trimethylhexane, not 2,4,5-trimethylhexane, because the correct name has a "3" at the first point of difference in the set of locants while the other has a "4."

Here is another example to illustrate the principle:

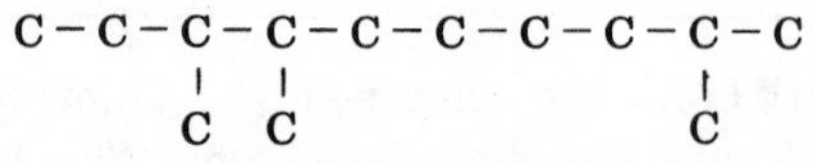

```
      C - C - C - C - C - C - C - C - C - C
          |   |                   |
          C   C                   C
```

2,7,8-trimethyldecane
(not 3,4,9-trimethyldecane)

It doesn't matter that the sum of the locants in the incorrect set is 16, as compared to 17 in the correct set. At the first point of difference it's 2 vs. 3. A final example is

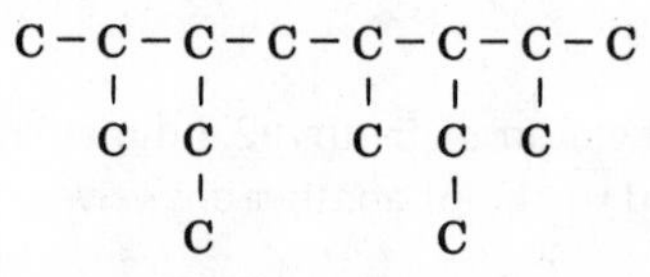

```
      C - C - C - C - C - C - C - C
          |   |       |   |   |
          C   C       C   C   C
              |           |
              C           C
```

3,6-diethyl-2,4,7-trimethyloctane

It is not 3,6-diethyl-2,5,7-trimethyloctane because the set 2,3,4,6,7 is lower than 2,3,5,6,7.

What is the correct systematic name for the alkane with this skeleton?

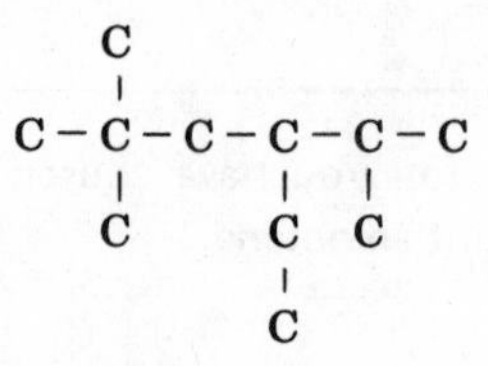

```
              C
              |
      C - C - C - C - C - C
          |       |   |
          C       C   C
                  |
                  C
```

A _____

3-ethyl-2,5,5-trimethylhexane

B _____

4-ethyl-2,2,5-trimethylhexane

C _____

4-isopropyl-2,2-dimethylhexane

A _____

This is certainly an adequate systematic name for the compound

$$
\begin{array}{cccccc}
 & \overset{\text{C}}{|} & & & & \\
6 & 5\ 4 & 3 & 2 & 1 & \\
\text{C} - \text{C} - \text{C} - \text{C} - \text{C} - \text{C} & & & & & \\
 & | & & | & | & \\
 & \text{C} & & \text{C} & \text{C} & \\
 & & & | & & \\
 & & & \text{C} & &
\end{array}
$$

3-ethyl-2,5,5-trimethylhexane

It's not the proper name according to the IUPAC rules, though. Compare it to the name in B to see that 2,2,4,5 is lower than 2,3,5,5.

B _____

Correct. 4-ethyl-2,2,5-trimethylhexane is the proper systematic name for this compound.

$$
\begin{array}{cccccc}
1 & 2 & 3 & 4 & 5 & 6 \\
 & \overset{\text{C}}{|} & & & & \\
\text{C} - \text{C} - \text{C} - \text{C} - \text{C} - \text{C} & & & & & \\
 & | & & | & | & \\
 & \text{C} & & \text{C} & \text{C} & \\
 & & & | & & \\
 & & & \text{C} & &
\end{array}
$$

4-ethyl-2,2,5-trimethylhexane

Go on to the next section.

C _____

Your answer is that 4-isopropyl-2,2-dimethylhexane is the correct name for this compound

$$
\begin{array}{cccc}
1 & 2 & 3 & 4 \\
 & \overset{\text{C}}{|} & & \\
\text{C} - \text{C} - \text{C} - \text{C} - \text{C} - \text{C} \\
 & | & & |\quad | \\
 & \text{C} & & 5\text{C}\quad \text{C} \\
 & & & | \\
 & & & 6\ \text{C}
\end{array}
$$

On the basis of what has been discussed so far, it would seem to be all right. There is another principle, however, that renders it incorrect. Go on to the next section to learn it.

Frequently there are several chains of equal length that might be chosen as the main chain on which to base the name of an alkane. Take, for example, the skeleton used for the last question. A chain of six carbon atoms could be found in these two ways:

$$
\begin{array}{cccc}
1\ 2 & 3 & 4 & \\
 & \overset{\text{C}}{|} & & \\
\text{C} - \text{C} - \text{C} - \text{C} - \text{C} - \text{C} \\
 & | & & |\quad | \\
 & \text{C} & & 5\text{C}\quad \text{C} \\
 & & & | \\
 & & & 6\ \text{C}
\end{array}
\qquad
\begin{array}{cccccc}
1\ 2 & 3 & 4 & 5 & 6 \\
 & \overset{\text{C}}{|} & & & \\
\text{C} - \text{C} - \text{C} - \text{C} - \text{C} - \text{C} \\
 & | & & |\quad | \\
 & \text{C} & & \text{C}\quad \text{C} \\
 & & & | \\
 & & & \text{C}
\end{array}
$$

In this situation the chain selected is the one with the largest number of side chains. The one on the right has four side chains and is chosen over the one on the left, which has three.

Here's another example:

```
C – C – C – C – C – C – C          C – C – C – C – C – C – C
        |   |   |   |                        |   |   |   |
        C   C   C   C                        C   C   C   C
        |                                    |
        C – C                                C – C
```

4 side chains   *2 side chains*

Now, what is the name of this substituted heptane?

A _____

2,3,5-trimethyl-4-propylheptane

B _____

2-ethyl-4,5-dimethyl-3-propylhexane

C _____

3,5,6-trimethyl-4-propylheptane

• •

A _____

Right. You used the correct main chain, numbered it from the end nearer a side chain, and arrived at 2,3,5-trimethyl-4-propylheptane as your answer. Continue on to the next section.

B _____

Wrong. If you were to draw a skeleton to match your name of 2-ethyl-4,5-dimethyl-3-propylhexane, you'd get the same one shown with the question. Your mistake lies in not finding the longest continuous chain to use as the main chain. Go back and work out another answer.

C _____

Not quite. Your choice of 3,5,6-trimethyl-4-propylheptane does describe the compound with this skeleton

```
C – C – C – C – C – C – C
        |   |   |   |
        C   C   C   C
        |
        C – C
```

However, you've numbered the main chain from the wrong end. If you start from the other end, you should find the name to be 2,3,5-trimethyl-4-propylheptane. If you do, go on to the next section.

Following are some drill exercises for you. The correct answers are given together following the exercises.

Where a formula is given, write the name of the alkane. Where a name is given, write the skeleton formula.

1. 2-methylhexane

2.

```
              C
              |
         C — C — C
              |
              C
```

3.

```
     C — C — C — C — C — C — C — C — C
         |   |                   |
         C   C                   C
```

4. 5-methyl-4-propylnonane

5. $CH_3(CH_2)_3CH(CH_3)_2$

6.

```
                   C
                   |
         C     C   C
         |     |   |
    C — C — C — C — C — C
         |         |
         C         C
```

●●●

1.

```
              C
              |
    C — C — C — C — C — C
```

2. 2,2-dimethylpropane, or neopentane

3. 2,3,8-trimethylnonane, *not* 2,7,8-trimethylnonane

4.

```
    C — C — C — C — C — C — C — C — C
                |   |
                C   C
                    |
                    C — C
```

5.

```
                                        C
                                        |
    2-methylhexane or isoheptane. The skeleton is C — C — C — C — C — C
```

6. 3-ethyl-2,2,5,5-tetramethylhexane, *not* 4-isopropyl-2,2-dimethylhexane, *and not* 4-ethyl-2,2,5,5-tetramethylhexane

Branched alkyl groups are named in a manner similar to branched alkanes. An important restriction is that the carbon atom with the available bond must be numbered 1. For example,

```
              CH₃
              |
CH₃CH₂CH₂CH₂CH —                    1-methylpentyl
```

$$CH_3CH_2CH_2CH_2\overset{\displaystyle CH_3}{\underset{|}{CH}}- \qquad \text{1-methylpentyl}$$

$$CH_3\overset{\displaystyle CH_3}{\underset{|}{CH}}CH_2\overset{\displaystyle CH_3}{\underset{|}{CH}}- \qquad \text{1,3-dimethylbutyl}$$

Trivial names are retained under the IUPAC rules for the following groups so long as they remain unsubstituted:

Structure	Trivial name	Systematic name
CH_3CH- $\quad\vert$ $\quad CH_3$	isopropyl	1-methylethyl
CH_3CHCH_2- $\quad\vert$ $\quad CH_3$	isobutyl	2-methylpropyl
CH_3CH_2CH- $\quad\quad\vert$ $\quad\quad CH_3$	*sec*-butyl	1-methylpropyl
$\quad\quad CH_3$ $\quad\quad\vert$ CH_3-C- $\quad\quad\vert$ $\quad\quad CH_3$	*tert*-butyl or *t*-butyl	1,1-dimethylethyl
$CH_3CHCH_2CH_2-$ $\quad\vert$ $\quad CH_3$	isopentyl	3-methylbutyl
$\quad\quad CH_3$ $\quad\quad\vert$ CH_3-C-CH_2- $\quad\quad\vert$ $\quad\quad CH_3$	neopentyl	2,2-dimethylpropyl
$\quad\quad CH_3$ $\quad\quad\vert$ CH_3CH_2C- $\quad\quad\vert$ $\quad\quad CH_3$	*tert*-pentyl or *t*-pentyl	1,1-dimethylpropyl
$CH_3CHCH_2CH_2CH_2-$ $\quad\vert$ $\quad CH_3$	isohexyl	4-methylpentyl

The end of the line for "iso-" branched alkyl groups is isohexyl, just as isohexane is the largest permitted "iso-" alkane. Take a few moments to study these names. Although you won't see them much in this book, your organic chemistry textbook is apt to use them frequently.

You should easily be able to give the correct name for the group whose trivial name is isoheptyl:

$$CH_3CH(CH_2)_3CH_2-$$
$$\vert$$
$$CH_3$$

A _____

5,5-dimethylpentyl

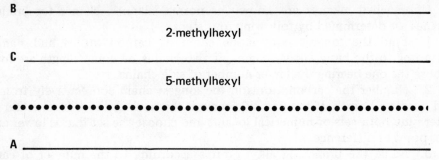

B

2-methylhexyl

C

5-methylhexyl

• •

A

Wrong. The name 5,5-dimethylpentyl would surely lead you to this formula,

$$CH_3 CH(CH_2)_3 CH_2-$$
$$|$$
$$CH_3$$

but it ignores the first rule in naming alkyl groups. Find the longest continuous chain beginning with the carbon atom having the available bond. Look the chain over carefully before you go back to choose another answer.

B

Incorrect. This is a methylhexyl group all right,

$$CH_3 CH(CH_2)_3 CH_2-$$
$$|$$
$$CH_3$$

but you started numbering from the wrong end of the chain. For branched alkyl groups the carbon atom with the available bond must be numbered 1. The correct answer is 5-methylhexyl. Go on to the next section.

C

You are right. The correct systematic name for this branched alkyl group is 5-methylhexyl.

$$CH_3 CH(CH_2)_3 CH_2-$$
$$|$$
$$CH_3$$

Summary: Naming of Alkanes

Alkanes are saturated aliphatic hydrocarbons. When two or more different compounds have the same molecular formula, they are called isomers. Alkanes with four or more carbon atoms have isomers that differ in the sequence of atoms in the molecule. These are known as constitutional isomers. Each has its own characteristic chemical and physical properties.

The general formula for the alkanes is $C_n H_{2n+2}$. The names of the first 10 normal isomers of the homologous series are methane, ethane, propane, butane, pentane, hexane, heptane, octane, nonane, and decane. All alkanes have names ending with the suffix *-ane*.

When one hydrogen atom is removed from an alkane, the resulting univalent group is named by substituting the suffix *-yl* for the *-ane* in the name of the alkane. Collectively, these are known as alkyl groups.

Unbranched alkanes are known as normal alkanes. Names for branched alkanes are determined by following four steps:

1. Find the longest continuous chain of carbon atoms and name it according to the basic alkane names. If two chains of equal length are found, choose the one bearing the larger number of side chains.

2. Number the carbon atoms in the longest chain consecutively from the end nearer a branch. If the first side chain is the same distance from either end, determine both sets of numerical locants and choose the set that is lower at the first point of difference.

3. Name the branching alkyl groups according to the number of carbon atoms each contains, list the names alphabetically, and specify the location of each group by the number of the atom to which it is attached.

4. When two or more of the same alkyl groups appear, indicate the fact by means of a prefix; the location of *each* group is shown by a number.

ALKANES

General formula: C_nH_{2n+2} Series name: *-ane*

When you are ready, take this short test.

Answer these questions one at a time and check your answers on pages 53 to 55.

1. One formula for 2,2,4-trimethylpentane is

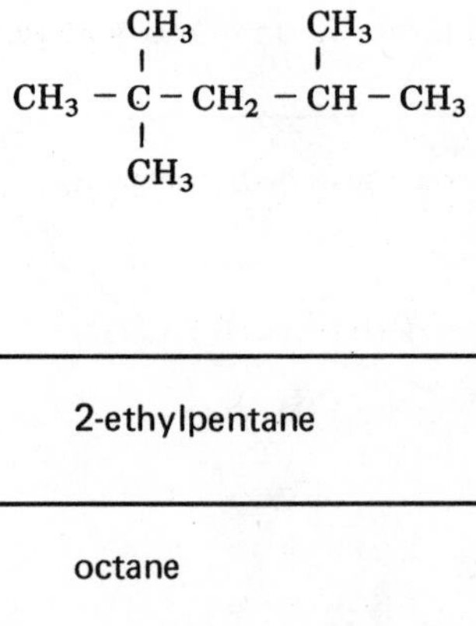

$$CH_3 - \overset{\overset{\displaystyle CH_3}{|}}{\underset{\underset{\displaystyle CH_3}{|}}{C}} - CH_2 - \overset{\overset{\displaystyle CH_3}{|}}{CH} - CH_3$$

It is an isomer of

1A _____

2-ethylpentane

1B _____

octane

1C _____

2,2,3-trimethylhexane

2. What is the name of the alkane represented by this skeleton:

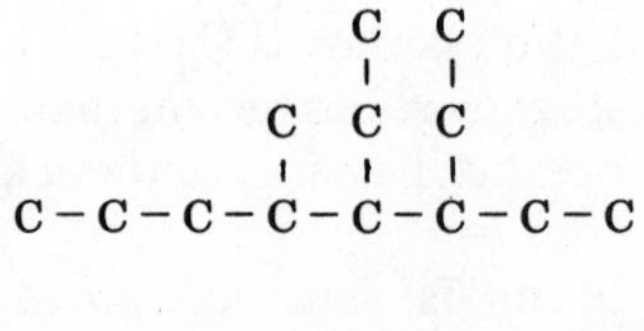

$$C-C-\underset{\underset{\displaystyle C}{|}}{C}-\underset{\underset{\displaystyle C}{|}}{\overset{\overset{\displaystyle C}{|}}{C}}-\underset{\underset{\displaystyle C}{|}}{\overset{\overset{\displaystyle C}{|}}{C}}-C-C$$

2A _____

5,6-ethyl-4-methyloctane

2B _____

3,4-diethyl-5-methyloctane

2C _____

5-methyl-3,4-diethyloctane

3. Name the alkane that has this skeleton:

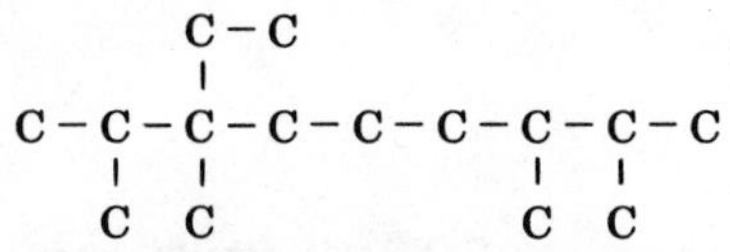

```
            C – C
              |
C – C – C – C – C – C – C – C – C
    |   |           |   |
    C   C           C   C
```

3A _____

3-ethyl-7-isopropyl-2,3-dimethyloctane

3B _____

3-ethyl-2,3,7,8-tetramethylnonane

3C _____

7-ethyl-2,3,7,8-tetramethylnonane

4. Write as many different kinds of formulas for 2,2,3,3-tetramethylbutane as you can. You should be able to write at least three.

•••

1A _____

You are incorrect. The isomer you seek must have the same number of carbon atoms as 2,2,4-trimethylpentane. Let's check the number.

Fragment of name	Number of carbons
2,2,4-trimethyl	3
pentane	$\frac{5}{8}$

You chose 2-ethylpentane as an isomer. Checking it, we find:

2-ethyl	2
pentane	$\frac{5}{7}$

Go back and choose another answer.

1B _____

You are right. Both 2,2,4-trimethylpentane and octane are alkanes containing eight carbon atoms. Their molecular formulas are C_8H_{18}. Therefore they are isomers.

1C _____

Wrong. Any isomer must have the same number of carbon atoms as the compound in question, 2,2,4-trimethylpentane. Let's check:

Fragment of name	Number of carbons
2,2,4-trimethyl	3
pentane	5
	8

Now let's check your answer, 2,2,3-trimethylhexane

2,2,3-trimethyl	3
hexane	6
	9

These cannot be isomers. Go back and choose another answer.

2A _____

You are wrong. The skeleton does represent an octane. But you have numbered the chain from the wrong end. Always start your numbers at the end nearer to a branch. One more thing: if there are more than one of the same branching group, show this by means of a prefix (*di-, tri-,* etc.). Go back and choose another answer.

2B _____

You are right. The skeleton represents 3,4-diethyl-5-methyloctane.

2C _____

Incorrect. You have named the compound correctly as an octane and identified the substituent groups. Your only error was not to name the groups alphabetically. Choose another answer.

3A _____

Incorrect. You are still having trouble finding the longest continuous chain. Go back and find one with more than eight carbon atoms.

3B _____

You are right. The skeleton represents 3-ethyl-2,3,7,8-tetramethylnonane. You chose the lower set of locants and named the substituents in alphabetical order.

3C _____

Incorrect. You are still having trouble assigning numbers to the carbon atoms in the longest chain. Go back and choose another answer carefully.

4. Here are several kinds of formulas for 2,2,3,3-tetramethylbutane:

Molecular formula $\qquad$ C_8H_{18}

Structural formula

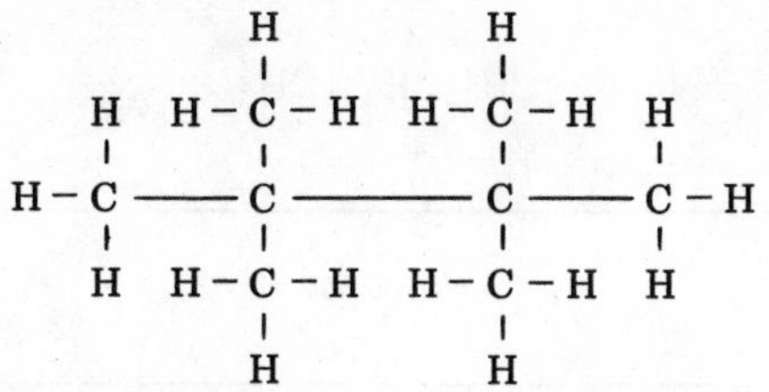

```
          H               H
          |               |
  H   H - C - H   H - C - H   H
  |       |           |       |
H - C ——— C ————————— C ——— C - H
  |       |           |       |
  H   H - C - H   H - C - H   H
          |               |
          H               H
```

Condensed structural formula

$$CH_3\,C(CH_3)_2\,C(CH_3)_2\,CH_3$$

Skeleton formula

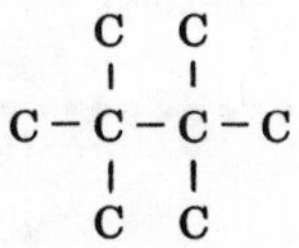

```
      C   C
      |   |
  C - C - C - C
      |   |
      C   C
```

One of several possible Newman projections

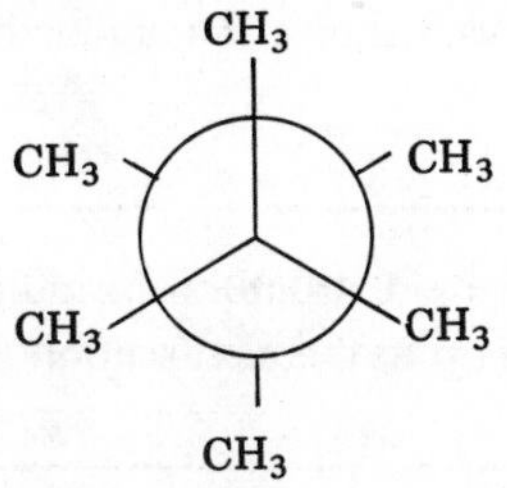

1.3 ALKENES

Once you have learned to name alkanes, the rest of your introduction to the IUPAC system of organic nomenclature is easy. Many of the common organic compounds are derived from alkanes. The next family we will consider is the *alkenes*. Simple alkenes are unsaturated aliphatic hydrocarbons that contain one carbon-carbon double bond. They are sometimes called olefins [ō' lĕ fīnz]. You can think of them as the product resulting from the removal of a molecule of hydrogen from adjacent carbon atoms of an alkane. For example

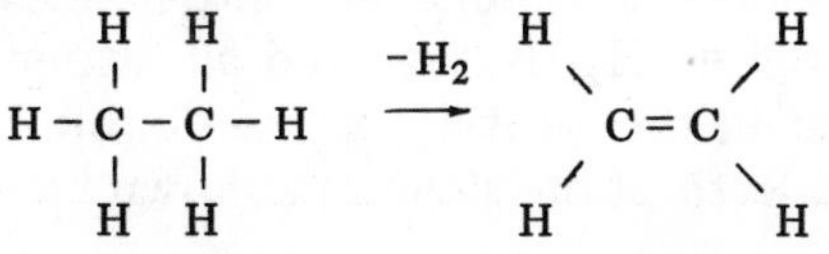

ethane   *ethene*

The names of the alkenes are formed by changing the *-ane* ending of the parent alkane to *-ene*. As you can see above, ethane becomes ethene. Obviously, there is no alkene corresponding to the simplest of the alkanes, methane.

What is the name of an alkene derived from this alkane?

$$CH_3(CH_2)_2CH_3$$

A _____

propene

B _____

butene

C _____

pentene

D _____

hexene

• •

A _____

Incorrect. Perhaps you did not look carefully at the condensed structural formula: $CH_3(CH_2)_2CH_3$. When written out more fully, it is $CH_3CH_2CH_2CH_3$, butane. If hydrogen is removed to give it a double bond, it becomes butene. Go on to the next section.

B _____

You are right. The condensed structural formula is butane. Alkenes derived from butane are butenes. Go on to the next section.

C _____

You are wrong. Look carefully at the condensed structural formula again. Does it have five carbon atoms? If not, an alkene derived from it will not be called pentene. Go back and choose another answer.

D _____

Wrong. Does the structural formula have a carbon chain of six carbon atoms? If not, an unsaturated derivative will not be a hexene. Go back and choose another answer.

Since they result from the removal of hydrogen from alkanes, the alkenes have the general formula C_nH_{2n}. Of course, n must have a value of 2 or more.

There is no chance for ambiguity in naming the first two members of the homologous series of alkenes. Ethene and propene can only correspond to $CH_2 = CH_2$ and $CH_3 - CH = CH_2$. Butene and all higher members of the series have isomers that differ in the location of the double bond. Let's look at the possibilities for butene. Both of the skeletons shown here represent butene:

$$C - C - C = C \qquad\qquad C - C = C - C$$

The solution to the problem again lies with numbering the carbon atoms. Assign numbers consecutively from the end of the chain nearer the double bond. In this case

$$C - C - C = C \qquad\qquad C - C = C - C$$
$$4 \quad 3 \quad 2 \quad 1 \qquad\qquad 4 \quad 3 \quad 2 \quad 1$$

The locant for the double bond is given by writing the lower numeral of the two carbon atoms joined by the double bond, as shown here.

$$C - C - C = C \qquad\qquad C - C = C - C$$

1-butene   *2-butene*

Which of these skeletons represents 3-heptene?

A _____

$$C - C = C - C - C - C - C$$

B _____

$$C - C - C = C - C - C$$

C _____

$$C - C - C = C - C - C - C$$

• •

A _____

Incorrect. The skeleton does represent a heptene. Look at it again.

$$C - C = C - C - C - C - C$$
$$1 \quad 2 \quad 3 \quad 4 \quad 5 \quad 6 \quad 7$$

The double bond is between the carbon atoms numbered 2 and 3. Positions of double bonds are always indicated by giving the *lower* number of the atoms which it joins. Hence, this skeleton represents 2-heptene. Go back and choose another answer.

B _____

You are wrong. Your answer is that this skeleton represents 3-heptene:

$$C - C - C = C - C - C$$

The double bond is in the 3-4 position and should be designated by the numeral 3, but how long is the chain? Only six carbons. Doesn't that make it the skeleton of 3-hexene? Sure. Now go back and choose another answer.

C _____

Right. The skeleton

$$C - C - C = C - C - C - C$$

represents a heptene since there are seven carbon atoms in the chain. The double bond joins the third and fourth atoms and is designated by the numeral 3 in the name 3-heptene. Go on to the next section.

Hydrocarbons that contain more than one double bond are known as alkadienes, alkatrienes, alkatetraënes, etc. The prefixes *di-*, *tri-*, *tetra-*, etc., are used to specify the number of double bonds. The location of each double bond is specified by giving the lower number of the two carbon atoms joined by each double bond. The chain is numbered to give the double bond(s) the lowest possible set of locants. A couple of examples will serve to make this clear:

$$C = C - C = C$$
$$1 \quad 2 \quad 3 \quad 4$$
1,3-butadiene (an ingredient of some synthetic rubbers)

$$C = C - C - C = C$$
$$1 \quad 2 \quad 3 \quad 4 \quad 5$$
1,4-pentadiene

$$C - C = C = C - C - C$$
$$1 \quad 2 \quad 3 \quad 4 \quad 5 \quad 6$$
2,3-hexadiene (*not* 3,4-hexadiene)

Compounds with one or more double bonds exhibit a kind of isomerism known as *cis-trans* or *E-Z* isomerism. This subject will be discussed in a later section (see p. 72).

Branched (substituted) alkenes are named in a manner similar to the branched alkanes. The carbon atoms are numbered, the main chain is named, and then substituent groups are named and located followed by location of the double bonds. For example,

$$\begin{array}{c} C \\ | \\ C - C - C - C = C \end{array}$$
2-methyl-1-pentene

$$\begin{array}{c} C \\ | \\ C - C = C - C = C \end{array}$$
2-methyl-2,3-pentadiene

What is the correct name for the alkene represented by this skeleton?

$$\begin{array}{c} C - C - C - C = C - C - C \\ | \\ C \end{array}$$

A ──────────────────────────────────

5-methyl-3-heptene

B ──────────────────────────────────

3-methyl-4-heptene

C ──────────────────────────────────

Help!

• •

A _____

You are right. The carbon chain is numbered from the end nearer the double bond. Then, any substituent groups are located. Consequently

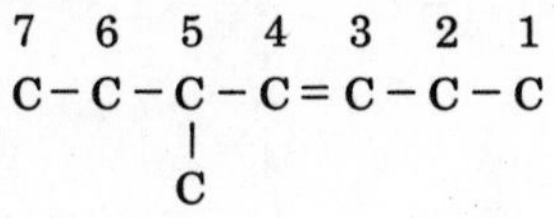

$$
\begin{array}{ccccccc}
7 & 6 & 5 & 4 & 3 & 2 & 1 \\
\end{array}
$$
$$
C-C-C-C=C-C-C
$$
$$
\underset{C}{\overset{|}{}}
$$

is the skeleton for 5-methyl-3-heptene. Numbering of the chain is shown. Go on to the next section.

B _____

Incorrect. You have numbered the carbon atoms incorrectly. Always number the chain from the end nearer the double bond as shown here.

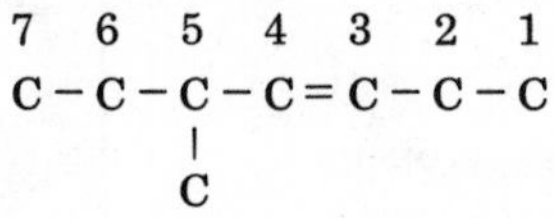

$$
\begin{array}{ccccccc}
7 & 6 & 5 & 4 & 3 & 2 & 1 \\
\end{array}
$$
$$
C-C-C-C=C-C-C
$$
$$
\underset{C}{\overset{|}{}}
$$

Then name and locate the substituent groups. Go back and choose another answer.

C _____

So you need help in naming the compound by the skeleton

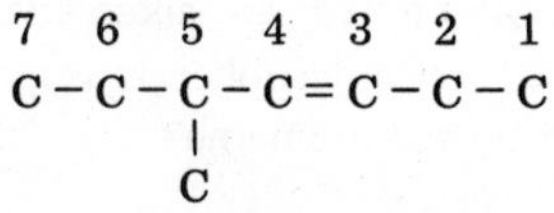

$$
\begin{array}{ccccccc}
7 & 6 & 5 & 4 & 3 & 2 & 1 \\
\end{array}
$$
$$
C-C-C-C=C-C-C
$$
$$
\underset{C}{\overset{|}{}}
$$

First of all, you can see that this is an alkene since it has a double bond. The last syllable of the name will be *-ene.* There are seven carbon atoms in the longest chain which includes the double bond, so it is a derivative of heptane. First number the chain from the end nearer the double bond and then locate the double bond and substituent group. The double bond joins the number 3 and 4 carbon atoms and its location is specified by the 3 in the name. Finally, there is a methyl group on carbon number 5. Put all of these together and you have the name. Go back and see if the name you have chosen is one of the answers on page 58.

Branched unsaturated hydrocarbons are named as derivatives of the unbranched hydrocarbons that contain the maximum number of double and triple bonds. Occasionally the longest continuous chain of carbon atoms will not be the main chain. For example,

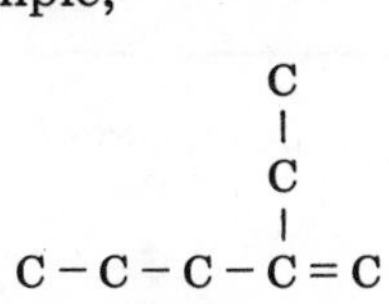

$$
\begin{array}{c}
C \\
| \\
C \\
| \\
C-C-C-C=C
\end{array}
$$

2-ethyl-1-pentene

This compound is correctly named as a pentene even though there is a continuous chain of six carbon atoms.

Whenever the longest carbon chains are of equal length, choose the one containing the maximum number of double bonds as the basis of the name. This skeleton should be named as a pent*ene* rather than a pentane:

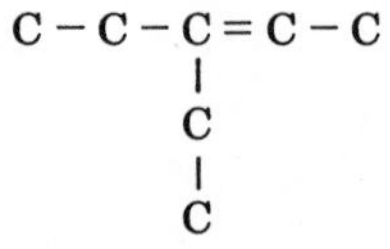

$$C - C - C = C - C$$
$$\overset{\displaystyle |}{\underset{\displaystyle |}{C}}$$
$$C$$

3-ethyl-2-pentene

There are only a few alkenes that have trivial names worth remembering. They are shown here.

Formula	Systematic name	Trivial name
$CH_2 = CH_2$	ethene	ethylene (also an acceptable systematic name)
$CH_3 CH = CH_2$	propene	propylene
$\overset{\displaystyle CH_3}{\underset{}{CH_3 C = CH_2}}$	methylpropene	isobutylene
$CH_3 CH_2 CH = CH_2$	1-butene	
$CH_3 CH = CHCH_3$	2-butene	

Natural rubber is a polymer of an alkadiene. The trivial name of the monomer, or basic unit, of natural rubber is isoprene. Its skeleton is shown below. What is its preferred systematic name?

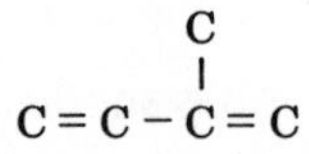

$$\overset{\displaystyle C}{\underset{}{C = C - C = C}}$$

A _____

2-methylbutadiene

B _____

2-methyl-1,3-butadiene

C _____

3-methyl-1,3-butadiene

• •

A _____

Incorrect. Your answer is that the skeleton

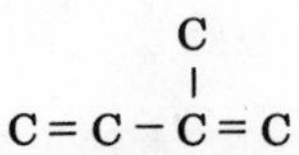

$$\overset{\displaystyle C}{\underset{}{C = C - C = C}}$$

represents 2-methylbutadiene. Everything you have written is correct, but you do not have enough. Couldn't you also call the compound represented by this skeleton by the same name?

$$
\begin{array}{c}
C \\
|\\
C = C = C - C
\end{array}
$$

The positions of the double bonds need to be specified. When you have done this, go back and choose another answer.

B _____

You are right. The systematic name for isoprene is 2-methyl-1,3-butadiene. Isoprene as well as 1,3-butadiene is used in some synthetic rubbers. Go on to the next section.

C _____

You are wrong. It appears that you numbered the carbon chain from the wrong end. Here is the skeleton again:

$$
\begin{array}{c}
C \\
|\\
C = C - C = C
\end{array}
$$

As far as the double bonds are concerned, it makes no difference about the numbers. It would be 1,3-butadiene either way. The presence of the methyl group, however, requires that you number from the end nearer to the methyl group. When you have determined the correct name, go back and choose it as your answer.

Condensed formulas for alkenes commonly show double bonds by means of a double dash (=). Although it is rarely seen, a colon is sometimes used for the same purpose. Consequently,

$$CH_2 = CHCH(CH_3)CH_2CH_3, \quad CH_2 : CHCH(CH_3)CH_2CH_3, \quad \text{and} \quad \begin{array}{c} C \\ | \\ C = C - C - C - C \end{array}$$

all represent 3-methyl-1-pentene. As always, the key is to make sure that each carbon atom has four covalent bonds.

What is the correct condensed structural formula for 3-ethyl-1,4-pentadiene?

A _____

$$CH_3 = CHCH(C_2H_5)CH = CH_3$$

B _____

$$CH_2 = CHCH(C_2H_5)CH = CH_2$$

C _____

$$CH_2CHCH(C_2H_5)CHCH_2$$

●●

A _____

Incorrect. Perhaps you did not examine the formula carefully enough. The answer you chose is $CH_3 = CHCH(C_2H_5)CH = CH_3$. It fits all the requirements for 3-ethyl-1,4-pentadiene except one. Check the number of bonds on the number 1 and number 5 carbon atoms. They would be like this:

$$
\begin{array}{c}
H \\
| \\
H - C = \\
| \\
H
\end{array}
$$

This is one bond too many on each of these carbon atoms. Go back and choose another answer.

B _____

Right. $CH_2 = CHCH(C_2H_5)CH = CH_2$ is the condensed "on-line" formula for 3-ethyl-1,4-pentadiene. Pay particular attention to the fact that each carbon atom has four and only four covalent bonds. This can be shown more clearly by writing the full structural formula.

$$
\begin{array}{c}
H \\
| \\
H - C - H \\
| \\
H - C - H \\
| \\
H - C = C - C - C = C - H \\
\quad |\quad |\quad |\quad |\quad | \\
\quad H\quad H\quad H\quad H\quad H
\end{array}
$$

C _____

You are wrong. You want the condensed structural formula for 3-ethyl-1,4-pentadiene. The *-diene* portion of the name is a signal that there should be two double bonds in the formula. The answer you have chosen, $CH_2CHCH(C_2H_5)CHCH_2$, has none at all. You can see the location of the double bonds more clearly if you write the full structural formula.

$$
\begin{array}{c}
H \\
| \\
H - C - H \\
| \\
H - C - H \\
| \\
H - C = C - C - C = C - H \\
\quad |\quad |\quad |\quad |\quad | \\
\quad H\quad H\quad H\quad H\quad H
\end{array}
$$

Where should the double bonds be placed in order to have four bonds on each carbon atom? When you have decided, go back and choose another answer.

Groups derived from alkenes are named in the same fashion as the groups derived from alkanes. The *-ene* ending of the name is changed to *-enyl*. This list illustrates the procedure:

Systematic name	Formula
ethenyl	$CH_2 = CH -$
2-propenyl	$CH_2 = CHCH_2 -$
1-propenyl	$CH_3 CH = CH -$
1-methylethenyl	$CH_2 = C -$ $\qquad\ \ \|$ $\qquad\ \ CH_3$
2-butenyl	$CH_3 CH = CHCH_2 -$
1,3-butadienyl	$CH_2 = CHCH = CH -$

It is interesting to note that propene, $CH_3 CH = CH_2$, gives rise to three alkenyl groups: 1-propenyl, 2-propenyl, and 1-methylethenyl.

The IUPAC rules permit the use of trivial names for three of the groups shown above. These trivial names were used by *Chemical Abstracts* prior to 1972, but now the systematic names are used exclusively. The three are:

Systematic name	Formula	Trivial name
ethenyl	$CH_2 = CH -$	vinyl
2-propenyl	$CH_2 = CHCH_2 -$	allyl
1-methylethenyl	$CH_2 = C -$ $\qquad\ \ \|$ $\qquad\ \ CH_3$	isopropenyl

Although the trivial names should not be found in recent literature, you will see them frequently.

The carbon atoms must be numbered for all but the ethenyl group. The carbon atom with the free valence is always number 1.

Even though you know how to name unsaturated groups, you should name hydrocarbons in such a way as to include any double bonds in the base name. For example, the compound represented by this skeleton should be named 3-ethyl-1-pentene rather than 3-ethenylpentane.

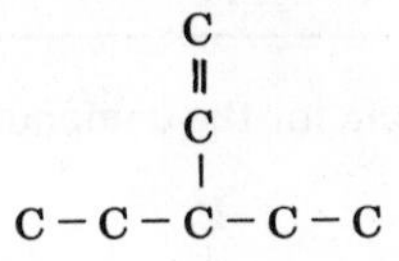

$$C$$
$$\|$$
$$C$$
$$\|$$
$$C - C - C - C - C$$

What is the preferred systematic name for this compound?

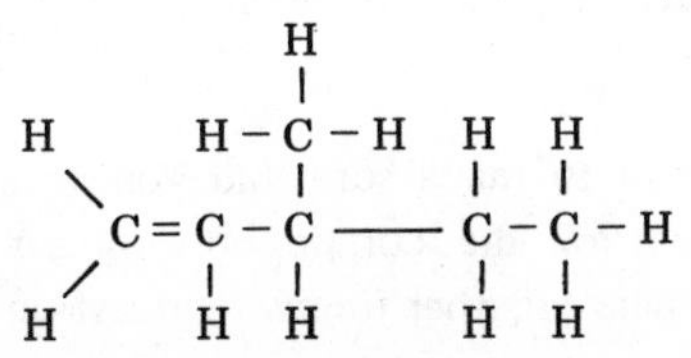

$$
\begin{array}{c}
\qquad\quad H \\
\qquad\quad | \\
H \qquad H - C - H \quad H \quad H \\
\ \ \backslash \qquad\qquad | \qquad\ \ |\quad | \\
\quad C = C - C \underline{\quad\quad} C - C - H \\
\ / \qquad | \quad | \qquad\ | \quad | \\
H \qquad H \quad H \qquad H \quad H
\end{array}
$$

A _____

ethenylbutane

B _____

2-ethenylbutane

C _____

3-methyl-1-pentene

•••

A _____

Incorrect. The preferred name for the compound represented by the skeleton

$$C = C - \overset{\overset{\textstyle C}{|}}{C} - C - C$$

is not ethenylbutane, although this name does describe it adequately. The name violates our first rule: find the longest continuous carbon chain. There is a carbon chain that includes more than four carbon atoms and contains the double bond as well. Look at the skeleton above until you see it. Then go back and choose another answer.

B _____

You are incorrect. The skeleton for the compound in question is

$$C = C - \overset{\overset{\textstyle C}{|}}{C} - C - C$$

The name you have chosen is 2-ethenylbutane. You have failed to apply the first rule for naming hydrocarbons: always look for the longest continuous carbon chain. If you look carefully, you will find a chain that has more than four carbon atoms and includes the double bond as well. You should also note that the 2 is redundant. "1-Ethenylbutane" would be simply 1-hexene. When you have found the longest carbon chain that includes the double bond, go back and choose another answer.

C _____

Correct. The skeleton formula for the compound in question is

$$C = C - \overset{\overset{\textstyle C}{|}}{C} - C - C$$

The longest carbon chain contains five carbon atoms and includes the double bond. The preferred systematic name is 3-methyl-1-pentene. Go on to the next section.

Now you need practice to make sure that you have the system down pat. Give the preferred name for the compound represented by each of these formulas. The answers appear together following question 7.

1.

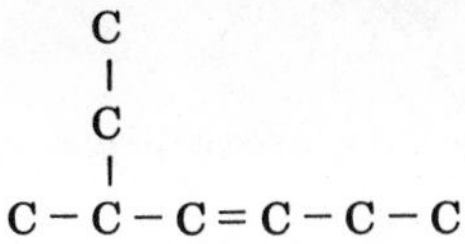

$$
\begin{array}{c}
\mathrm{C} \\
| \\
\mathrm{C} \\
| \\
\mathrm{C-C-C=C-C-C}
\end{array}
$$

2. $CH_2 = CHCH_2 -$

3.

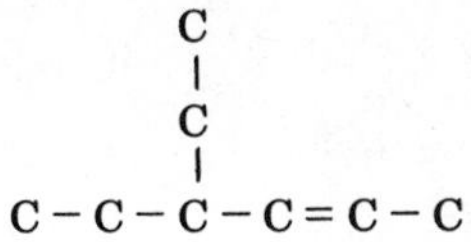

$$
\begin{array}{c}
\mathrm{C} \\
| \\
\mathrm{C} \\
| \\
\mathrm{C-C-C-C=C-C}
\end{array}
$$

4.

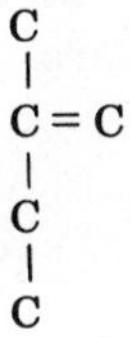

$$
\begin{array}{c}
\mathrm{C} \\
| \\
\mathrm{C = C} \\
| \\
\mathrm{C} \\
| \\
\mathrm{C}
\end{array}
$$

5.

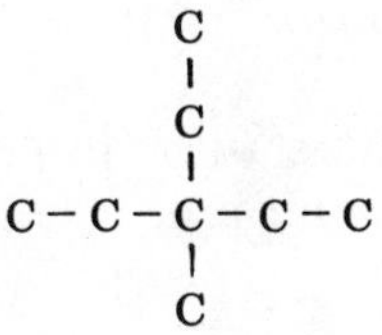

$$
\begin{array}{c}
\mathrm{C} \\
| \\
\mathrm{C} \\
| \\
\mathrm{C-C-C-C-C} \\
| \\
\mathrm{C}
\end{array}
$$

6.

$$
\begin{array}{c}
\mathrm{C} \\
\| \\
\mathrm{C-C=C-C-C-C}
\end{array}
$$

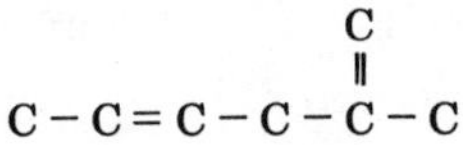

7. $C = C -$

• •

1. 5-methyl-3-heptene (See page 66 if you chose any other.)

2. 2-propenyl group, commonly called by its trivial name, the allyl group (See page 66 if you chose a different answer.)

3. 4-ethyl-2-hexene (See page 66 if you chose 3-ethyl-4-hexene.)

4. 2-methyl-1-butene (See page 66 if you chose 1-methyl-1-ethylethene.)

5. 3-ethyl-3-methylpentane

6. 2-methyl-1,4-hexadiene (See page 66 for explanation of answer.)

7. ethenyl group, also called by its trivial name, the vinyl group (see page 63.)

If you named all of these correctly, go on to page 66. If you made more than one error, you should review the section on alkenes.

Explanation of Answers

1.   The skeleton with the proper numbering is shown here. The meanings of the different parts of the name are:

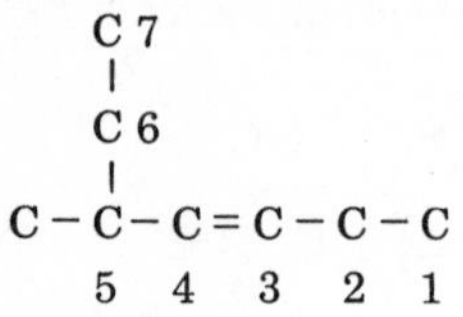

```
     C 7
     |
     C 6
     |
 C - C - C = C - C - C
 5   4   3   2   1
```

-ene	one double bond
hept-	seven C's in chain
-3-	double bond between numbers 3 and 4
5-methyl-	methyl group on number 5

3.   The skeleton is shown here with the carbon atoms numbered properly. The meanings of the name fragments are:

```
         C 6
         |
         C 5
         |
 C - C - C - C = C - C
 4   3   2   1
```

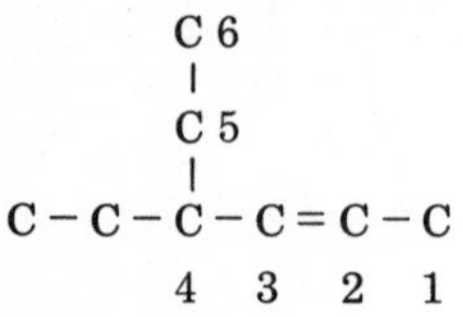

-ene	one double bond
hex-	six C's in chain
-2-	double bond between numbers 2 and 3
4-ethyl-	ethyl group on number 4

4.   The skeleton is shown here with the carbon atoms numbered properly. The meanings of the name fragments are:

```
       C
       |  1
   2 C = C
       |
   3 C
       |
   4 C
```

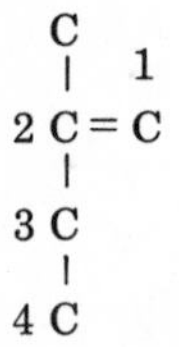

-ene	one double bond
but-	four C's in chain
-1-	double bond between numbers 1 and 2
2-methyl	methyl group on number 2

6.   The skeleton is shown here with the carbon atoms numbered properly. The meanings of the name fragments are:

```
             C 1
             ‖
 C - C = C - C - C - C
 6   5   4   3   2
```

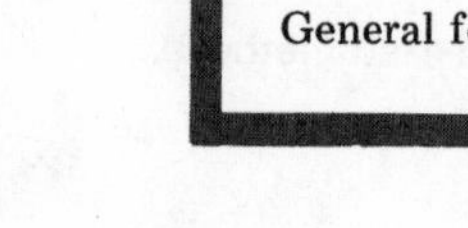

-diene	two double bonds
hexa-	six C's in chain
-1,4-	double bonds in 1-2 and 4-5 positions
2-methyl	methyl group on number 2

ALKENES

General formula: C_nH_{2n} Series name: *-ene*

1.4 ALKYNES

A third family of hydrocarbons is the *alkynes* [ăl′ kīnz]. These are unsaturated hydrocarbons containing a carbon-carbon triple bond. The simplest and most common alkyne is ethyne (ĕth′ ine). Its formula is

$$H - C \equiv C - H$$

You may recognize its trivial name, acetylene. Since the alkynes could result from the removal of two molecules of hydrogen from the alkanes, their general formula is C_nH_{2n-2}.

The names of the alkynes are formed by changing the *-ane* ending of the parent alkane to *-yne*. All of the rules you have learned for alkenes apply also to alkynes. Number the carbon atoms, if necessary, from the end of the chain nearer the triple bond. Locate the triple bond by the lower number of the two carbon atoms joined.

What is the name of the alkyne represented by this skeleton?

$$\begin{array}{c} \text{C} \\ | \\ \text{C} - \text{C} - \text{C} \equiv \text{C} - \text{C} \end{array}$$

A _____

4-methyl-2-pentyne

B _____

2-methyl-3-pentyne

C _____

4-methyl-3-pentyne

D _____

Help!

● ●

A _____

You are correct. This skeleton represents 4-methyl-2-pentyne.

$$\begin{array}{c} \text{C} \\ | \\ \text{C} - \text{C} - \text{C} \equiv \text{C} - \text{C} \end{array}$$

Since it contains a triple carbon-carbon bond, the suffix on the name is *-yne.* Write the complete structural formula and check to be certain that it obeys the general formula C_nH_{2n-2}. Go on to the next section.

B _____

Incorrect. You numbered the carbon chain from the wrong end. The carbon chain in alkynes must be numbered from the end nearer the triple bond. Go back and study the skeleton again carefully before you choose another answer.

C _____

You are so close to right that your error may only be a careless one. Remember that the location of double and triple bonds is specified by the lower number of the two carbon atoms joined. Study the skeleton again before you go back to pick another answer.

$$\begin{array}{c} \text{C} \\ | \\ \text{C} - \text{C} - \text{C} \equiv \text{C} - \text{C} \end{array}$$

D _____

Here is a bit of help for you. Hydrocarbon names are most readily put together from back to front. The steps are:

1. Determine suffix (*-yne* here because of triple bond).
2. Find base name by locating longest carbon chain.
3. Number chain from end nearer triple bond and specify the location of the triple bond.
4. Name and locate, by means of a number, any substituent groups. Go back and work out an answer.

Hydrocarbons having both double and triple bonds are named by replacing *-ane* by *-en-yne*, *-adien-yne*, *-ene-diyne*, etc. For example,

$$CH_2 = CHC \equiv CCH = CH_2$$

1,5-hexadien-3-yne

The smallest set of locants is used even if it results in *-yne* with a lower number than *-ene* as in

$$CH_3CH = CHC \equiv CH$$

3-penten-1-yne
(not 2-penten-4-yne)

When there is a choice in numbering, double bond(s) are given the lowest numbers.

$$CH \equiv CCH = CHCH = CH_2$$

1,3-hexadien-5-yne
(not 3,5-hexadien-1-yne)

Regardless of the locants, the part of the name resulting from double bonds precedes the part resulting from triple bonds.

Univalent groups derived from alkynes are named in the same fashion as those derived from alkanes and alkenes. The final *-e* is replaced by *-yl*. Hence, $HC \equiv C-$ is the ethynyl group. Where necessary, locants are used to specify the position of the triple bond. The carbon atom with the available bond is always numbered 1. For example,

$CH \equiv C - CH_2 -$	2-propynyl
$CH_3CH_2C \equiv C -$	1-butynyl

As shown in these examples, groups containing more than one multiple bond are named according to the same principles:

$CH_2 = CH - CH = CH -$	1,3-butadienyl
$CH \equiv C - CH = CH - CH_2 - CH_2 -$	3-hexen-5-ynyl
$CH_2 = CH - C \equiv C - CH_2 - CH_2 -$	5-hexen-3-ynyl

Here are five univalent groups. The correct names are listed together following the fifth structural formula.

1. $CH_2 = CH - CH_2 -$

2. $CH_2 = CH - CH = CH - CH_2 -$

3.
$$CH_3$$
$$|$$
$$CH_3 - C = CH -$$

4. $CH_2 = CH - C \equiv C -$

5.
$$CH_3$$
$$|$$
$$CH_3 - CH - CH_2 - CH_2 -$$

● ●

1. 2-propenyl (trivial: allyl)

2. 2,4-pentadienyl

3. 2-methyl-1-propenyl

4. 3-buten-1-ynyl

5. 3-methylbutyl

If you answered all these questions correctly, go on to the next section. If you made errors or are unsure of yourself, review page 63 and pages 66 through 68.

> ### ALKYNES
> General formula: C_nH_{2n-2} Series name: *-yne*

1.5 STEREOISOMERISM IN HYDROCARBONS AND THEIR DERIVATIVES

Anyone who studies organic chemistry discovers quickly that chemistry in space — stereochemistry — is important in explaining the properties and behavior of organic substances. Concepts and terminology in stereochemistry are expanding rapidly, not just in organic chemistry, but also in biochemistry, inorganic chemistry, and other fields. Because the topics of particular interest vary from one area to another, different specialized vocabularies have developed. Fortunately, there is reasonable agreement on the basic ideas and terms presented in this book.

The authors of one textbook of organic chemistry close their chapter on stereochemistry with this paragraph, which is appropriate to the beginning of this section:*

An Apology To The Student: Most students find that stereochemistry is the most difficult topic covered in this course. No way is known to make this topic "easy." Nature has chosen the tetrahedron as the basic geometric element of organic chemistry and the complications that follow are the unavoidable result of the geometric properties of a tetrahedron. On the encouraging side, if you are able to master the material in this chapter, things are not likely to get any worse.

Isomers have already been defined as compounds that have the same molecular formula but differ in some respect. Isomers may differ in the nature or sequence of bonding of their atoms or in the arrangement of their atoms in space. Isomers that differ in the nature or sequence of bonding have been designated as *constitutional isomers* (see p. 17). For example, CH_3OCH_3 is a constitutional isomer of C_2H_6O. By the same token pentane, isopentane, and neopentane are constitutional isomers of C_5H_{12}.

$$CH_3CH_2CH_2CH_2CH_3 \qquad\qquad CH_3CH_2\underset{\underset{\displaystyle CH_3}{|}}{CH}CH_3 \qquad\qquad H_3C-\underset{\underset{\displaystyle CH_3}{|}}{\overset{\overset{\displaystyle CH_3}{|}}{C}}-CH_3$$

pentane   *isopentane*   *neopentane*

At one time these were called structural isomers, but this term has been abandoned as insufficiently specific.

When two isomers have the same constitution and differ only in the arrangement of their atoms in space, they are *stereoisomers*. They are said to have different *configurations* and may be called *configurational isomers*.

This classification of the types of isomerism may serve as a guide through the kinds of stereoisomerism.

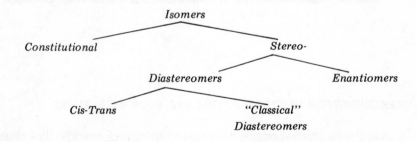

Don't worry if some of the terms are unfamiliar. We'll get to them soon.

Here are two pairs of formulas. The first pair represents constitutional isomers while the second shows configurational isomers.

---

*N. L. Allinger, *et al.*, *Organic Chemistry*. New York: Worth Publishers, 1971, page 126.

$$\underset{\underset{\text{H}}{|}}{\overset{\overset{\text{H}}{|}}{\text{Cl} - \text{C}}} - \text{O} - \underset{\underset{\text{H}}{|}}{\overset{\overset{\text{H}}{|}}{\text{C}}} - \text{H} \quad \text{and} \quad \text{Cl} - \underset{\underset{\text{H}}{|}}{\overset{\overset{\text{H}}{|}}{\text{C}}} - \underset{\underset{\text{H}}{|}}{\overset{\overset{\text{H}}{|}}{\text{C}}} - \text{O} - \text{H}$$

Now consider this pair. What do they represent?

A _____

Constitutional isomers.

B _____

Configurational isomers.

C _____

I need help.

• •

A _____

Right you are. The molecules represented by these formulas differ in the sequence of bonding. This is clearly shown when they are changed to line formulas: $CH_3CHClCH_3$ and $CH_3CH_2CH_2Cl$. Go on to the next section.

B _____

You are wrong. You believe these figures represent configurational isomers.

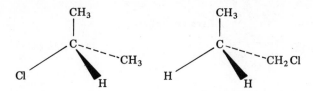

To be configurational isomers they must have the same constitution and differ only in the arrangement of the atoms in space. You should be able to see your error if you draw regular structural formulas or line formulas for both molecules. When you see that they are constitutional isomers, go on to the next section.

C _____

So you need help. Let's see if this will do it. The question is whether these two molecules have the same constitution or not. Suppose we draw conventional structural formulas.

$$
\begin{array}{ccc}
\text{H} & \text{Cl} & \text{H} \\
| & | & | \\
\text{H}-\text{C}-\text{C}-\text{C}-\text{H} \\
| & | & | \\
\text{H} & \text{H} & \text{H}
\end{array}
\quad\text{and}\quad
\begin{array}{ccc}
\text{H} & \text{H} & \text{H} \\
| & | & | \\
\text{H}-\text{C}-\text{C}-\text{C}-\text{Cl} \\
| & | & | \\
\text{H} & \text{H} & \text{H}
\end{array}
$$

Now what do you think? Go back and choose another answer.

At first glance, 2-butene does not seem worthy of special attention. Nonetheless, it is the simplest example of an important type of stereoisomerism known as *cis-trans* isomerism. The 2-butene molecule, as explained in terms of the sp^2 hybridization of the carbon atoms joined by the double bond, is planar and can exist in two forms:

$$
\begin{array}{cc}
\text{H}_3\text{C} & \quad\quad \text{CH}_3 \\
 \diagdown & \diagup \\
 & \text{C}=\text{C} \\
 \diagup & \diagdown \\
\text{H} & \quad\quad \text{H}
\end{array}
\quad\quad\quad\quad
\begin{array}{cc}
\text{H} & \quad\quad \text{CH}_3 \\
 \diagdown & \diagup \\
 & \text{C}=\text{C} \\
 \diagup & \diagdown \\
\text{H}_3\text{C} & \quad\quad \text{H}
\end{array}
$$

cis   *trans*

cis-trans Isomerism occurs only in structures that are rigid by virtue of double bonds or rings. Atoms or groups at opposite ends of a double bond are considered to be *cis* or *trans* to one another when they are respectively on the same or opposite side of a specified reference plane. For alkenes such as 2-butene the reference plane is perpendicular to a plane containing the $C-C$ and $C=C$ bonds. These drawings illustrate the point.

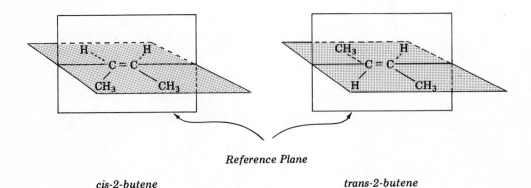

Reference Plane

cis-2-butene   *trans-2-butene*

These *cis-trans* isomers belong to a large class of stereoisomers known as *diastereoisomers* or *diastereomers*. Diastereomers are stereoisomers that are not mirror images of one another. The meaning of this definition will become clear later in this section when enantiomers, the stereoisomers that *are* mirror images of one another, are discussed.

Neglecting names for a moment, which of the following represents a *trans* isomer?

A _____

$$\begin{matrix} Cl & & Cl \\ \backslash & & / \\ & C = C & \\ / & & \backslash \\ H & & H \end{matrix}$$

B _____

$$\begin{matrix} & H & Cl & \\ & | & | & \\ H - & C - & C & - H \\ & | & | & \\ & Cl & H & \end{matrix}$$

C _____

$$\begin{matrix} H & & Cl \\ \backslash & & / \\ & C = C & \\ / & & \backslash \\ Cl & & H \end{matrix}$$

• •

A _____

Incorrect. You say that this formula represents a *trans* isomer.

$$\begin{matrix} Cl & & Cl \\ \backslash & & / \\ ----C & = C & ---- \\ / & & \backslash \\ H & & H \end{matrix}$$

If we think of the page as the plane containing the $C = C$ or $C - C$ bonds, then the reference plane is perpendicular to the page and intersects it along the dashed line. By definition, the *trans* isomer must have the Cl or H atoms on opposite sides of the reference plane. Here they are on the same side, so this must be a *cis* isomer. Go back and choose another answer.

B _____

You are wrong. You may have missed the first point about *cis-trans* isomerism: the structure must be rigid by virtue of a double bond or ring structure. Your choice has neither. Go back and choose another.

C _____

Right. If the page is thought of as the plane containing the $C = C$ or $C - C$ bonds, then the reference plane is perpendicular to the page and intersects it along the $C = C$ bond. The methyl groups are on opposite sides of the reference plane, so this formula represents a *trans* isomer. Go on to the next section.

For compounds of the type $aCH = CHb$ (where a and b represent an atom or a group other than H) it is easy to see which isomer is *cis* and which *trans*.

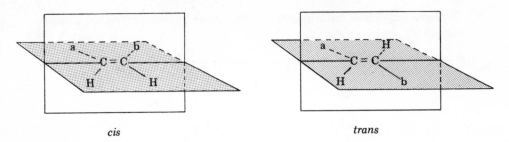

cis   trans

The distinction between *cis* and *trans* is less clear for those with the formula abC = CHc and impossible to draw for those of the type abC = Ccd.

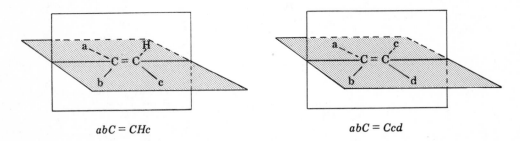

abC = CHc   abC = Ccd

In order to avoid ambiguity, the steric relations around double bonds can be more precisely described by use of the affixes Z and/or E. Their use requires knowledge of the sequence rule formulated by Robert S. Cahn, Christopher K. Ingold, and Vladimir Prelog. The sequence rule provides a way to rank the substituent atoms or groups in order of preference based principally on atomic number. For ease of expression the order can be written $a > b > c > d$ where ">" has the meaning "is preferred to" or "ranks above."

In the example abC = Ccd above, we need to know the relative rank of the pairs a,b and c,d. If the preferred groups from each pair are on the same side of

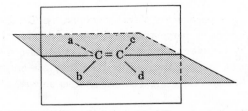

the reference plane, the designation Z (from German, *zusammen*, together) is used. If they are on opposite sides of the reference plane, E (from German, *entgegen*, opposite) is used. To illustrate, let's suppose $a > b$ and $c > d$. Then

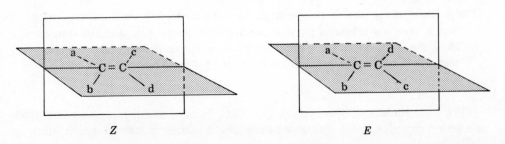

Z   E

Before going to real compounds, let's see if you have the hang of it. Consider these two compounds where a > b > c > d.

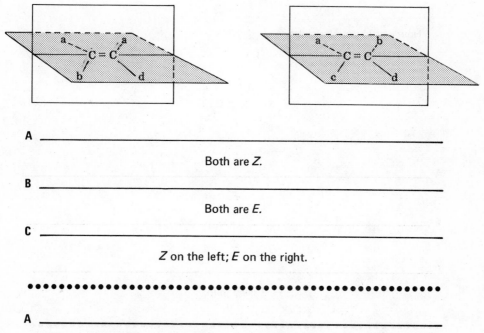

A _____

Both are Z.

B _____

Both are E.

C _____

Z on the left; E on the right.

••

A _____

You are right. For the compound on the left, the pairs in question are a,b and a,d. The preferred group in both pairs is a. The a's lie on the same side of the reference plane so the designation is Z. On the right it's a > c and b > d. The two preferred substituents, a and b, are on the same side so it's Z again. Go on to the next section.

B _____

Wrong. Let's discuss the compound on the left. It has the formula abC = Cad. We are concerned over the relative rank in the pairs a,b and a,d. Since a > b and a > d and the two a's are on the same side of the reference plane, the designation is Z. Apply the same line of reasoning to the compound on the right when you go back to choose another answer.

C _____

You are only partly right. The compound on the left is Z, but the one on the right is not E. Go back and take another look.

The first sequence subrule states that substituent atoms or groups rank in order of decreasing atomic number. For example, Br > Cl > F > H. This simple principle suffices for many cases. Using it, we find

Cl   Cl   Cl   Cl
 \   /   \   /
  C = C   C = C
 /   \   /   \
H   Br   F   H

E   *Z*

Look at the two formulas carefully. Both appear to be *Z* (or *cis*) insofar as the Cl alone is concerned. Since Br > Cl, however, the one on the left is *E (trans)*.

How would you designate

$$\begin{array}{ccc} H & & Cl \\ \diagdown & & \diagup \\ & C = C & \\ \diagup & & \diagdown \\ F & & F \end{array}$$

A _____

E.

B _____

Z.

C _____

I need help.

• •

A _____

You are correct. The pairs of substituents are F > H and Cl > F. Since the preferred members, F and Cl, lie on opposite sides of the reference plane, the designation is *E.* Go on to the next section.

B _____

Incorrect. Perhaps you were misled by the appearance of the two F's on the same side of the reference plane. If you look again, you'll see that F is the preferred member of the pair F > H, but not in the pair Cl > F. Consequently, the designation is *E.* Keep this subrule in mind as you go on to the next section.

C _____

So you need help. Let's see if we can find some. We need to look at the formula and determine the pairs of substituents on the two carbon atoms joined by the double bond. They are F,H and Cl,F. Because the atomic number of F is greater than that of H, F is the preferred member of the first pair. By the same reasoning, Cl is the preferred member of the second pair. Where do these atoms lie with respect to the reference plane? If they are on the same side, the desig-

$$\begin{array}{ccc} H & & Cl \\ \diagdown & & \diagup \\ & C = C & \\ \diagup & & \diagdown \\ F & & F \end{array}$$

nation is *Z*; if opposite, *E.* Go back now and choose one of the other answers.

If a substituent is a group of atoms, we deal first with the atom attached directly to the double-bonded carbon. For example,

$$
\begin{array}{c}
\text{Cl} \qquad \qquad \text{CH}_3 \\
\diagdown \qquad \diagup \\
\text{C} = \text{C} \\
\diagup \qquad \diagdown \\
\text{H}_3\text{C} \qquad \qquad \text{H}
\end{array}
\qquad\qquad
\begin{array}{c}
\text{Cl} \qquad \qquad \text{CH}_3 \\
\diagdown \qquad \diagup \\
\text{C} = \text{C} \\
\diagup \qquad \diagdown \\
\text{H} \qquad \qquad \text{Cl}
\end{array}
$$

Z (because Cl > C and C > H)   *E (because Cl > H and Cl > C)*

The hydrogen atoms in the methyl group are ignored because the order of preference can be established without them.

Do you remember the compound that started this entire business? It is 2-butene. Here are the two isomers again. It's easy to see that application of the sequence rule leads to the proper designations.

$$
\begin{array}{c}
\text{H}_3\text{C} \qquad \qquad \text{CH}_3 \\
\diagdown \qquad \diagup \\
\text{C} = \text{C} \\
\diagup \qquad \diagdown \\
\text{H} \qquad \qquad \text{H}
\end{array}
\qquad\qquad
\begin{array}{c}
\text{H} \qquad \qquad \text{CH}_3 \\
\diagdown \qquad \diagup \\
\text{C} = \text{C} \\
\diagup \qquad \diagdown \\
\text{H}_3\text{C} \qquad \qquad \text{H}
\end{array}
$$

(Z)-2-butene   *(E)-2-butene*
(cis-2-butene)   *(trans-2-butene)*

Alkenes with non-terminal double bonds exhibit *cis-trans* isomerism. Frequently the atoms bonded directly to the double-bonded carbons are identical — for example, the C's on the right side of this formula:

$$
\begin{array}{c}
\text{Cl} \qquad \qquad \text{CH}_2\text{OH} \\
\diagdown \qquad \diagup \\
\text{C} = \text{C} \\
\diagup \qquad \diagdown \\
\text{H} \qquad \qquad \text{CH}_3
\end{array}
$$

We can easily establish that $Cl > H$ on the left. To find the order of preference on the right we look at the atoms attached to the two C's and find

$$
\begin{array}{c}
\text{H} \\
| \\
-\,\text{C}-\text{H} \\
| \\
\text{H}
\end{array}
\qquad \text{and} \qquad
\begin{array}{c}
\text{H} \\
| \\
-\,\text{C}-\text{O}- \\
| \\
\text{H}
\end{array}
$$

which can be written C(H,H,H) and C(O,H,H). When the bonding sequences are written in this manner, atoms listed within parentheses are cited in order of preference. Because $O > H$, $CH_2OH > CH_3$ and the compound is designated Z.

Now suppose the highest priority atom in each set of three is O as on the right side of

$$
\begin{array}{c}
\text{Cl} \qquad \qquad \text{CH}_2\text{OH} \\
\diagdown \qquad \diagup \\
\text{C} = \text{C} \\
\diagup \qquad \diagdown \\
\text{H} \qquad \qquad \text{CH(CH}_3)\text{OH}
\end{array}
$$

To resolve this situation the second highest priority atom in each set is used and, if necessary, the third. Here C(O,C,H) > C(O,H,H) and the designation is E.

How would you classify these two?

$$H_3C \diagdown \qquad CH_2OH$$
$$C = C$$
$$ClH_2C \diagup \qquad \diagdown CH_3$$

$$HO \diagdown \qquad CH_2OH$$
$$C = C$$
$$H \diagup \qquad \diagdown CH_2CH_2OH$$

A _____

E on the left; *Z* on the right.

B _____

Z on the left; *E* on the right.

C _____

Both *E*.

● ●

A _____

Right. For the compound on the left, C(Cl,H,H) > C(H,H,H) and C(O,H,H) > C(H,H,H). Because the preferred groups, ClH_2C- and $-CH_2OH$ are on opposite sides of the reference plane, the designation is *E*. Similar analysis shows the one on the right to be *Z*. Go on to the next section.

B _____

You are incorrect. Let's analyze the formula on the left to learn why. The task is to establish the relative priority between H_3C- and ClH_2C- and between $-CH_2OH$ and $-CH_3$. The first two can be rewritten as follows:

$$H_3C - \qquad C(H,H,H)$$
$$ClH_2C - \qquad C(Cl,H,H)$$

According to the principle stated above, C(Cl,H,H) > C(H,H,H). The second pair can be rewritten:

$$-CH_2OH \qquad C(O,H,H)$$
$$-CH_3 \qquad C(H,H,H)$$

By the same principle, C(O,H,H) > C(H,H,H).
The preferred groups, ClH_2C- and $-CH_2OH$, are on opposite sides of the reference plane so the designation is *E*. Analyze the compound on the right and then choose another answer.

C _____

You are half right. The compound on the left is designated *E*, but the one on the right is *Z*. Convince yourself of this before you go on to the next section.

Branched groups are treated in the same fashion. Although this formula is complex, the ultimate decision is easy. You can quickly see that $Cl > CH_3$ on the left side.

$$\begin{array}{c} Cl \\ \diagdown \\ \\ H_3C \diagup \end{array} C = C \begin{array}{c} \diagup CH(CH_3)(CH_2Cl) \\ \\ \diagdown CH(CH_2OH)(CHOHCH_3) \end{array}$$

Consideration of the two substituents on the right side shows them both to be C(C,C,H) so we must go on. Remember always to list atoms within the parentheses in order of preference. The three branches attached to the initial carbon atom of the upper substituent can be ordered as follows: C(Cl,H,H) > C(H,H,H) > H. Similarly, for the lower substituent it is C(O,C,H) > C(O,H,H) > H. On comparison of the preferred branches from the top and the bottom we find C(Cl,H,H) > C(O,C,H) and designate the compound Z.

So, as you can see, it all boils down to establishing the order of preference for substituent groups. Examine this pair and rank them.

$$-CH \begin{array}{c} \diagup CH_2OH \\ \\ \diagdown CH_2CH_2Cl \end{array} \qquad -CH \begin{array}{c} \diagup C(OH)(CHClOH)CH_2Cl \\ \\ \diagdown CH_2Cl \end{array}$$

$$A \qquad\qquad\qquad\qquad\qquad B$$

A _____

A > B.

B _____

B > A.

•••

A _____

Wrong. You believe that substituent A above ranks higher than B. If we look carefully, we can see that both of the initial C's can be described by C(C,C,H). We must, therefore, examine the next groups in the chain. In A we find C(O,H,H) > C(C,H,H). In B we find C(Cl,H,H) > C(O,C,C). The next step is to compare the preferred set from each. We find C(Cl,H,H) > C(O,H,H) and, in consequence, B > A. Review carefully to insure your understanding before you go on to the next section.

B _____

Right. Your answer is B > A. The initial C's of both groups can be described as C(C,C,H). Examining the next groups in A we find C(O,H,H) > C(C,H,H). In B we find C(Cl,H,H) > C(O,C,C). Comparing the preferred set from each gives C(Cl,H,H) > C(O,H,H) and, in consequence, B > A. Go on to the next section.

Perhaps you've wondered what happens when a double (or triple) bond is encountered. Each one is expanded into two (or three) bonds until both atoms joined by the multiple bond have the maximum number of single bonds, and hypothetical duplicate representations of the atoms at the other end are added.

One illustration is worth a lot of words. In these three examples the hypothetical duplicates are enclosed in parentheses.

$$
\text{A.} \quad
\underset{\displaystyle H}{\overset{\displaystyle O}{\underset{\vert}{\overset{\Vert}{-C}}}}
\qquad \text{is expanded to} \qquad
\underset{\displaystyle H}{\overset{\displaystyle O-(C)}{\underset{\vert}{\overset{\vert}{-C-(O)}}}}
$$

$$
\text{B.} \quad
\underset{\displaystyle OH}{\overset{\displaystyle O}{\underset{\vert}{\overset{\Vert}{-C}}}}
\qquad \text{is expanded to} \qquad
\underset{\displaystyle OH}{\overset{\displaystyle O-(C)}{\underset{\vert}{\overset{\vert}{-C-(O)}}}}
$$

$$
\text{C.} \quad -C \equiv CH
\qquad \text{is expanded to} \qquad
\underset{(C)\ (C)}{\overset{(C)\ (C)}{-C-CH}}
$$

How would these three be ranked? Do you agree with $B > A > C$? If you do, skip the next paragraph. If you don't, read it.

When expanded, the three groups become:

      A.   C(O,O,H)

      B.   C(O,O,O)

      C.   C(C,C,C)

From these expansions you should conclude that $C(O,O,O) > C(O,O,H) > C(C,C,C)$ and, therefore, $B > A > C$.

Appendix A.1 contains lists of the common substituent groups in order of sequence rule preference and alphabetically.

The sequence rule is useful also with another group of stereoisomers known as *chiral* [kī′ rǎl] molecules. Such molecules are *not* superimposable on their mirror images. This condition is equivalent to handedness and is called *chirality* [kī·rǎl′ ĭ tē]. Chiral and chirality are derived from the Greek χειρ, for hand.

If you compare your left and right hands, assuming they're normal, you'll find they're mirror images of one another. Although they have the same constitution (four fingers and a thumb), they have different configurations. Stereoisomers that have this relationship to one another — that is, are nonsuperimposable mirror images — are called *enantiomers* [ĕn ǎn′ tǐ·ō·mĕrz].

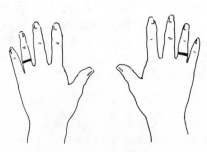

nonsuperimposable mirror images
enantiomers

A century ago the Dutch chemist Jacobus H. van't Hoff suggested that the bonds of a carbon atom might be conceived as directed toward the apices of a tetrahedron. An understanding of the meaning of chirality and the relationship between enantiomers may be easier if you make two paper tetrahedra like those of van't Hoff. They are found on pages 83 and 85 (Figs. 1.2 and 1.3). One of the commercial model kits will do as well. Or you can use toothpicks and gumdrops. An advantage of the latter is that you can eat the gumdrops after a hard night's study.

Whatever aid you use, you'll quickly find that, while the molecules represented by the tetrahedra have the same structural formula (C is replaced by the tetrahedron),

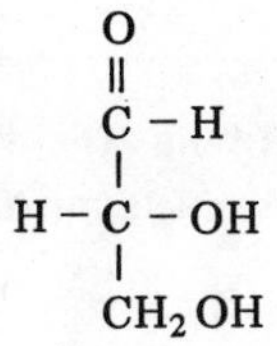

$$
\begin{array}{c}
O \\
\parallel \\
C - H \\
\mid \\
H - C - OH \\
\mid \\
CH_2\,OH
\end{array}
$$

they are definitely not the same in all respects. Experimentation with a mirror will reveal that they are mirror images of one another. Since they are *not* superimposable, they are enantiomers.

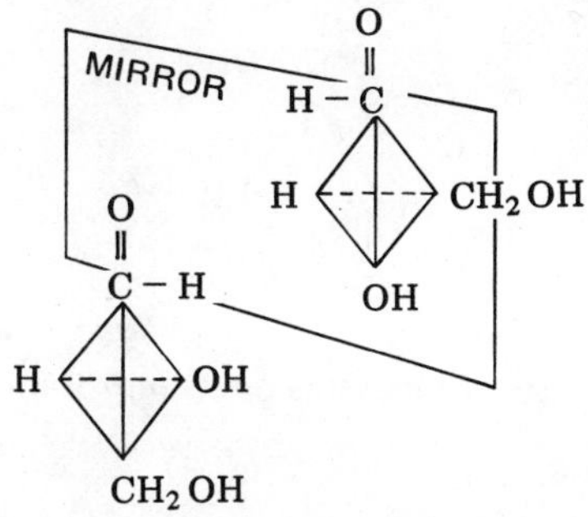

If you replace $-\overset{\displaystyle O}{\overset{\parallel}{C}} - H$ with $-H$, the molecules are still mirror images but are now identical and can be superimposed on each other. They are no longer chiral, but are now *achiral* [ā′ kī·răl].

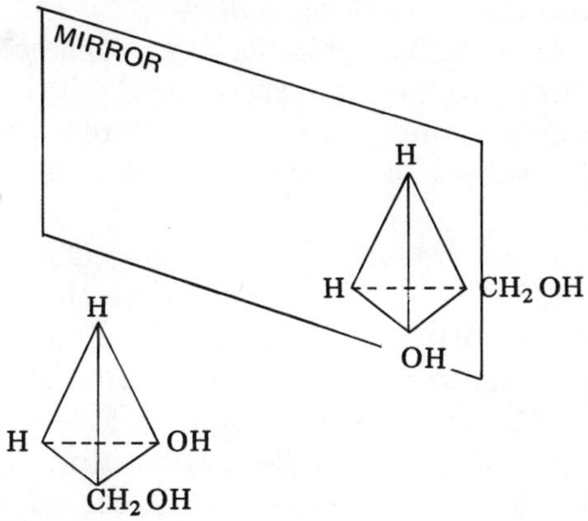

Achiral means without chirality. All objects have mirror images, but only chiral objects have nonsuperimposable mirror images.

Which of these structural formulas represents a chiral molecule? (Use your tetrahedra and a mirror if you need them.)

$$
\begin{array}{ccc}
\underset{\displaystyle A}{\overset{\displaystyle H}{\underset{|}{HO-\underset{|}{\overset{|}{C}}-CH_2OH}}}
& \qquad &
\end{array}
$$

$$
\begin{array}{ccc}
& H & O \\
& | & \| \\
HO-C-CH_2OH & H-C-C-OH & HO-C-C-OH \\
| & | & | \\
H & H & CH_3 \\
A & B & C
\end{array}
$$

A _____

You are wrong. Structural formula A is the same molecule as shown in the last drawing. It is superimposable on its mirror image and represents an achiral molecule. Try another answer.

B _____

Incorrect. This is the structural formula for acetic acid, a compound without stereoisomers. If you build a tetrahedral molecule, it should look like this:

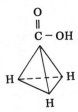

You can see that it will be superimposable on its mirror image. Go back and pick another answer.

C _____

You are right. This structural formula represents a chiral molecule, one not superimposable on its mirror image.

The most common chiral molecules are those with one or more carbon atoms bonded to four different substituents. Carbon atoms bearing four different substituents are called *chiral centers*.

It is worth emphasizing that chirality is a property of the space and nature of substituents on the central atom (carbon) and not of the atom itself. So long as the central atom is bonded to four different groups in a tetrahedral configuration, the resulting molecule is chiral. Some ammonium salts, for example, are chiral.

All chiral molecules are molecules of optically active compounds, and molecules of all optically active compounds are chiral. Optical activity refers to the rotation of the plane of polarization of plane-polarized light and is especially important in natural products (see Chapt. 5).

The sequence rule is used to designate the configuration of chiral molecules, too. Let's consider a single carbon atom with four achiral groups attached to it and call them A, B, C, and D for convenience. The resulting

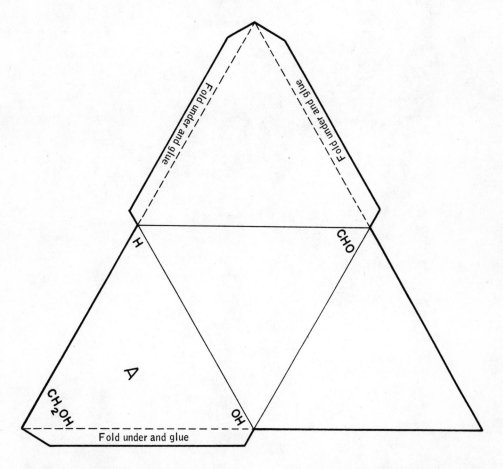

Figure 1.2 Tetrahedral model of L-(+)-glyceraldehyde. Cut out and assemble in the same way as Figure 1.1.

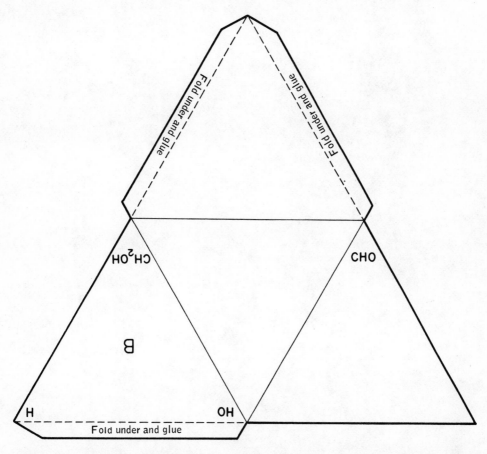

Figure 1.3 Tetrahedral model of D-(−)-glyceraldehyde. Cut out and assemble in the same way as Figure 1.1.

molecule can exist in two, and only two, configurations. They are non-superimposable mirror images, or enantiomers.

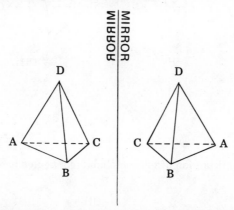

If you are skeptical, experiment with your tetrahedra. You'll find that every model you build will be superimposable on one or the other of the above structures.

The sequence rule is applied by arranging the four groups in order of preference. Suppose it's A > B > C > D in this instance. Now picture the two enantiomeric forms as shown here with the lowest ranking group away from the eye of the viewer. Looking squarely at the ABC face of the tetrahedron, your line of sight passes through the central carbon.

R   S

If the path traced from A to B to C is clockwise, the configuration is designated *R* (from Latin *rectus*, right). If the path is counterclockwise, the designation is *S* (from Latin *sinister*, left).

Conversion of one enantiomer into the other always changes *R* to *S*, or *S* to *R*.

When real groups are substituted for the letters, the rule is applied just as you learned it for *E* and *Z*. In these examples $Cl > OH > CH_3 > H$.

Notice that the tetrahedron on the left of each pair has been reoriented at the right to make it easier to trace around the path on the face opposite the lowest-ranking substituent group. To determine the configuration of a chiral center you must view it properly, either in your mind's eye or by redrawing the tetrahedron. Models are useful in gaining the proper perspective.

R

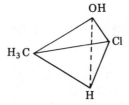

S

What is the configuration of the compound represented here?

A _____

R.

B _____

S.

C _____

I don't know.

•••

A _____

Incorrect. Your answer is that the compound represented by this tetrahedron has R configuration. The order of preference for the four substituent
groups is $Cl > OH > CH_3 > H$. If you place your mind's eye so it looks through
the face opposite H, this is what you'll see.

Now you should be able to verify that the configuration is S. Go on to the
next section.

B _____

You are right. When viewed toward the lowest-ranking substituent, H, the
sequence in descending order of rank for the remaining three groups is
counterclockwise. The configuration is, therefore, S. Go on to the next section.

C _____

Let's see if this will help. The order of preference for the four substituent groups is $Cl > OH > CH_3 > H$. If you place your mind's eye so it looks through the tetrahedron toward H, this is what you'll see.

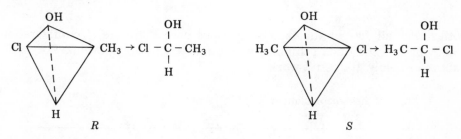

How does the sequence in descending order go? Clockwise or counterclockwise? Go back and choose another answer.

Fischer projections offer a convenient way to represent molecules with chiral centers. Recall that they show the four bonds from carbon at right angles to one another on the page. Actually, the substituents on the left and right of the chiral center are in front of the page while those above and below it project behind the page.

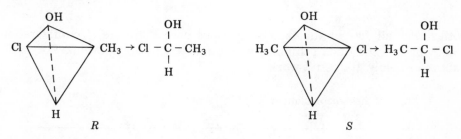

If the substituent of lowest rank is above or below the chiral center in the Fischer projection, then the configuration is R if the sequence of the remaining three describes a clockwise array and S if it is counterclockwise.

Far and away the best method to determine configuration of a chiral center is to perceive a mental picture and move your mind's eye around until it's looking through the tetrahedron toward the lowest-ranking substituent. Because that's sometimes a difficult feat of mental gymnastics, and because many textbooks still use Fischer projections, it's worthwhile to learn to manipulate them.

If a given Fischer projection doesn't have the substituent of lowest rank above or below the chiral center, it can be rewritten with no change in configuration in either of two ways:

(1) Move the group of lowest rank and any two others to new positions, placing the one of lowest rank either above or below the chiral center. For example, using $a > b > c > d$ for simplicity:

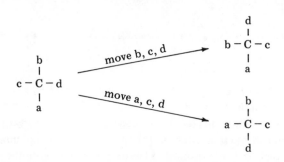

(2) Exchange a pair to put the group of lowest rank either above or below the chiral center. Then exchange any other pair that does not include the lowest ranking group. For example,

```
                                                                    b
                                                                    |
                                                                a - C - c
                                                                    |
                                                                    d
                                                  exchange a & c  ↗

      b                       b              exchange a & b         a
      |                       |          ──────────────────→       |
  c - C - d   exchange    c - C - a                             c - C - b
      |       ────────→       |                                     |
      a        a & d          d                                     d
                                                  exchange b & c  ↘
                                                                    c
                                                                    |
                                                                b - C - a
                                                                    |
                                                                    d
```

Notice that all five of the final projections have the same R configuration. Don't fall into the trap of thinking you can merely rotate a formula 90° to put d above or below. It doesn't work.

Which of these manipulations is *not* correct?

A _____

```
      d                           b
      |                           |
  a - C - b       ──────→      a - C - c
      |                           |
      c                           d
```

B _____

```
      b                           b
      |                           |
  a - C - c       ──────→      c - C - a
      |                           |
      d                           d
```

C _____

```
      b                           b
      |                           |
  c - C - d       ──────→      a - C - c
      |                           |
      a                           d
```

• •

A _____

Sorry. This one is all right. Look at the first projection again. The lowest ranking substituent is above the chiral center and the other three describe a clockwise array in descending order. Although the lowest ranking group is below the chiral center in the projection on the right, the array is still clockwise. Both projections depict an R configuration. Go back and choose another answer.

B _____

You are right. There's something wrong with this manipulation. The lowest ranking substituent is below the chiral center in both projections. The remaining three descend in a clockwise order on the left but in a counterclockwise order on the right. Since the configurations are different, the manipulation must be faulty. Go on to the next section.

C _____

Incorrect. The manipulation shown is all right. Let's follow it through.

$$\begin{array}{c} b \\ | \\ c-C-d \\ | \\ a \end{array} \xrightarrow[\text{a \& d}]{\text{exchange}} \begin{array}{c} b \\ | \\ c-C-a \\ | \\ d \end{array} \xrightarrow[\text{a \& c}]{\text{exchange}} \begin{array}{c} b \\ | \\ a-C-c \\ | \\ d \end{array}$$

Go back and choose another answer.

The example used earlier is 2,3-dihydroxypropanal. Prior to 1972, *Chemical Abstracts* used the trivial name glyceraldehyde, a name that is still acceptable under the IUPAC rules. It was used by Emil Fischer to assign chiral

$$\begin{array}{c} O \\ \| \\ C-H \\ | \\ H-C-OH \\ | \\ CH_2OH \end{array}$$

molecules, into one of two families that he designated D and L. Until 1951 only these relative configurations were known. When a procedure was found to determine the real, or absolute, configurations, D-2,3-dihydroxypropanal turned out to have the R configuration.

Another common chiral molecule, the amino acid L-serine, was used for many years as the reference substance in assigning relative configurations to amino acids. Here is its Fischer projection. Is its configuration R or S?

$$\begin{array}{c} O \\ \| \\ C-OH \\ | \\ H_2N-C-H \\ | \\ CH_2OH \end{array} \longrightarrow \begin{array}{c} O\ \ NH_2 \\ \| \ \ | \\ HO-C-C-CH_2OH \\ | \\ H \end{array} \quad \text{or} \quad \begin{array}{c} b \\ | \\ a-C-d \\ | \\ c \end{array} \longrightarrow \begin{array}{c} a \\ | \\ b-C-c \\ | \\ d \end{array}$$

It's S. Fischer's D and L families correspond, commonly, to our modern R and S configurations. Occasionally there's a complication with amino acids that are related to carbohydrates. All of this will appear again (see Chapt. 5).

If a molecule has two or more chiral centers, the configuration of each is determined separately. Tartaric acid, which figures prominently in the history of stereochemistry, is an example. The Fischer projections for the two enantiomers are

$$
\begin{array}{ll}
1 & \quad \overset{\displaystyle O}{\overset{\displaystyle \|}{C}} - OH \\
2 & H - C - OH \\
3 & HO - C - H \\
4 & C - OH \\
& \quad \underset{\displaystyle O}{\overset{\displaystyle \backslash\backslash}{}}
\end{array}
\qquad\qquad
\begin{array}{l}
\overset{\displaystyle O}{\overset{\displaystyle \|}{C}} - OH \\
HO - C - H \\
H - C - OH \\
C - OH \\
\quad \underset{\displaystyle O}{\overset{\displaystyle \backslash\backslash}{}}
\end{array}
$$

(2R,3R)-tartaric acid   (2S,3S)-tartaric acid

Here is the detailed proof for the R configuration of the carbon 2 atom in the ($2R,3R$)-enantiomer.

The others can be done in a similar manner.

There is still a third stereoisomer of tartaric acid called *meso*-tartaric acid. Here is its Fischer projection.

$$
\begin{array}{ll}
\overset{\displaystyle O}{\overset{\displaystyle \|}{C}} - OH & 1 \\
H - C - OH & 2 \\
H - C - OH & 3 \\
C - OH & 4 \\
\underset{\displaystyle O}{\overset{\displaystyle \backslash\backslash}{}}
\end{array}
$$

The configurations of the two chiral centers in *meso*-tartaric acid are opposite. For this reason it is sometimes called *RS*-tartaric acid. It would be a good idea for you to check the configurations for yourself before going on.

Although *meso*-tartaric acid has two chiral centers, it is achiral and optically inactive. The molecule has a plane of symmetry between the two chiral carbons. It is superimposable on its mirror image, and therefore is not part of an enantiomeric pair. All three forms of tartaric acid are diastereomers, often called a *dl* pair and the *meso* form

The chiral centers in *meso* compounds need not be adjacent to one another. For example,

$$
\begin{array}{c}
\overset{\displaystyle O}{\overset{\displaystyle \|}{C}} - OH \\
| \\
H - C - OH \\
| \\
H - C - H \\
| \\
H - C - OH \\
| \\
C - OH \\
\overset{\displaystyle \|}{\underset{\displaystyle O}{}}
\end{array}
$$

Special terminology is sometimes used for molecules of the type xABC–CABy. If x and y are different, the two chiral centers are different and there are two pairs of enantiomers. If x and y are the same, the two chiral centers are alike and the result is one pair of enantiomers and an inactive *meso* isomer. Common examples of the first type are the four-carbon sugars, erythrose and threose. Their Fischer projections are:

O ‖ C – H | H – C – OH | H – C – OH | CH₂OH	O ‖ C – H | HO – C – H | HO – C – H | CH₂OH	O ‖ C – H | HO – C – H | H – C – OH | CH₂OH	O ‖ C – H | H – C – OH | HO – C – H | CH₂OH
D-erythrose	*L-erythrose*	*D-threose*	*L-threose*

Oxidation of the terminal carbons leads to the isomers of tartaric acid:

COOH | H – C – OH | H – C – OH | COOH	COOH | HO – C – H | H – C – OH | COOH	COOH | H – C – OH | HO – C – H | COOH
meso-tartaric acid *(erythraric acid)*	*SS-tartaric acid* *(D-threaric acid)*	*RR-tartaric acid* *(L-threaric acid)*

Enantiomers that bear a formal resemblance to erythrose are often called *erythro* while those that resemble threose are called *threo*. The example most often cited is 3-bromo-2-butanol:

CH₃ | H – C – Br | H – C – OH | CH₃	CH₃ | Br – C – H | HO – C – H | CH₃	CH₃ | H – C – Br | HO – C – H | CH₃	CH₃ | Br – C – H | H – C – OH | CH₃
erythro enantiomers		*threo enantiomers*	

If the bromine is replaced by a hydroxy group, to yield 2,3-butanediol, the result is a pair of *threo* enantiomers and a single inactive *erythro* form.

$$
\begin{array}{ccc}
& CH_3 & \\
& | & \\
H- & C-OH & \\
& | & \\
H- & C-OH & \\
& | & \\
& CH_3 &
\end{array}
$$

$$
\begin{array}{cc}
CH_3 & CH_3 \\
| & | \\
H-C-OH & HO-C-H \\
| & | \\
HO-C-H & H-C-OH \\
| & | \\
CH_3 & CH_3
\end{array}
$$

erythro (meso) form   *threo enantiomers*

Threo and *erythro* isomers are often used in stereochemical studies.

Another situation in which no optical activity is observed occurs when equal numbers of enantiomeric molecules of a substance are present. Such an aggregation can occur only on the macroscopic scale and is known as a *racemic modification.* In a racemic modification the optical activity of every molecule present is exactly offset by the activity of its enantiomer.

Since starting materials that are not optically active cannot produce optically active products, ordinary synthesis leads to racemic modifications. Thus, although many of the compounds encountered in the next three chapters will possess chiral centers, they are commonly found in racemic modifications. Their chirality is of little importance because enantiomers have identical physical properties, except for the direction of rotation of the plane of polarized light, and identical chemical properties except toward optically active reagents.

Although pure enantiomers are difficult to isolate and to prepare in the laboratory, nature has less trouble. Many substances found in nature — D-tartaric acid, sugars such as glucose and sucrose, steroids such as cholesterol, vitamins, alkaloids, proteins, and others — occur in the form of a single enantiomer. We'll see some of them in Section 3.9 and Chapter 5.

Here are some questions to check your understanding of the terms and principles presented in this section. The answers follow.

1. The simplest hydrocarbon having a chiral center is 3-methylhexane. Draw a Fischer projection for (*R*)-3-methylhexane.

2. Give a systematic name for each of the following hydrocarbons:

(a)

$$
\begin{array}{c}
C \\
| \\
C-C-C-C-C-C \\
| \\
C \\
| \\
C \\
| \\
C
\end{array}
$$

(b)

$$
\begin{array}{cc}
H & CH_2CH_2CH_3 \\
\diagdown & \diagup \\
C=C & \\
\diagup & \diagdown \\
H_3C & CH_2CH_3
\end{array}
$$

(c)

$$
\begin{array}{c}
C-C-C-C-C-C \\
| \\
C \\
| \\
C
\end{array}
$$

(d)

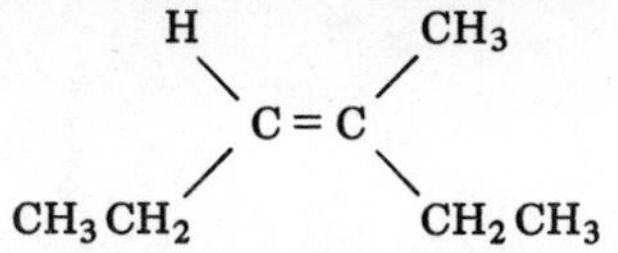

$$
\begin{array}{ccc}
H & & CH_3 \\
 & C=C & \\
CH_3CH_2 & & CH_2CH_3
\end{array}
$$

3. Give *E, Z, R,* or *S* as appropriate for each of these structures:

(a)

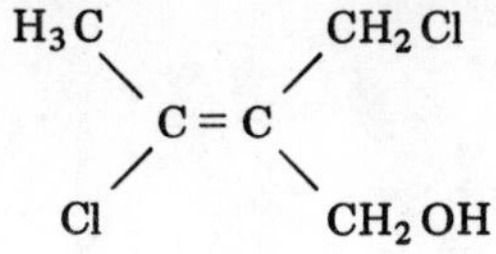

$$
\begin{array}{ccc}
H_3C & & CH_2Cl \\
 & C=C & \\
Cl & & CH_2OH
\end{array}
$$

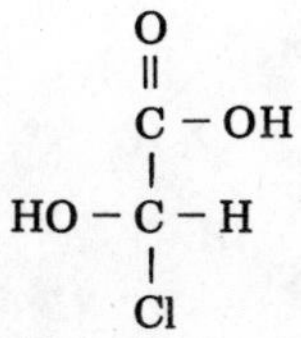

(b)

$$
\begin{array}{c}
O \\
\| \\
C-OH \\
| \\
HO-C-H \\
| \\
Cl
\end{array}
$$

(c)

$$
\begin{array}{ccc}
H_3C & & H \\
 & C=C & \\
Cl & & Br
\end{array}
$$

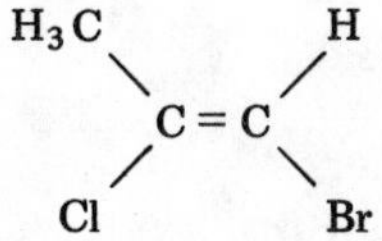

(d)

$$
\begin{array}{c}
O \\
\| \\
C-OH \\
| \\
H_3C-C-OH \\
| \quad\quad O \\
\quad\quad \| \\
CH_2C-OH
\end{array}
$$

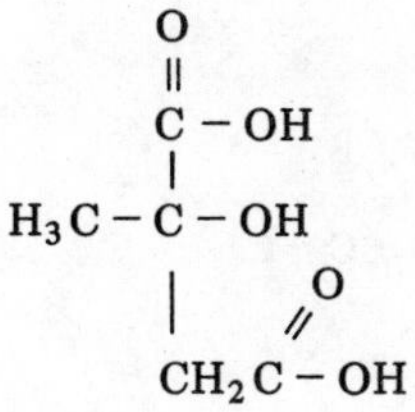

4. Draw unambiguous skeleton formulas or Fischer projections for each of the following:

   (a) (*Z*)-4-ethyl-3-methyl-3-heptene

   (b) (*E*)-2-pentene

   (c) (*3R,4R*)-4-ethyl-3-methylheptane

   (d) (*S*)-4-methyloctane

• •

1.

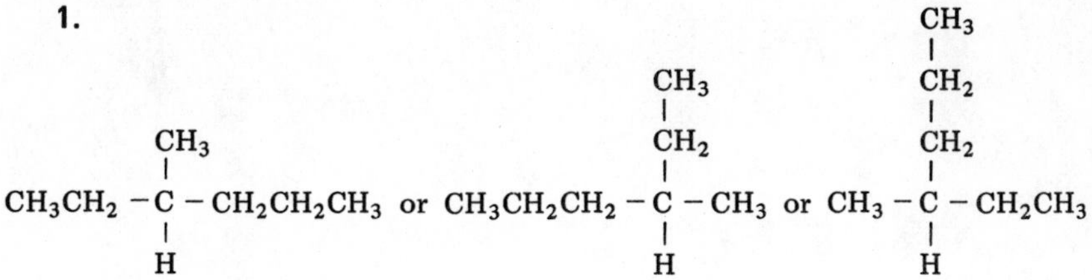

$$
CH_3CH_2-\underset{\underset{H}{|}}{\overset{\overset{CH_3}{|}}{C}}-CH_2CH_2CH_3 \quad \text{or} \quad CH_3CH_2CH_2-\underset{\underset{H}{|}}{\overset{\overset{CH_2}{\overset{|}{\overset{CH_3}{|}}}}{C}}-CH_3 \quad \text{or} \quad CH_3-\underset{\underset{H}{|}}{\overset{\overset{CH_2}{\overset{|}{\overset{CH_2}{\overset{|}{\overset{CH_3}{|}}}}}}{C}}-CH_2CH_3
$$

2. (a) (*R*)-4-ethyl-4-methyloctane

   (b) (*E*)-3-ethyl-2-hexene

   (c) (*S*)-3-methylheptane

   (d) (*Z*)-3-methyl-3-hexene

3. (a) *E*

   (b) *S*

   (c) *Z*

   (d) *R*

4. (a) 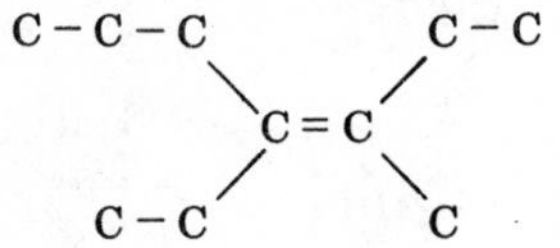
$$
\begin{array}{cc}
C-C-C & \quad C-C \\
& C=C \\
C-C & \quad C
\end{array}
$$

   (b)
$$
\begin{array}{ccc}
C & & \\
\quad C=C & \text{or} & C \qquad\qquad H \\
\quad\quad C-C & & \quad C=C \\
& & H \qquad\qquad C-C
\end{array}
$$

   (c)
$$
\begin{array}{c}
H \\
| \\
CH_3CH_2 - C - CH_3 \\
| \\
H - C - CH_2CH_3 \\
| \\
CH_2CH_2CH_3
\end{array}
$$

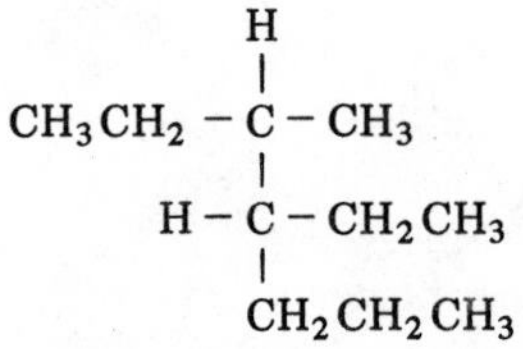

   (d) 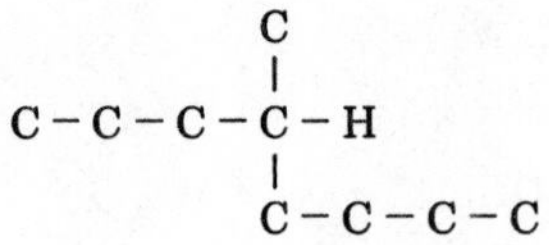
$$
\begin{array}{c}
C \\
| \\
C-C-C-C-H \\
| \\
C-C-C-C
\end{array}
$$

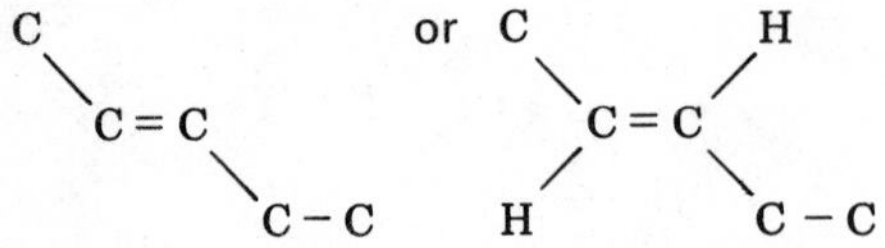

Chapter Two

CYCLIC HYDROCARBONS AND SUBSTITUTED HYDROCARBONS

From a structural point of view the two major classes of organic compounds are acyclic (aliphatic) and cyclic. Chapter One was concerned with acyclic compounds, principally hydrocarbons. Acyclic compounds with elements other than carbon and hydrogen will be found in this and the following chapters. Cyclic compounds may be further subdivided into alicyclic and aromatic classes. Alicyclic substances have many of the same chemical properties as their aliphatic counterparts. Aromatic compounds have distinctive chemical reactions that can be explained in terms of delocalized π electron bonds. Both classes may have rings consisting entirely of carbon atoms, or of carbon and other elements. The latter, called heterocyclic compounds, will be discussed in Chapter 4.

Our complete classification scheme is, then:

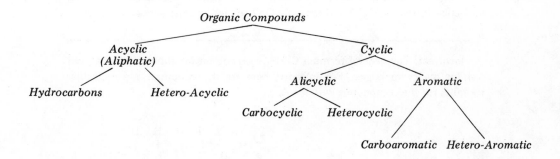

Any classification scheme is arbitrary to some degree. Some workers apply the term heterocyclic more generally to include both the saturated heterocyclic compounds and the unsaturated aromatic ones.

2.1 MONOCYCLIC HYDROCARBONS – CYCLOALKANES, CYCLOALKENES, AND CYCLOALKYNES

Alicyclic hydrocarbons resemble the corresponding aliphatic, or open-chain, hydrocarbons in many ways, including their names. The names are derived from the systematic names of the aliphatic hydrocarbons which have the same number of carbon atoms. The prefix *cyclo-* [sī′ klō] is attached. Here are some examples:

$$
\begin{array}{c}
CH_2 \\
/ \ \backslash \\
H_2C \text{———} CH_2
\end{array}
\qquad
\begin{array}{c}
H_2C - CH_2 \\
| \qquad | \\
H_2C - CH_2
\end{array}
\qquad
\begin{array}{c}
CH_2 - CH_2 \\
/ \qquad\quad \backslash \\
CH_2 \qquad CH_2 \\
\backslash \qquad / \\
CH_2 - CH_2
\end{array}
$$

cyclopropane   *cyclobutane*   *cyclohexane*

What is the general formula for cycloalkanes?

A _____

$$C_n H_{2n}$$

B _____

$$C_n H_{2n+2}$$

C _____

$$C_n H_{2n} \text{ with } n \geqslant 3$$

• •

A _____

You are close, but not entirely correct. Cycloalkanes do have two fewer hydrogen atoms than the aliphatic alkane with the same number of carbon atoms. Consequently, if the general formula for aliphatic alkanes is $C_n H_{2n+2}$, the general formula for the cycloalkanes ought to be $C_n H_{2n}$. But shouldn't there be a further limitation? Suppose that n = 1. Is there a cycloalkane with the molecular formula CH_2? No, there isn't. Go back for another answer.

B _____

Incorrect. The general formula $C_n H_{2n+2}$ is correct for aliphatic alkanes, but not for the cycloalkanes. For instance, here are the structural and molecular formulas for the compounds with n = 4.

$$
\begin{array}{c}
H \quad H \\
| \quad\ | \\
H - C - C - H \\
| \quad\ | \\
H - C - C - H \\
| \quad\ | \\
H \quad H
\end{array}
\qquad\qquad
\begin{array}{c}
H \quad H \quad H \quad H \\
| \quad\ | \quad\ | \quad\ | \\
H - C - C - C - C - H \\
| \quad\ | \quad\ | \quad\ | \\
H \quad H \quad H \quad H
\end{array}
$$

cyclobutane, $C_4 H_8$ *butane*, $C_4 H_{10}$

If comparison of these two formulas doesn't lead you to the correct general formula for the cycloalkanes, try writing one or two more similar pairs. Then go back and choose another answer.

C _____

Right. A cycloalkane has two fewer hydrogen atoms than the aliphatic alkane with the same number of carbon atoms. Therefore, the general formula is $C_n H_{2n}$. Since there are no cycloalkanes with 1 or 2 carbon atoms, there must be the further statement that *n* is equal to or greater than 3.

Cyclopropane is a widely used anesthetic. Most important of the cycloalkanes, by far, is cyclohexane, which is used as a solvent and as an intermediate in organic synthesis. Cyclohexane exists in several conformations. Although they all bear the same name, they are noteworthy because of the insight they give into the shapes and names of substituted derivatives of cyclohexane.

There are three conformations in which the bond angles are normal for sp^3-hybridized carbon; that is, the tetrahedral angle, 109.5°. Such conformations are said to be free of angle strain. Because of their shapes, they are called chair, boat, and twist-boat (twist).

The figure below shows three different ways to represent each of the conformations. In the top row each carbon atom is represented by a circle. Smaller circles for the twelve hydrogen atoms are added in the second row. The bottom row shows only the lines that represent the axes of the carbon-carbon bonds.

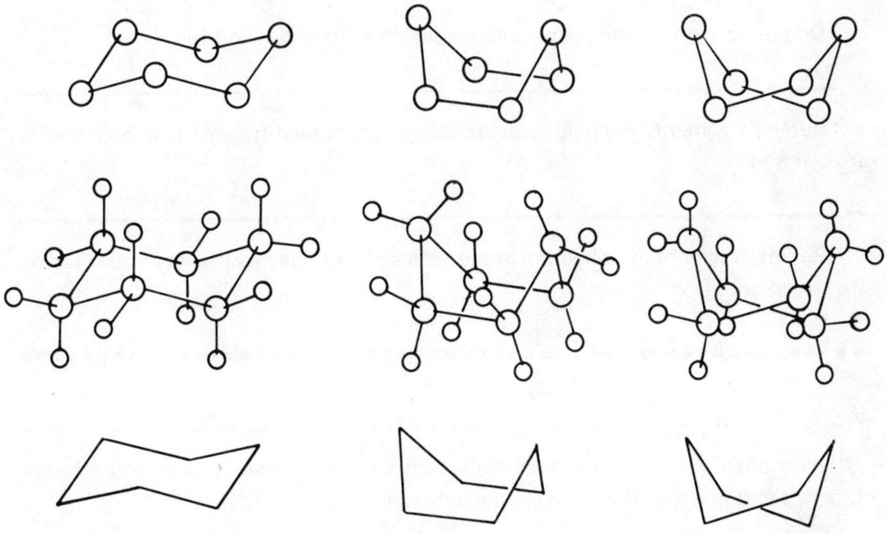

Chair   Boat   Twist-boat

The Newman projection of the chair form appears like this:

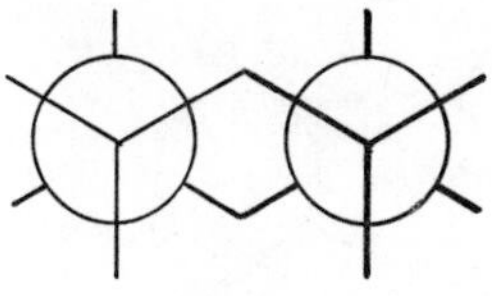

Chair
cyclohexane

Sighting along any one of the carbon-carbon bonds shows the same relationship among the carbon-hydrogen bonds.

If the foot of the chair is "flipped" up to make the boat form, the Newman projection becomes

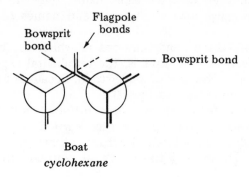

Boat
cyclohexane

Notice the different relationship among the carbon-hydrogen bonds.

Do you remember the terms used to describe these relationships?

A _____

Butane fragments in chair form are staggered; butane fragments in boat form are eclipsed.

B _____

Butane fragments in chair form are eclipsed; butane fragments in boat form are staggered.

• •

A _____

You are right. The Newman projection for a staggered conformation looks like this and matches that of chair cyclohexane:

The Newman projection of an eclipsed conformation looks like this and matches that of boat cyclohexane:

Go on to the next section.

B _____

You have it backwards. See page 10 for the explanation of Newman projections and then read *A* above.

In general, staggered conformations are thermodynamically more stable than eclipsed. In fact, the chair form is the most stable conformation of cyclohexane, most derivatives of cyclohexane, and most other six-membered rings as well.

The various representations of the cyclohexane ring clearly show that it is not planar. As shown here, a plane that bisects the axes of all of the carbon-carbon bonds will leave three carbon atoms above the plane and three below it.

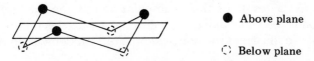

● Above plane

◯ Below plane

Plane bisecting C − C bonds

Six of the carbon-hydrogen bonds are roughly perpendicular to the plane shown in the figure above. They are called *axial* hydrogens. The remaining six are roughly parallel to the plane, although three point up and three down. They are shown here, too, and are called *equatorial* hydrogens. Note that they all point toward the plane.

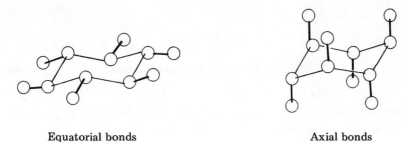

Equatorial bonds   Axial bonds

Depending on whether the methyl group is in an equatorial or axial position, methylcyclohexane can exist in two forms.

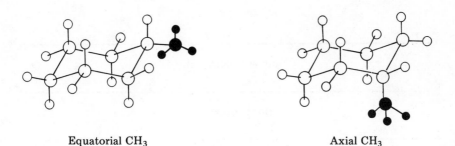

Equatorial CH$_3$ Axial CH$_3$

The equilibrium at room temperature favors the equatorial conformation about 19 to 1 because of the greater distance between the methyl group and the neighboring hydrogen atoms in the equatorial form. In the equatorial position the methyl group points away from its nearest neighbors, the two hydrogens on the adjacent carbon atoms. The methyl group in the axial position is held by a bond that is parallel to its nearest neighbors, two axial hydrogen atoms.

Equatorial methyl group   Axial methyl group

As if all this weren't confusing enough, one chair form can be converted into another. As a result of this chair-chair conversion, or "ring-flipping," all equatorial bonds become axial and all axial bonds become equatorial. This figure attempts to illustrate the conversion. Your model will do it better.

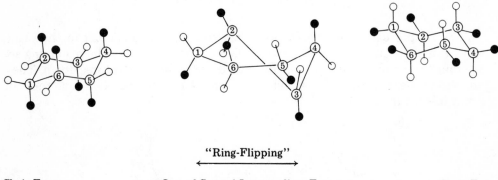

"Ring-Flipping"

Chair Form   One of Several Intermediate Forms   Chair Form

Here is a representation of 1,2-dimethylcyclohexane in the chair conformation. What positions do the two methyl groups occupy?

A _____

Both are equatorial.

B _____

Both are axial.

C _____

One is axial and one is equatorial.

• •

A _____

Right. Both methyl groups are in equatorial positions relative to the cyclohexane ring. The carbon-carbon bonds between the ring carbons and the methyl carbons are roughly parallel to the plane that bisects the C – C bonds of the ring. These positions are designated equatorial. Go on to the next section.

B _____

Wrong. You may have misread the discussion of axial and equatorial bonds. Those that are roughly perpendicular to the plane that bisects the C – C bonds of the ring are called axial while those that are roughly parallel to the plane are called equatorial. Take another look at the formula and choose another answer.

C _____

You are incorrect. According to the earlier discussion, axial bonds are the ones that are roughly perpendicular to the plane that bisects the C – C bonds in the ring. Equatorial bonds are roughly parallel to the plane. One of each is labeled in this diagram. Look at it carefully and then choose another answer.

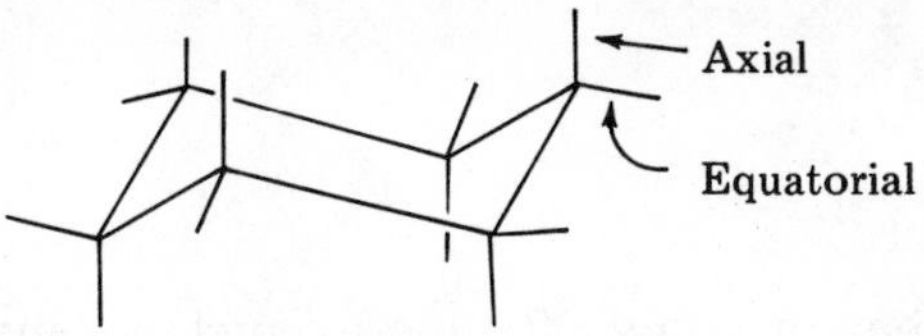

Let's continue to examine this diequatorial chair conformation of 1,2-dimethylcyclohexane:

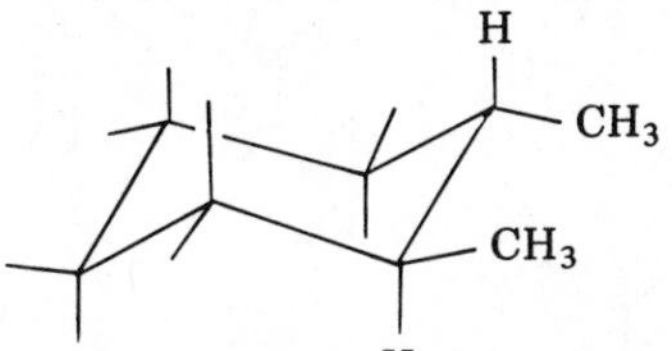

Recall that *cis-trans* isomerism designates groups which lie on the same side or opposite sides of a plane. Using the plane shown on page 101 for reference, this diequatorial form is a *trans* isomer.

Consider the diaxial chair conformation of 1,2-dimethylcyclohexane. Is it a *cis-* or *trans-* isomer?

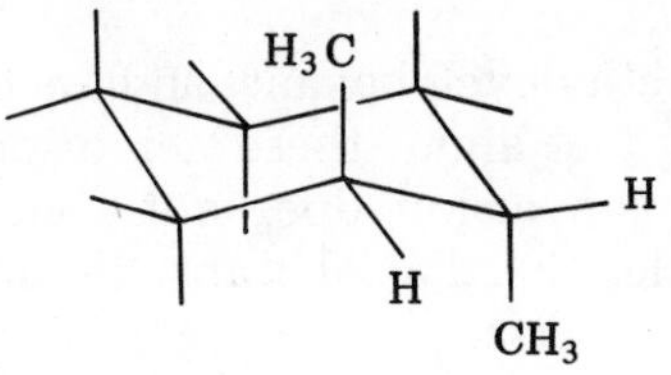

A _____

cis-

B _____

trans-

••

A _____

You are wrong. You should recall that the substituent groups in a *cis-* isomer lie on the same side of a reference plane. For cyclohexane and its derivatives the reference plane is the plane that bisects the $C-C$ bonds in the ring (see p. 101). It should be easy for you to see that the two methyl groups here are on opposite sides of the plane. Go on to *B* below.

B _____

You are right. The two axial methyl groups are on opposite side of the plane that bisects the $C-C$ bonds in the ring. The formula represents a *trans-* isomer. Go on to the next section.

These two isomers of *trans*-1,2-dimethylcyclohexane are diastereomers. They are not mirror images of one another. Because of less crowding between the methyl groups and the axial hydrogens, the diequatorial form is more stable.

Two other possibilities for 1,2-dimethylcyclohexane are equatorial-axial and axial-equatorial substitution, as shown here:

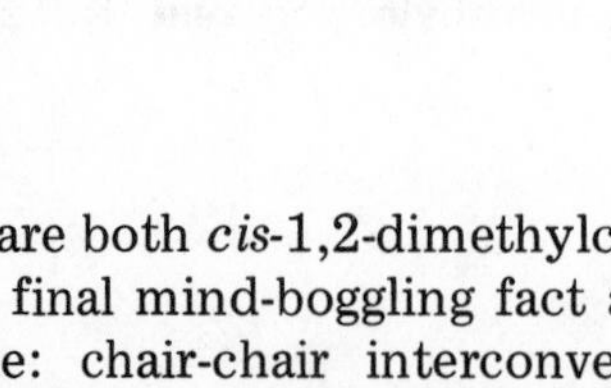

Equatorial-axial 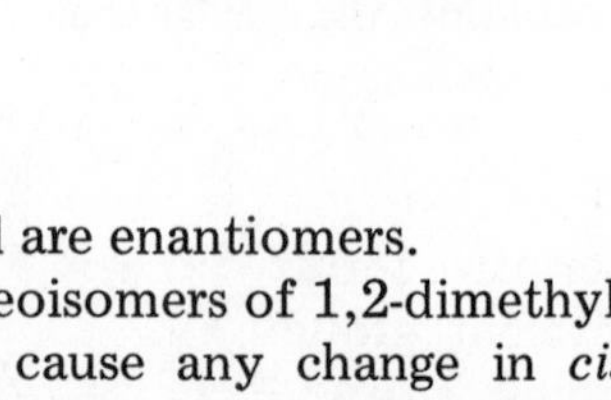   Axial-equatorial

They are both *cis*-1,2-dimethylcyclohexane, and are enantiomers.

A final mind-boggling fact about these stereoisomers of 1,2-dimethylcyclohexane: chair-chair interconversion does not cause any change in *cis-trans* relationships. For example, the diaxial *trans-* isomer converts to the diequatorial *trans-* isomer and vice versa.

Is there a possibility of chirality in 1,2-dimethylcyclohexane? The drawing below represents *trans*-1,2-dimethylcyclohexane in its more stable diequatorial conformation and its mirror image. Because they are not superimposable, they are enantiomers. Chair-chair interconversion leads to the diaxial conformers that are enantiomers and are not superimposable on the diequatorial pair. Therefore, *trans*-1,2-dimethylcyclohexane should, and does, exist as two optically active enantiomers.

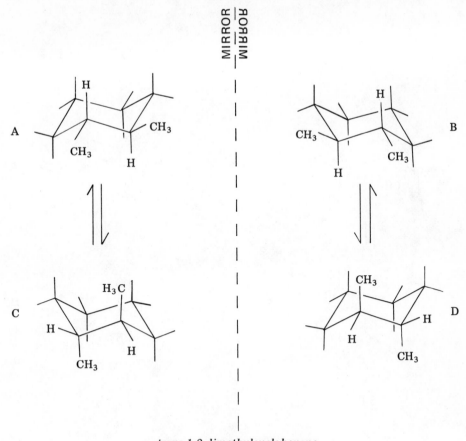

trans-1,2-dimethylcyclohexane
a resolvable racemic modification
(chair-chair interconversion of A and B leads to new enantiomers C and D)

On the next page is a representation of *cis*-1,2-dimethylcyclohexane similar to the figure above. Because the molecule and its mirror image are not superimposable, they are also enantiomers. An important difference, however, is that chair-chair interconversion does *not*lead to a new pair of enantiomers, but to the mirror image. In other words, *cis*-1,2-dimethylcyclohexane and its mirror image are conformational enantiomers. At ordinary temperatures the interconversion is too rapid to permit resolution (separation) and measurement of optical activity.

The foregoing is by no means a complete discussion of the stereochemistry of the conformations of cyclic compounds. It should, however, explain most of the descriptive terms you will encounter.

Univalent groups derived from cycloalkanes are named by replacing the *-ane* ending with *-yl*. The carbon atom with the available bond is designated 1. For example:

cyclopropyl group   *cyclohexyl group*

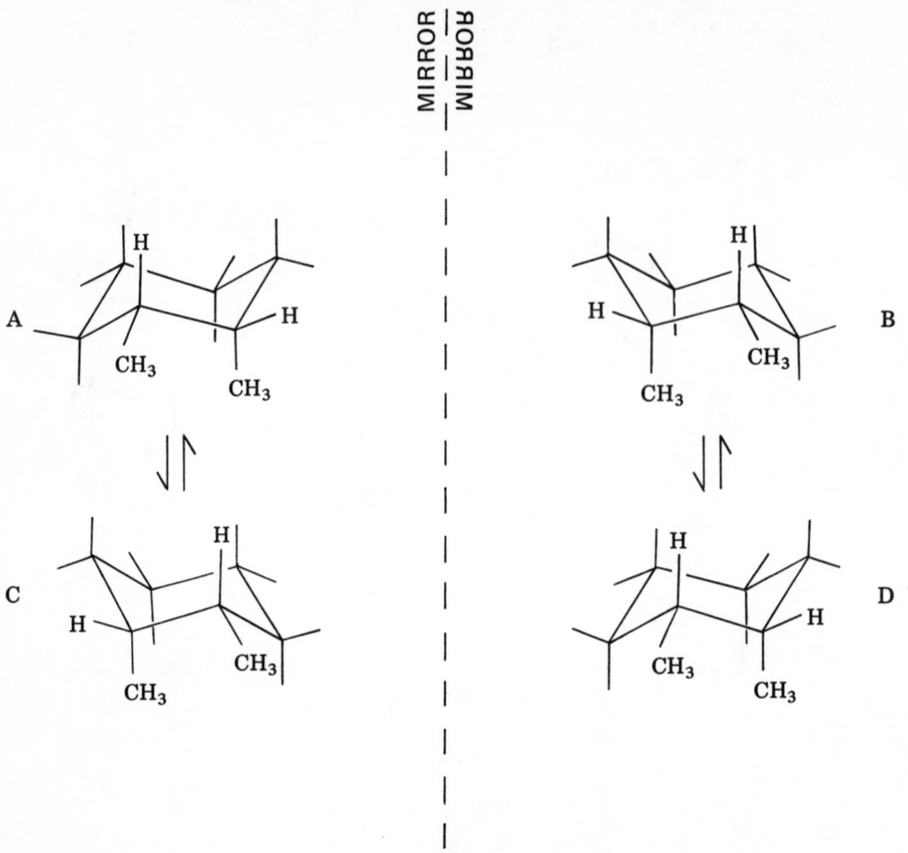

cis-1,2-dimethylcyclohexane
a non-resolvable racemic modification
(chair-chair interconversion of A leads to mirror image C; chair-chair interconversion of B
leads to mirror image D. A is identical to D; B is identical to C)

The unsaturated alicyclic hydrocarbons are named by adding the prefix *cyclo-* to the name of the corresponding alkene or alkyne. Two examples are:

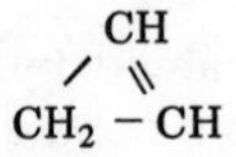

cyclopropene

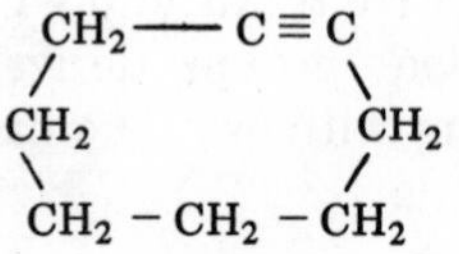

cyclooctyne

Cyclooctyne is the smallest stable cyclic hydrocarbon that can accommodate the triple bond.

If there is more than one multiple bond, locants are needed to specify their locations. Begin the numbering so that one of the double bonds joins the number 1 and 2 carbon atoms. Study this example:

$$
\begin{array}{ccc}
 & CH & \\
 & \diagup \ \diagdown\diagdown & \\
CH_2 & & CH \\
| & & | \\
CH_2 & & CH \\
 & \diagdown \ \diagup\diagup & \\
 & CH &
\end{array}
$$

1,3-cyclohexadiene

Cis-trans isomerism is possible in cycloalkenes just as in the acyclic alkenes. When a double bond occurs in a ring smaller than cyclooctene, the ring bonds must be *cis-* to one another and the unsubstituted hydrocarbon is always designated *Z*.

$$
\begin{array}{ccc}
CH_2\text{-} CH_2 & & \\
\diagup \quad | & & \\
CH_2 \quad CH_2 & & \\
\diagdown \quad | & & \\
\quad C = C & & \\
\diagup \quad \diagdown & & \\
H \qquad H & &
\end{array}
\qquad
\begin{array}{c}
(CH_2)_4 \\
\diagup \qquad \diagdown \\
CH_2 \qquad CH_2 \\
\diagdown \qquad \diagup \\
C = C \\
\diagup \qquad \diagdown \\
H \qquad\qquad H
\end{array}
\qquad
\begin{array}{c}
\overparen{CH_2 \quad} \ H \ (CH_2)_4 \\
(\qquad\qquad | \\
C = C \qquad CH_2 \\
\diagup \qquad \underparen{\qquad} \\
H
\end{array}
$$

  (Z)-cyclohexene   *(Z)-cyclooctene*   *(E)-cyclooctene*

The names of univalent groups derived from cycloalkenes and cycloalkynes have the endings *-en-yl*, *-yn-yl*, *-dien-yl*, etc. The carbon atom with the available bond is numbered 1 and the position(s) of the multiple bond(s) indicated in the usual manner. For example,

$$
\begin{array}{c}
CH - CH - \\
|| \quad\quad | \\
CH - CH_2
\end{array}
\qquad\qquad\qquad
\begin{array}{c}
| \\
CH \\
\diagup \ \diagdown \\
CH \quad CH \\
|| \quad\quad || \\
CH - CH
\end{array}
$$

   2-cyclobuten-1-yl group   *2,4-cyclopentadien-1-yl group*

Here are five questions concerning unsaturated alicyclic hydrocarbons. Keep the answers covered until you have worked out your own answer.

1. Write the carbon skeleton for cycloheptene.

2. What is the name of this compound?

$$
\begin{array}{c}
CH - CH \\
\qquad\qquad \diagdown\diagdown \\
|| \qquad\qquad\quad CH \\
\qquad\qquad \diagup \\
CH - CH_2
\end{array}
$$

3. What is the name of this compound?

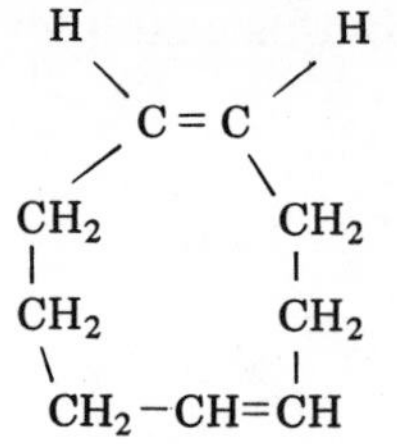

$$
\begin{array}{ccc}
H & & H \\
\diagdown & & \diagup \\
& C = C & \\
\diagup & & \diagdown \\
CH_2 & & CH_2 \\
| & & | \\
CH_2 & & CH_2 \\
\diagdown & & | \\
& CH_2 - CH = CH &
\end{array}
$$

4. Write a structural formula for (*E*)-1-methylcyclodecene.

5. What is the name of this group?

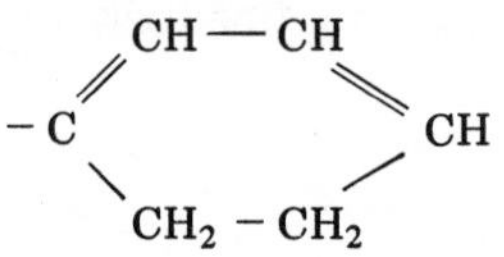

$$
\begin{array}{ccc}
& CH - CH & \\
\diagup\!\!\!= & & =\!\!\!\diagdown \\
- C & & CH \\
\diagdown & & \diagup \\
& CH_2 - CH_2 &
\end{array}
$$

• •

1.

$$
\begin{array}{ccc}
& C & \\
\diagup & & \diagdown \\
C & & C \\
| & & | \\
C & & C \\
\diagdown & & \diagup \\
& C = C &
\end{array}
$$

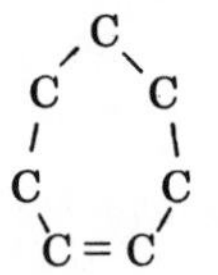

2. 1,3-cyclopentadiene

3. *(Z)-1,5-cyclononadiene*

4.

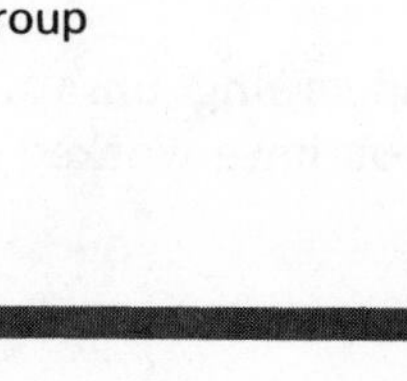

$$
\begin{array}{ccc}
& (CH_2)_6 & \\
CH_2 & & CH_3 \\
\diagdown & & \diagup \\
& C = C & \\
\diagup & & \diagdown \\
H & & CH_2
\end{array}
$$

5. 1,3-cyclohexadien-1-yl group

<table>
<tr><td colspan="2" align="center">*MONOCYCLIC HYDROCARBONS*</td></tr>
<tr><td>General formulas:</td><td>Series names:</td></tr>
<tr><td>C_nH_{2n} ($n \geqslant 3$)</td><td>*cyclo-* . . . . *-ane*</td></tr>
<tr><td>C_nH_{2n-2} ($n \geqslant 3$)</td><td>*cyclo-* . . . . *-ene*</td></tr>
<tr><td>C_nH_{2n-4} ($n \geqslant 8$)</td><td>*cyclo-* . . . . *-yne*</td></tr>
</table>

2.2 SPIRO, FUSED, AND BRIDGED HYDROCARBONS; RING ASSEMBLIES

When two rings have a single atom in common, the result is a *spiro system*. When two rings share two adjacent atoms, the system is *fused*. If two rings share non-adjacent atoms, the system is *bridged*. Rings joined by a single bond are called *ring assemblies*.

spiro   fused   bridged   ring assembly

Spiro systems, bridged systems, and ring assemblies will be discussed in this section. Because most common fused ring systems are aromatic, they will be covered in Section 2.3.

Spiro compounds consisting of only two alicyclic rings are named by placing *spiro* before the name of the hydrocarbon containing the same total number of carbon atoms. The number of carbon atoms in each ring linked to the spiro atom is placed in brackets and inserted between the prefix spiro and the hydrocarbon name. The numerals are arranged in ascending order and separated by a period. For example:

spiro[2.3]hexane   *spiro[3.4]octane*

Note that the sum of the bracketed numerals is one less than the number of carbon atoms.

When necessary, the carbon atoms are numbered consecutively, starting with an atom in the smaller ring next to the spiro junction; after passing around this ring and through the junction, numbering continues so as to afford lowest locants for substituents in the larger ring.

2-methylspiro[3.3]heptane   *1,4-dimethylspiro[2.4]heptane*
                                     (not *2,4-dimethylspiro[2.4]heptane*)

The final example above has three chiral centers, namely the carbons numbered 1, 3, and 7. It could, therefore, have eight stereoisomers. The same holds true for the compound shown below. How would you name it? Which three carbons are the chiral centers?

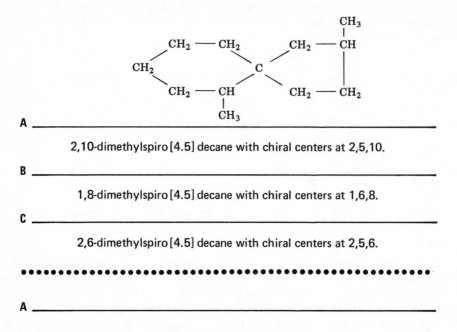

A _____

2,10-dimethylspiro[4.5]decane with chiral centers at 2,5,10.

B _____

1,8-dimethylspiro[4.5]decane with chiral centers at 1,6,8.

C _____

2,6-dimethylspiro[4.5]decane with chiral centers at 2,5,6.

• •

A _____

Incorrect. You have made an error in numbering the carbons in the rings. You began all right with the smaller ring but went the long way around the second ring. Go back and choose another answer.

B _____

You are wrong. It appears you began numbering with the larger rather than the smaller ring. You should start with the carbon in the smaller ring that is adjacent to the spiro carbon and closer to the substitutent methyl group. Go back and try again.

C _____

Right! The compound is 2,6-dimethylspiro[4.5]decane and the chiral centers are 2,5,6. You correctly began the numbering in the smaller ring and continued around the larger ring in the direction to give the lowest locant.

When unsaturation is present in a spiro hydrocarbon, the same approach is used for numbering with the direction chosen to give the double and triple bonds the lowest possible locants. For instance

spiro[4.5]deca-1,6-diene
(not *spiro[4.5]deca-1,9-diene*)

Here are three spiro compounds for you to name or to draw the skeletons of. The answers follow question 3. Keep them covered until you have your answer.

1. Draw the carbon skeleton for spiro[4.4]nonane.

2. Name this compound, neglecting possible stereoisomers.

$$CH_3 - CH \diagdown \qquad CH - CH_3$$
$$\Big| \qquad \diagdown C \diagup \qquad \Big|$$
$$CH_2 \qquad CH_2$$

3. Draw the carbon skeleton for spiro[3.4]oct-1-ene.

● ●

1.
$$C - C \diagdown \qquad C - C$$
$$\Big| \qquad \diagdown C \diagup \qquad \Big|$$
$$C - C \diagup \qquad \diagdown C - C$$

2. 1,4-dimethylspiro[2.2]pentane

3.
$$C - C \diagdown \qquad C$$
$$\Big| \qquad \diagdown C \diagup \diagdown$$
$$\qquad \diagup C \diagdown \qquad C$$
$$C - C \qquad C$$

As stated earlier, systems in which two or more rings share non-adjacent carbon atoms are said to be bridged. The simplest are those which have two rings sharing two carbon atoms. For example:

$$\begin{array}{c} H \\ C \\ \diagup | \diagdown \\ H_2C \quad | \quad CH_2 \\ \diagdown | \diagup \\ C \\ H \end{array}$$

The bridges are the chain(s) of atoms or the valence bond(s) connecting the different parts of the molecule. The two tertiary carbon atoms at the ends of the bridges are called *bridgeheads*. In the example above, the two methylidyne groups are the bridgeheads. The bridges are the two methylene groups and the valence bond between the bridgeheads. While the two bridgeheads might be considered adjacent to one another along the bond, they are non-adjacent along the other two bridges.

Saturated bridged carbocyclic compounds consisting of two rings only take the name of the alkane with the same total number of carbon atoms preceded by *bicyclo-*. The number of carbon atoms in each of the three bridges is placed in brackets and inserted between the prefix and hydrocarbon name. The numerals are arranged in descending order and separated by periods. Two examples, including the one previously used, are:

$$CH_2 \diagdown \begin{array}{c} CH \\ | \\ CH \end{array} \diagup CH_2$$

bicyclo[1.1.0]butane

$$\begin{array}{ccc} & CH_2 - CH & \\ CH_2 & CH_2 & CH_2 \\ CH_2 & | & CH_2 \\ & CH & \end{array}$$

bicyclo[3.2.1]octane

Note that the sum of the bracketed numerals is two less than the total number of carbon atoms.

Norbornane, a relative of camphor, has this structure. What is its systematic name?

$$\begin{array}{ccc} & CH & \\ CH_2 & & CH_2 \\ & CH_2 & \\ CH_2 & & CH_2 \\ & CH & \end{array}$$

A _____

bicyclo[1]heptane

B _____

bicyclo[2.2.1]heptane

C _____

bicyclo[3.3.2]heptane

• •

A _____

Wrong. Perhaps you view the norbornane structure as a six-membered ring with a single methylene group bridging two of the ring carbons. The systematic name, however, must cite *all* of the paths between the bridgeheads. Go back and pick another answer.

B _____

You are correct. The correct systematic name for norbornane is bicyclo[2.2.1]heptane. The numerals reflect the number of carbon atoms in each bridge and are cited in descending order. Go on to the next section.

C _____

Incorrect. You seem to have counted one of the bridgeheads as part of the bridges. The numerals should reflect the number of carbon atoms *between* the bridgeheads. Go back and choose another answer.

The presence of the bridges in norbornane forces the cyclohexane ring to have the boat conformation. Here are three representations with the bridgehead atoms circled:

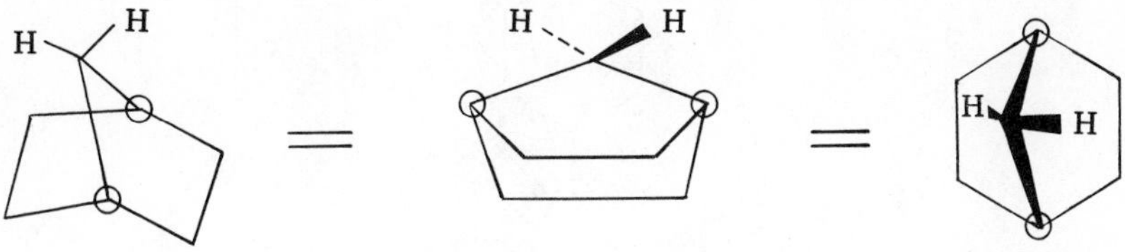

As usual, locants are required to locate substituent groups or unsaturation. Bicyclic systems are numbered by assigning 1 to one of the bridgehead atoms, following the longest path to the second bridgehead atom, the longest remaining path back to the first bridgehead atom, and finally the remaining path. Study these examples and the accompanying remarks:

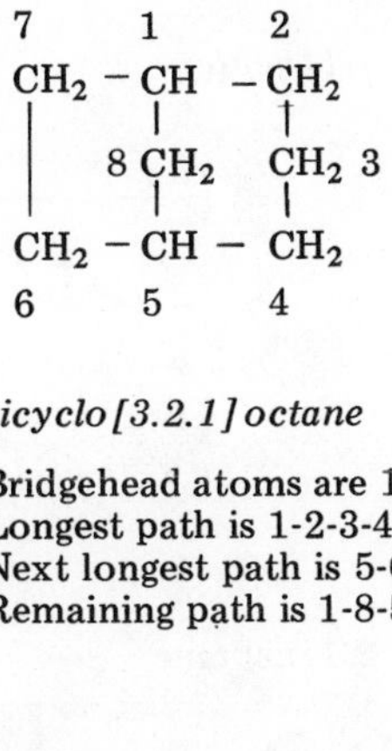

bicyclo[3.2.1]octane

Bridgehead atoms are 1 and 5
Longest path is 1-2-3-4-5
Next longest path is 5-6-7-1
Remaining path is 1-8-5

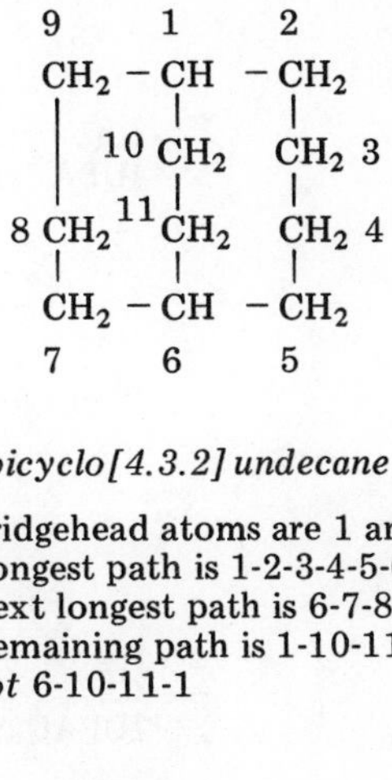

bicyclo[4.3.2]undecane

Bridgehead atoms are 1 and 6
Longest path is 1-2-3-4-5-6
Next longest path is 6-7-8-9-1
Remaining path is 1-10-11-6,
not 6-10-11-1

Note carefully in the second example that the numbering continues from 9 through the bridgehead to 10 and 11.

Unsaturated bridged hydrocarbons are named in the usual manner. If there is a choice in numbering, unsaturation is given the lowest number(s). Look at this example closely:

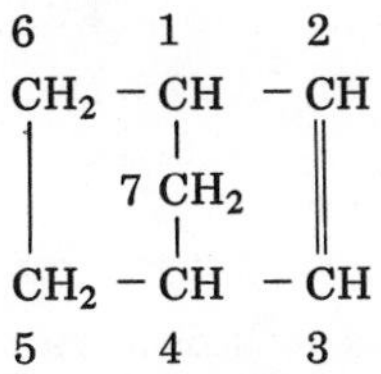

bicyclo[2.2.1]hept-2-ene

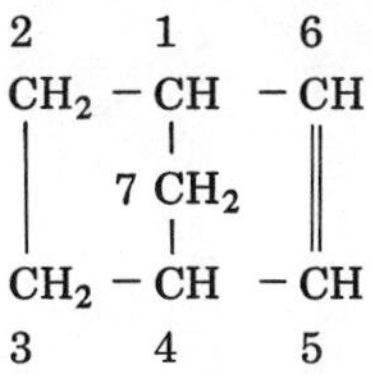

incorrect

Some of the terpenes (see Section 5.4) are derivatives of these saturated bicyclic hydrocarbons:

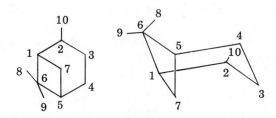

CA: *4-methyl-1-(1-methylethyl)bicyclo[3.1.0]hexane*
IUPAC: *thujane*

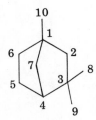

CA: *2,6,6-trimethylbicyclo[3.1.1]heptane*
IUPAC: *pinane*

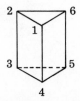

CA: *1,3,3-trimethylbicyclo[2.2.1]heptane*
IUPAC: *fenchane*

Cyclic hydrocarbon systems of three or more rings are known. Their nomenclature, while not overly difficult, is complicated and will not be discussed in this book. The following two examples are given for your amusement and to show that organic chemists do have their lighthearted moments.

CA: *tetracyclo[2.2.0.0^{2,6}.0^{3,5}]hexane*
Trivial: *prismane*

CA: pentacyclo[4.2.0.0^{2,5}.0^{3,8}.0^{4,7}]octane
Trivial: *cubane*

An assembly of two *identical* cyclic hydrocarbon groups may be named in one of two ways under the IUPAC rules: (1) by placing the prefix *bi-* in front of the name of the corresponding univalent group; or (2) by placing the prefix *bi-* in front of the name of the corresponding hydrocarbon. *CA* restricts the use of the first method to two-component assemblies of monocyclic hydrocarbons and hetero systems with "cyclo" names, and uses the second method for all others. For instance:

CA: 1,1'-*bicyclopropyl*
IUPAC: above, and 1,1'-*bicyclopropane*

Notice that the numbering is that of the corresponding hydrocarbon, with one group given an unprimed number and the other a primed one.

Assemblies of three or more groups are named by the second method. For example:

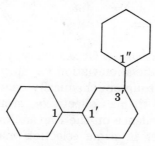

1,1':3',1"-*tercyclohexane*

SPIRO HYDROCARBONS

   Series name: *spiro*[   ]

BRIDGED HYDROCARBONS

   Series name: *bicyclo*[   ]

HYDROCARBON RING ASSEMBLIES

   Series name: *bi-, ter-, quater,* etc.

2.3 AROMATIC HYDROCARBONS

The aromatic series of compounds has special characteristics that distinguish it from the alicyclic series, and completes the classification scheme on page 97. The most common aromatic hydrocarbon is benzene, and the term aromatic is applied to all stable cyclic compounds having delocalized π electron bonds. Most of them contain six-membered rings but the ring size can vary from three to twenty-two. The rings are unsaturated, but do not undergo the same reactions as the cycloalkenes and cycloalkynes. Benzene and the other aromatics are examples of resonant compounds that cannot be represented adequately by a single structural formula. The generic term for this class of compounds is *arene*.

In formula writing, benzene is usually represented by a hexagon with the six hydrogen atoms not shown.

benzene   *benzene*

The three double bonds or the circle designate the aromatic character of the ring; the circle will be used in this book.

When a hydrogen atom is removed from one of the carbon atoms, the group formed has the formula C_6H_5- and may be shown as:

It is called a *phenyl* [fĕn' ĭl] *group.* Occasionally, you may still see it shown by the Green letter phi, ϕ, or the abbreviation Ph. Since all of the carbon atoms in the unsubstituted phenyl group are chemically equivalent, the free valence can be shown on any one of them. Collectively, groups derived from aromatic compounds with the free valence on the ring are known as *aryl* [âr' ĭl] *groups.*

The substantial changes in name selection made in 1972 for the ninth collective index period of *Chemical Abstracts* are quite apparent in the aromatic hydrocarbons. In this section it is important to distinguish between the names still acceptable under the IUPAC rules and those used by *Chemical Abstracts.* For example, when a simple alkyl group is joined to a phenyl group, the IUPAC rules name the resulting compound as either a substituted benzene or a substituted alkane, depending upon the chemical intent. *CA* names them as substituted benzenes.

CH_3

CA: *methylbenzene*
IUPAC: *methylbenzene* or *phenylmethane*

Some alkyl-substituted benzenes with trivial names you need to remember are toluene [tŏl' ū ēn], styrene [stī' rēn], and cumene [kū' mēn]. These names are accepted for use in the IUPAC system, but *Chemical Abstracts* uses the systematic names as shown below.

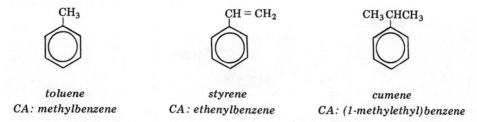

toluene	styrene	cumene
CA: methylbenzene	*CA: ethenylbenzene*	*CA: (1-methylethyl)benzene*

The location of substituent groups on a benzene ring becomes significant when two or more groups are substituted on the benzene ring. There are three isomeric dimethylbenzenes. Their formulas are:

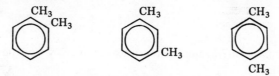

How can they be distinguished? There are two systems. One involves numbers and the other uses a special set of prefixes. In the first system a carbon atom bearing a substituent is given the number 1 and the rest are numbered consecutively around the ring in the direction that will give the smallest set of locants (see p. 46) when all of the substituents are located. For example

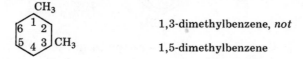

1,3-dimethylbenzene, *not*

1,5-dimethylbenzene

Unlike the dimethylhexanes, the dimethylbenzenes are planar and there are no *cis-trans* isomers.

What is the structural formula for 1-ethyl-4-methylbenzene?

A _____

CH_2CH_3 — benzene ring — CH_3

B _____

benzene ring — CH_2CH_3 / CH_3

C _____

CH_3CH_2 — benzene ring — CH_3

A _____

You are incorrect. You may have selected this answer hurriedly. Look at the formula again:

Do you see anything wrong? Look at the ethyl group carefully. Its structural formula would be

That is wrong, isn't it? Examine all formulas thoroughly from now on. Go on to the next section.

B _____

You are wrong. The formula you have selected for 1-ethyl-4-methylbenzene is

Number the carbon atoms again. The carbon bearing the ethyl group is number 1. The one with the methyl group is number 3. This is the formula for 1-ethyl-3-methylbenzene. Go back and choose another answer.

C _____

You are correct. The structural formula for 1-ethyl-4-methylbenzene is

In this particular instance the numbering of the ring can go in either direction.

The trivial name for all three dimethylbenzenes is xylene [zī′ lēn]. Xylene is accepted by the IUPAC rules. The formulas and names of the three xylenes are:

1,2-dimethylbenzene   *1,3-dimethylbenzene*

1,4-dimethylbenzene

The second system for locating substituents involves prefixes and is used in the IUPAC rules but not by *Chemical Abstracts*. The prefixes are *ortho-*, *meta-*, and *para-*. They represent the positions of carbon atoms in the ring relative to the carbon atom bearing the first substituent group. In the xylenes the first substituent is a CH_3- group. This diagram shows the prefixes for the other five positions in the benzene ring:

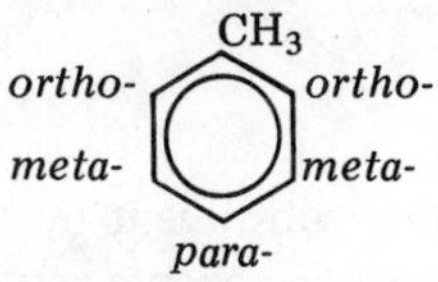

Note: The two *ortho-* positions are adjacent to the first substituent. The two *meta-* positions are separated by one carbon atom from the first substituted carbon atom. The single *para-* position is separated by two carbon atoms from the first.

The prefixes are abbreviated *o-*, *m-*, and *p-* for *ortho-*, *meta-*, and *para-*, respectively. Neither the words nor their abbreviations are considered in establishing the alphabetical order for listing substituent groups.

What is the commonly seen name for 1,3-dimethylbenzene?

A _____

ortho-xylene

B _____

meta-xylene

C _____

para-xylene

●●

A _____

Incorrect. Here is the diagram from above.

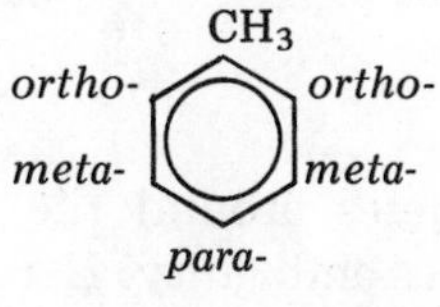

Notice that the two *ortho-* positions are adjacent to the first substituent. *Ortho*-xylene is 1,2-dimethylbenzene. Go back and choose another answer.

B _____

You are right. The common name of 1,3-dimethylbenzene is *meta*-xylene. Here is the diagram from above again.

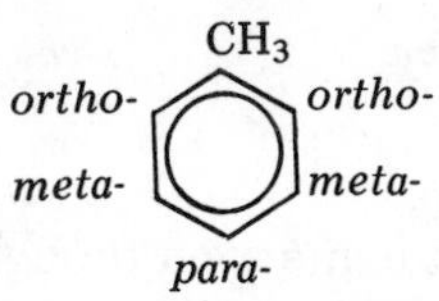

Remember that the 1,2-combination is *ortho-*, the 1,3-combination is *meta-*, and the 1,4-combination is *para-*. Go on to the next section.

C _____

You are wrong. Here is the diagram from the previous page.

Notice that the *para-* position is separated from the first substituent by two other carbons. *Para*-xylene is 1,4-dimethylbenzene. Go back and choose another answer.

Although it is unlikely that *ortho-, meta-, para-* nomenclature will ever disappear completely from textbooks and periodicals, it has been eliminated from names used for indexing by *Chemical Abstracts* since 1972.

Two more derivatives of benzene with trivial names accepted by the IUPAC rules are mesitylene [mĕ sĭt′ ĭl ĕn] and cymene [sĭ′ mĕn]. Cymene has 1,2-, 1,3-, and 1,4- isomers.

IUPAC:, *mesitylene*
CA: *1,3,5-trimethylbenzene*

p-cymene
1-methyl-4-(1-methylethyl)benzene

Notice the use of parentheses around the compound substituent group 1-methylethyl. Their use removes ambiguity. Here are two more examples:

1-chloro-2-(chloromethyl)benzene

1-chloro-2-(2-chloropropyl)benzene

Parentheses are used in the first name even though the chloromethyl group can be only $-CH_2Cl$. They are used in the second to make clear that the 2-chloropropyl group is clearly identified as the substituent group in the 2 position on the benzene ring.

The lowest set of locants is used in numbering the benzene ring. If there is a choice of identical sets, the group that comes first in alphabetical order is numbered one. Two examples are:

1-butyl-4-ethylbenzene

1-butyl-3-ethyl-2-propylbenzene
preferred over
3-butyl-1-ethyl-2-propylbenzene

Here are the structural formulas for five benzene derivatives. Work out a systematic name for each one. Be careful; they've been selected to illustrate the points explained above. The answers follow.

1.

2.

3.

4.

5.

1. 2-ethyl-1,3-dimethylbenzene (IUPAC accepts 2-ethyl-*m*-xylene)

2. 1,4-diethenylbenzene (IUPAC accepts *p*-divinylbenzene)

3. 1-(2-chloroethyl)-3,5-dimethyl-2-(1-methylethyl)benzene

4. 1-methyl-3-(1-methylethyl)benzene (IUPAC accepts *m*-cymene)

5. 4-butyl-1-ethyl-2-methylbenzene (If the butyl group were considered to be in the number 1 position, the locant set would be 1,3,4. The set 1,2,4 is lower and must be used.)

Univalent groups derived from these monocyclic aromatic hydrocarbons having the available bond at a ring atom have the names listed here. The carbon with the available bond is always numbered 1.

CA: *phenyl*
IUPAC: *phenyl*

$CH(CH_3)_2$

2-(1-methylethyl)phenyl
(also *3-* and *4-*)
o-cumenyl (also *m-* and *p-*)

H_3C CH_3

CH_3

2,4,6-trimethylphenyl
mesityl

CH_3

CA: *2-methylphenyl*
        (also *3-* and *4-*)
IUPAC: *o-tolyl*
        (also *m-* and *p-*)

CH_3
CH_3

2,3-dimethylphenyl
(also *2,4-, 2,5-, 2,6-, 3,4-* and *3,5-*)
2,3-xylyl
(also *2,4-, 2,5-, 2,6-, 3,4-* and *3,5-*) .

Other groups with the free valence at a ring atom are named as substituted phenyl groups under the IUPAC rules and by *Chemical Abstracts.* For example,

CH_2CH_3

2-ethylphenyl

$CH=CH_2$

3-ethenylphenyl

CH_3
CH_2CH_3

3-ethyl-2-methylphenyl

Some groups with the free valence in a side chain have trivial names that are accepted by the IUPAC rules. Only the systematic names are recognized by *Chemical Abstracts.* Following are six of them. The systematic name used by *Chemical Abstracts* is given first, followed by the trivial name acceptable to the IUPAC.

phenylmethyl
benzyl

CH_2-

or $C_6H_5CH_2-$

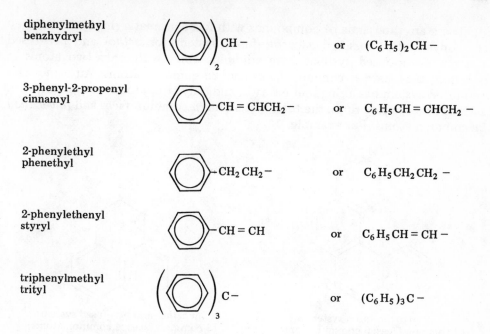

diphenylmethyl
benzhydryl

$(C_6H_5)_2CH-$

3-phenyl-2-propenyl
cinnamyl

$C_6H_5CH=CHCH_2-$

2-phenylethyl
phenethyl

$C_6H_5CH_2CH_2-$

2-phenylethenyl
styryl

$C_6H_5CH=CH-$

triphenylmethyl
trityl

$(C_6H_5)_3C-$

The name of the ring assembly consisting of two benzene rings is biphenyl. In the formula below, note the numbering with unprimed numbers for one ring and primed numbers for the other. The two points of attachment are always 1 and 1'.

biphenyl
CA: 1,1'-biphenyl

Substituted biphenyls are named with the use of the smallest set of numbers; in this connection, an unprimed number is considered lower than the same number when primed. When two identical sets of numbers are at hand, alphabetical order for the substituent groups prevails.

2,3,3',4',5'-pentamethylbiphenyl
(not 2',3,3',4,5-)
CA: 2,3,3',4',5'-pentamethyl-1,1'-biphenyl

2-ethyl-2'-propylbiphenyl
CA: 2-ethyl-2'-propyl-1,1'-biphenyl

In the example at the left, the ring with the methyl groups in the 2 and 3 positions is the unprimed ring because 2 is lower than 2'. At the right, the ring with the ethyl group is unprimed because ethyl comes before propyl alphabetically.

There are thousands of compounds with fused aromatic ring systems. They are conveniently classified as *ortho*-fused systems or *ortho*- and *peri*-fused systems. *Ortho*-fused systems have adjoining rings with only two atoms in common; they have *n* common faces and 2*n* common atoms. An *ortho*- and *peri*-fused system has a ring that has two, and only two, atoms in common with each of two or more rings; the total system has *n* common faces and *fewer than* 2*n* common atoms. For example,

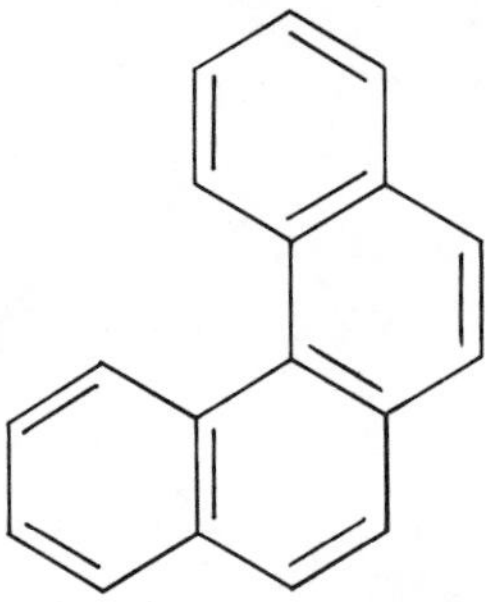

an *ortho*-fused system
(3 common faces; 6 common atoms)

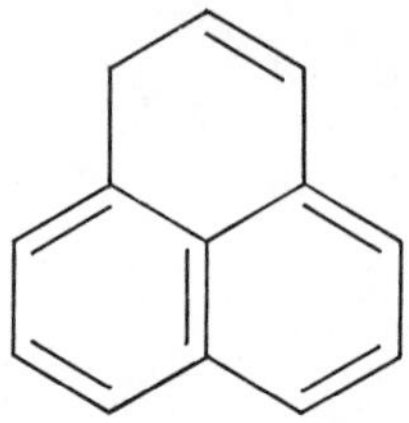

an *ortho*- and *peri*-fused system
(3 common faces; 4 common atoms)

A few ring systems, together with their numbering sequences, are shown below. In understanding these formulas, you should remember that there is one hydrogen on each numbered carbon atom unless more are indicated (cf. indene). Each numbered carbon atom is a member of only one ring. The numbers designate these carbon atoms and are used also to specify the location of substituents. Neither the names nor the numbering will be used later in this book. They are presented solely for reference. All are *ortho*-fused systems, except pyrene, which is *ortho*- and *peri*-fused.

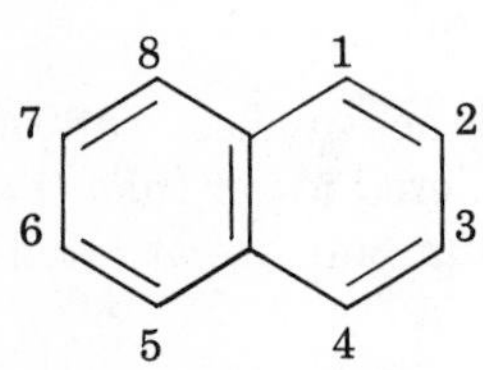

naphthalene, $C_{10}H_8$

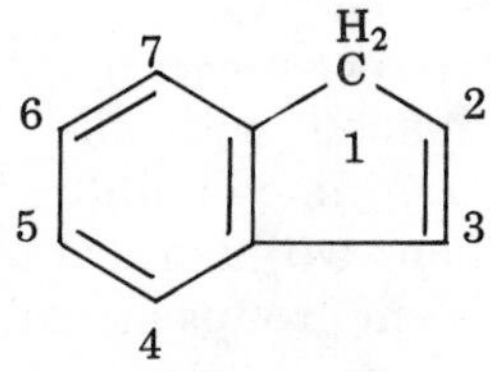

indene, C_9H_8

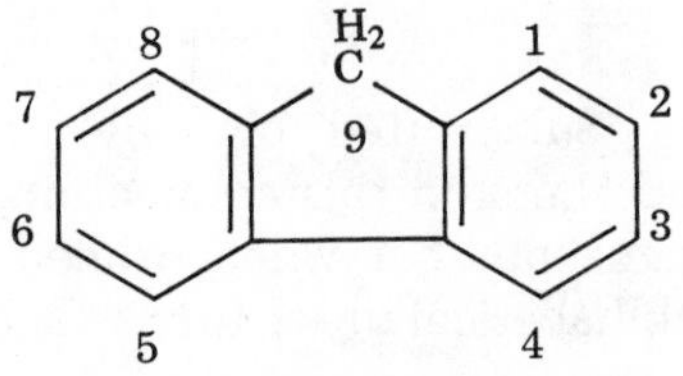

fluorene, $C_{13}H_{10}$

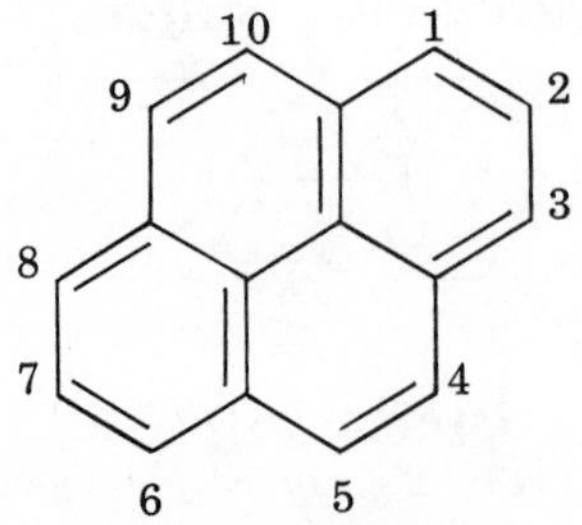

pyrene, $C_{16}H_{10}$

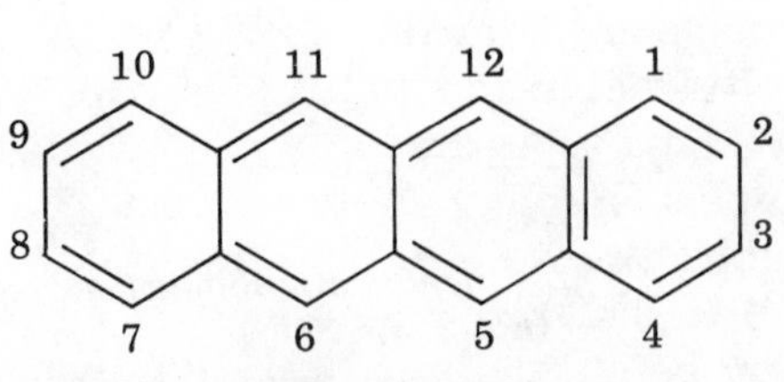

naphthacene, $C_{18}H_{12}$

Two common fused aromatic systems have the same molecular formula, but differ in constitution. They are anthracene and phenanthrene.

anthracene, $C_{14}H_{10}$ phenanthrene, $C_{14}H_{10}$

Occasionally you will see the circle used to represent the π electrons in these fused systems as it is for benzene. Naphthalene, for example, can be pictured as

Some authors object to this on the grounds that the circle does not always stand for the same number of electrons. Others admit to the objection but rationalize that the circles do represent the overlapping clouds of electrons that are characteristic of aromatic rings. This book will restrict the circle to the benzene ring only.

Our discussion of the naming of hydrocarbons is now finished. The next sections will deal with the names of organic compounds containing elements other than carbon and hydrogen.

AROMATIC HYDROCARBONS

General formula: One ring   Series name: *-benzene*
                Two or more rings   Depends on constitution

2.4 FREE RADICALS AND IONS

Neutral organic chemical structures that contain one or more unpaired electrons are called *free radicals*. Species that carry a net positive or negative charge are termed ions. The class names *carbocation* and *carbanion* are used for ion structures in which the charge is on a carbon atom. Some radicals and ions of these types are known to exist and are often called "reactive intermediates;" others are hypothetical. In either instance, a readily understood and unambiguous systematic nomenclature is needed.

In a structural formula, the unpaired electron of a free radical is shown as a dot beside the appropriate atom. A superscript dot following the molecular formula serves the same purpose although it is less specific in showing the location of the unpaired electron. For example:

$(CH_3)_2\overset{\cdot}{C}H$ or $C_3H_7^{\cdot}$ $C_6H_5^{\cdot}$ or

Names of free radicals generally coincide with the prefix name of the structure as a substituent group, followed by the word *radical*. For those prefixes that end in *o* or *y*, the ending is changed to *-yl*. Here are two examples:

$CH_3\dot{C}H_2$ or $C_2H_5^{\boldsymbol{\cdot}}$ ethyl radical

⬡-$\dot{C}H_2$ phenylmethyl radical

Free radicals rank highest in the order of precedence of compound classes and all functional groups that may be present are named as substituents.

$HO\dot{C}HCH_3$ 1-hydroxyethyl radical

[⬡]$^{\boldsymbol{\cdot}}$ with NH_2 aminobenzene radical

Methylene, $H_2C\colon$, is a neutral molecular structure in which a divalent carbon structure has only six electrons in its bonding shell. Corresponding structures with similarly electron-deficient nitrogen are amidogen, $H_2N\boldsymbol{\cdot}$, and imidogen, $HN\colon$. Two unshared electrons are designated by two dots, often printed as a colon. Groups attached to the structures $H_2C\colon$, $H_2N\boldsymbol{\cdot}$, and $HN\colon$ are specified with the usual prefixes. For instance:

$(CH_3)_2\dot{C}\boldsymbol{\cdot}$ dimethylmethylene (triplet form)
       CA: 1-methylethylidene

⬡-$\overset{H}{\underset{}{C\colon}}$ phenylmethylene (singlet form)

$(CH_3CH_2)_2N\boldsymbol{\cdot}$ diethylamidogen

⬡-$\overset{COOH}{\underset{H}{C}}-N\colon$ (carboxyphenylmethyl)imidogen

The IUPAC rules permit the use of the name carbene as an alternative to methylene, but *Chemical Abstracts* now uses the names shown above.

Which of the following structures represents the 2-naphthyloxyl (*CA:* naphthalenyloxy) radical?

A _____

B _____

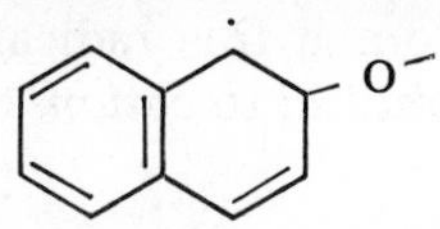

C _____

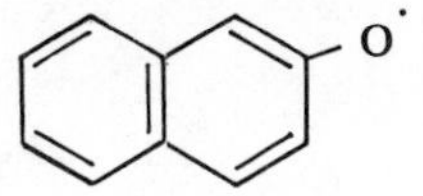

••

A _____

Incorrect. You seem to have forgotten the numbering scheme for naphthalene. This is the 1-naphthyloxyl radical that you have chosen:

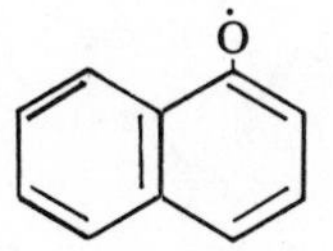

Go back and choose another answer.

B _____

You are wrong. The structure you have chosen is both a free radical and a substituting group since it has both an unpaired electron, shown by the dot, and an available bond, shown by the dash. The correct answer will be a structure that has only the dot to indicate an unpaired electron. Go back and look for it.

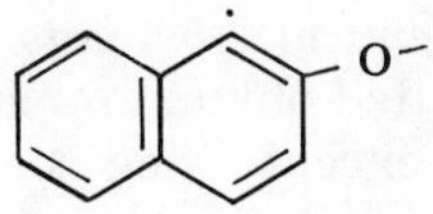

C _____

You are right. This is the structure for the 2-naphthyloxyl radical. There is an unpaired electron on the oxygen atom attached to the number two carbon atom in the naphthalene structure.

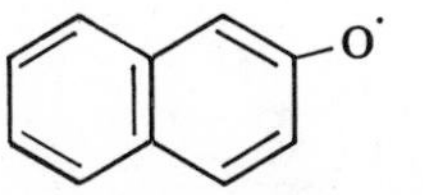

Formally, an organic cation can be considered as being formed by the loss of an unpaired electron from a free radical. The cation is often named by changing the terminal word *radical* to *cation.* Some examples are

$CH_3 CH_2 CH_2^+$ propyl cation

$CH_3 CH_2 NH^+$ ethylaminyl cation

$H_3C -\langle \text{cyclohexyl} \rangle +$ 4-methylcyclohexyl cation

While these names serve admirably to emphasize the ionic nature of the entity, they are not suitable when the ion is part of a compound. For this purpose, it is convenient to view the cation as having been formed by the conceptual gain of a proton by the neutral species, and adding the ending *-ium.* A final *e* may be elided as in the first of these examples:

$CH_3 NH_3^+ Br^-$ methanaminium bromide

$CH_3^+ Cl^-$ methylium chloride

$CH_3 \overset{+}{C}HCH_3 \ Cl^-$ 1-methylethylium chloride

On the rare occasions when it is necessary to name a carbocation derived by the addition of a proton to a saturated carbon atom, the ending *-ium* is used, dropping a final *e* if necessary. For instance,

CH_5^+ methanium ion

$C_6 H_{13}^+$ cyclohexanium ion

Onium ions are those in which the cationic atom, most commonly nitrogen, is not attached to hydrogen. Their suffix names end in *-onium* and the prefix name is *onio-*. For example, $R_4 N^+$ is called ammonium and ammonio, respectively. Since 1972 *Chemical Abstracts* has used *aminium* as a suffix in place of ammonium. Substituted onium ions are named substitutively as derivatives of the parent cations, except when the charged atom is part of a ring system. For instance,

$(CH_3)_4 N^+$

IUPAC: *tetramethylammonium ion*
CA: *N,N,N-trimethylmethanaminium*

$\langle \text{phenyl} \rangle - \overset{\overset{CH_3}{|}}{\underset{\underset{CH_3}{|}}{P^+}} - CH_3$ trimethylphenylphosphonium ion

$CH_3 CH_2 \overset{+}{Br} CH_2 CH_3$ diethylbromonium ion

Carbocations of the type R_3C^+ are often incorrectly called carbonium ions. Although such terminology may continue to be seen, they are better named as substituted carbenium ions. For example,

$(CH_3)_3C^+$ trimethylcarbenium ion

So-called quaternary ammonium salts, $RR'R''R'''N^+X^-$, are important as cationic detergents and as cleaning and antimicrobial agents. An example is:

$(CH_3)_3 \overset{+}{N}CH_2CH_2CH_2 \overset{+}{N}(CH_3)_3$ 2 Br$^-$ N,N,N,N',N',N'-hexamethyl-1,3-propanediaminium

Oxonium compounds are useful and exceptionally reactive alkylating agents. One example is:

$(CH_3CH_2)_3O^+ BF_4^-$ triethyloxonium fluoroborate

A special situation involves compounds having the general formula RN_2^+ X^-. They are called *diazonium compounds* and are named by adding the suffix *-diazonium* to the name of the corresponding compound, RH, followed by the name of the anion, X^-. Two examples are:

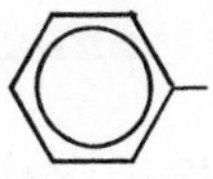

$-N_2^+$ Cl$^-$ benzenediazonium chloride

$CH_3NHN_2^+$ Br$^-$ (methylamino)diazonium bromide

An organic anion may be formed conceptually by the addition of an electron to a free radical and can be named by changing the word *radical* to *anion*. When the ion is a component of an ionic compound, the suffix *-ide* is used. For instance,

CH_3^- methyl anion   methanide (as ionic component)

$$CH_3\overset{\overset{\textstyle CH_3}{|}}{C}HCH_2CH_2^-$$ 3-methylbutyl anion   3-methylbutanide

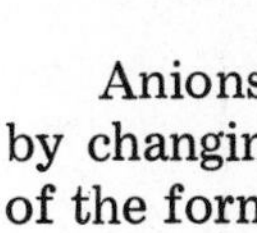

$-$$-$ 1,4-cyclohexylene dianion   1,4-cyclohexanediide

Anions formed by removal of a proton from an alcohol or thiol are named by changing the suffixes *-ol* and *-thiol* to *-olate* and *-thiolate*. The compounds of the former may also be given abbreviated names ending in oxide.

$CH_3CH_2CH_2O^-K^+$ potassium 1-propanolate
                                   potassium propoxide

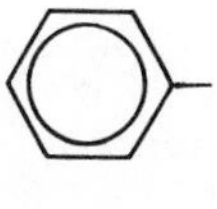

$-O^-K^+$ potassium phenolate
                                   potassium phenoxide

$(CH_3 S^-)_2 Ca^{2+}$ calcium dimethanethiolate

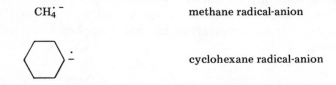

sodium benzenethiolate
not sodium thiophenolate

For the anions of carboxylic acids, see page 176.

Species that carry both a net charge and an unpaired electron are called *radical-ions.* An organic *radical-cation* may be formed conceptually by removal of an electron from a neutral non-radical molecule. It is named by adding the hyphenated word *radical-cation* to the name of the parent molecule. For example,

$CH_3 \dot{C}H_3^+$ or $C_2 H_6^{\cdot\,+}$ ethane radical-cation

$H_3 C$ ⬡ $\dot{+}$ 4-methylbenzene radical-cation

$C_3 H_7 N(C_2 H_5)_2^{\cdot\,+}$ *N,N*-diethylpropanamine *N*-radical-cation

Both the IUPAC rules and *Chemical Abstracts* use the suffix *-iumyl* for radical-cations that are components of ionic compounds. Thus, the ethane radical-cation, when associated with an anion, would be called 1-ethanium-1-yl.

An organic *radical-anion* may be formed conceptually by the addition of an electron to a neutral non-radical molecule. It is named by adding the hyphenated word *radical-anion* to the name of the parent. For instance,

$CH_4^{\cdot\,-}$ methane radical-anion

⬡ $\dot{-}$ cyclohexane radical-anion

The IUPAC rules use the suffix *-ylide* for radical-anions that are components of ionic compounds. A recent recommendation is to use *-idyl* to parallel the construction of *-iumyl.*

Here are some exercises to text your knowledge. If you see a name, answer with a structural formula. Answer with a name if you see a formula. Compare your answers with those that follow the final question.

1. 1-propanium ion

2. trimethylenediaminyl diradical

3. $(CH_3 CH_2)_2 C$:

4. cyclohexane radical-cation

5.

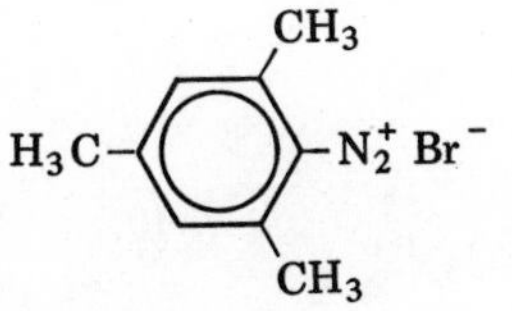

6. dimethylchloronium ion

7.

$$(C_6H_5)_2 \overset{\overset{\displaystyle CH_3}{|}}{\underset{\underset{\displaystyle CH_3}{|}}{N^+}} OH^-$$

8. sodium ethanide

9.

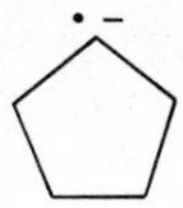

10. potassium 2-methylpropoxide

11. $CH_3CH_2^+$

12. disodium 1,4-cyclohexanediide

● ●

Here are the answers:

1. $CH_3CH_2\overset{+}{C}H_4$ or $C_3H_9^+$

2. $H\overset{\cdot}{N}CH_2CH_2CH_2\overset{\cdot}{N}H$

3. diethylmethylene or diethylcarbene

4.

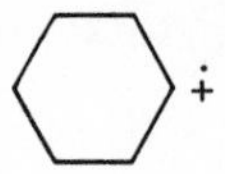

5. 2,4,6-trimethylbenzenediazonium bromide

6. $(CH_3)_2 Cl^+$

7. *N,N*-dimethyl-*N*-phenylbenzaminium hydroxide

8. $NaCH_2 CH_3$ or $Na^+ CH_2 CH_3^-$

9. cyclopentane radical-anion

10. $(CH_3)_2 CCH_2 O^- K^+$

11. ethyl cation or ethylium (as ionic component)

12.

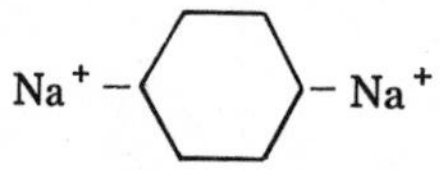

2.5 ALKYL AND ARYL HALIDES (HALOCARBONS)

When fluorine, chlorine, bromine, or iodine replaces one of the hydrogen atoms in an aliphatic or aromatic hydrocarbon, the resulting compound is an *alkyl* or *aryl halide*, respectively. They are useful in the synthesis of more complex organic compounds. Replacement of more than one hydrogen leads to a polyhalide.

Alkyl halides and polyhalides are named in the same way as branched hydrocarbons. The prefixes are *fluoro-*, *chloro-*, *bromo-*, and *iodo-*. Both the number of substituent groups and the location of each must be given. To illustrate this point, compare the two skeletons shown here:

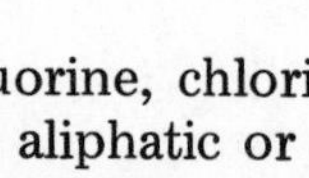

$$
\begin{array}{cc}
C\quad C \\
|\quad | \\
C-C-C-C
\end{array}
\qquad\qquad
\begin{array}{cc}
Cl\quad Cl \\
|\quad | \\
C-C-C-C
\end{array}
$$

2,3-dimethylbutane 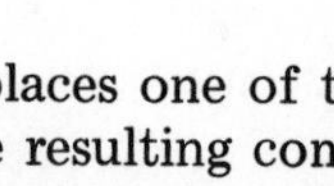   *2,3-dichlorobutane*

You should notice that 2,3-dichlorobutane has two chiral centers. Since they have the same substituent groups, there are three isomers: a *threo-* pair of enantiomers and an *erythro-* diastereomer. They are, however, not of any practical importance.

An inhalation anesthetic called halothane is a substituted alkane. What is the systematic name for halothane?

$$
\begin{array}{ccc}
& Cl & F \\
& | & | \\
Br - & C - & C - F \\
& | & | \\
& H & F
\end{array}
$$

halothane

A _____

2-bromo-2-chlorotrifluoroethane

B _____

1-bromo-1-chloro-2,2,2-trifluoroethane

C _____

2-bromo-2-chloro-1,1,1-trifluoroethane

• •

A _____

Incorrect. You chose 2-bromo-2-chlorotrifluoroethane as the systematic name for halothane:

$$
\begin{array}{ccc}
& Cl & F \\
& | & | \\
Br - & C - & C - F \\
& | & | \\
& H & F
\end{array}
$$

Because there might be another arrangement of the three fluorine atoms, their location needs to be specified, too. Go back and choose another answer.

B _____

Not quite. Your answer would certainly enable a person to recognize halothane, but it doesn't have the lowest set of locants. Go back and choose another answer.

C _____

Correct. The systematic name for halothane is 2-bromo-2-chloro-1,1,1-tri-fluoroethane. The locants are chosen to give the "lowest" number set, and the substituents are listed alphabetically in the name.

There are a number of unsaturated aliphatic halides that are important. One of them has the trivial name allyl chloride. Its formula is

$$CH_2=CHCH_2Cl$$

What is its systematic name? It is 3-chloro-1-propene, isn't it? Allyl chloride is used as an intermediate in organic synthesis.

Another important unsaturated halide is chloroethene, or ethenyl chloride. Many molecules of chloroethene can combine with each other to form a polymer known as polyvinyl chloride (PVC) or vinyl plastic.

$$3x \ CH_2 = CHCl \ \rightarrow \ [- CH_2 CHCl \ \vdots \ CH_2 CHCl \ \vdots \ CH_2 CHCl -]_x$$

The unsaturated chloride commonly called methallyl chloride has this formula:

$$CH_2 = CCH_2 Cl$$
$$|$$
$$CH_3$$

What is its correct name?

A _____

1-chloro-2-methyl-2-propene

B _____

chloromethylpropene

C _____

3-chloro-2-methyl-1-propene

• •

A _____

Incorrect. Your answer is nearly right. You made one small error. Remember that the carbon chain should be numbered from the end nearer the double bond. Here is the formula again:

$$CH_2 = CCH_2 Cl$$
$$|$$
$$CH_3$$

It should be named as a derivative of 1-propene. Work out the name again and then go back to see if there is a corresponding answer.

B _____

You are wrong. The compound represented by the formula

$$CH_2 = CCH_2 Cl$$
$$|$$
$$CH_3$$

is a derivative of propene. It does have a chlorine and a methyl group as substituents. You did not specify their locations. You must do this in order to have a unique name. Go back and try again.

C _____

You are correct. The systematic name for methallyl chloride is 3-chloro-2-methyl-1-propene. It is used as a raw material for the production of Lucite and Plexiglas.

The skeletons or names of a few more halocarbons are given below. If you see a skeleton, write the name; if you see a name, write the skeleton. Cover the answers until you have worked out your own.

1.

$$
\begin{array}{c}
\text{Cl} \qquad \text{Cl} \\
| \qquad\quad | \\
\text{C} - \text{C} - \text{C} - \text{C}
\end{array}
$$

2.

$$
\begin{array}{c}
\qquad\qquad\qquad \text{Br} \\
\qquad\qquad\qquad | \\
\text{C} = \text{C} - \text{C} - \text{C} - \text{C}
\end{array}
$$

3. 1-bromo-3-chlorobenzene

4.

5. 1,3,5-tribromo-2-methylbenzene

• •

1. 1,4-dichlorobutane

2. 5-bromo-1-pentene, *not* 1-bromo-4-pentene

3.

4. 1,3,5-trichlorobenzene

5.

Fluorinated hydrocarbons are widely used as refrigerants and propellants in aerosol-spray preparations. They are sold under the trademarked names Freon, Ucon, and Genetron. The generic term for all of them is halocarbon.

Halocarbon-12, used for aerosol bombs, is dichlorodifluoromethane, CCl_2F_2. Halocarbon-114 is the most widely used refrigerant for household refrigerators. Its structural formula is:

$$
\begin{array}{c}
\text{Cl} \quad \text{Cl} \\
| \qquad | \\
\text{F} - \text{C} - \text{C} - \text{F} \\
| \qquad | \\
\text{F} \quad\; \text{F}
\end{array}
$$

What is the preferred systematic name for halocarbon-114?

A _____

dichlorotetrafluoroethane

B _____

1,2-dichloro-1,2-tetrafluoroethane

C _____

1,2-dichloro-1,1,2,2-tetrafluoroethane

D _____

Freon-114

• •

A _____

Incorrect. The compound represented by the formula

$$\begin{array}{ccc} & Cl & Cl \\ & | & | \\ F - & C - & C - F \\ & | & | \\ & F & F \end{array}$$

is indeed a dichlorotetrafluoroethane. Isn't this one, too?

$$\begin{array}{ccc} & F & Cl \\ & | & | \\ F - & C - & C - F \\ & | & | \\ & F & Cl \end{array}$$

The preferred systematic name must specify the locations of the halogen atoms to make it a unique name. Go back and choose another answer.

B _____

You are wrong. The compound represented by the formula

$$\begin{array}{ccc} & Cl & Cl \\ & | & | \\ F - & C - & C - F \\ & | & | \\ & F & F \end{array}$$

is not 1,2-dichloro-1,2-tetrafluoroethane. The formula and the tetra-prefix both indicate that there are four fluorine atoms, and you have specified the locations of only two of them with numbers. Go back and choose another name.

C _____

Correct. The preferred systematic name for halocarbon-114 is as above. Go on to the next section.

D _____

You are incorrect. Freon-114 is a trademarked name. A refrigerator expert might be able to infer the formula from it, but most chemists would want something more. Go back and choose a satisfactory chemical name.

Numbers such as 12 and 114 are given to the halocarbons to avoid the confusion that might arise if laymen were to attempt to use the systematic names. To the initiated, the numbers imply the formula. You are about to be initiated.

The number consists of three digits, with the first being omitted if it is zero. The meanings are:

First digit: one less than the number of carbon atoms (zero omitted)
Second digit: one more than the number of hydrogen atoms
Third digit: number of fluorine atoms

All other atoms necessary to form a saturated molecule are chlorine.

Let's see how this works for halocarbon-114. The first digit tells us there are two carbons

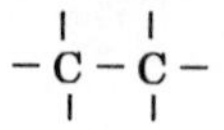

$$
-\overset{|}{\underset{|}{C}} - \overset{|}{\underset{|}{C}} -
$$

The second says there are no hydrogen atoms and the third that there are four fluorine atoms.

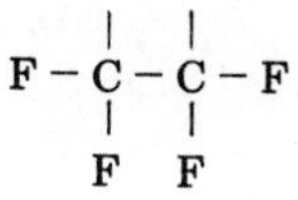

$$
F - \overset{|}{\underset{|}{C}} - \overset{|}{\underset{|}{C}} - F
$$
$$
\quad\quad F \quad F
$$

The remaining two atoms are chlorine.

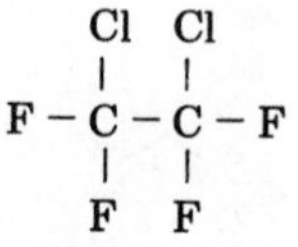

$$
\overset{Cl}{\underset{|}{}} \quad \overset{Cl}{\underset{|}{}}
$$
$$
F - \overset{|}{\underset{|}{C}} - \overset{|}{\underset{|}{C}} - F
$$
$$
\quad\quad F \quad F
$$

halocarbon-114

One final part of the code is that the fluorines are distributed as symmetrically as possible whenever constitutional isomerism is possible. This explains why halocarbon-114 is not

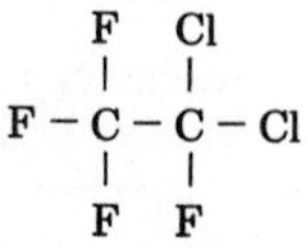

$$
\overset{F}{\underset{|}{}} \quad \overset{Cl}{\underset{|}{}}
$$
$$
F - \overset{|}{\underset{|}{C}} - \overset{|}{\underset{|}{C}} - Cl
$$
$$
\quad\quad F \quad F
$$

Now let's try halocarbon-12. The missing 0 tells us there's one carbon atom, the 1 that there are no hydrogens, and 2 that there are two fluorine atoms. Filling in with chlorine atoms gives us

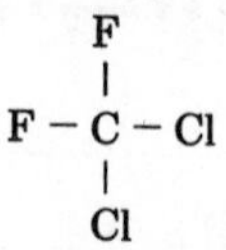

$$
\overset{F}{\underset{|}{}}
$$
$$
F - \overset{|}{\underset{|}{C}} - Cl
$$
$$
\overset{|}{Cl}
$$

halocarbon-12

A substituted phenyl group can itself be a substituent. For instance:

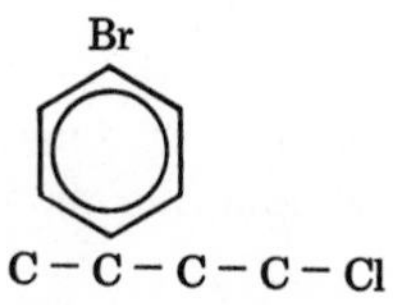

$$
C - C - C - C - Cl
$$

According to what you have learned so far, this compound should be named as a substituted butane. There is a chlorine on the number one carbon atom, but what about that aryl group on number three? It is a substituted phenyl group. Let's look at it alone.

The free valence, or point of attachment to the butane, is considered to be the number one carbon atom of the ring. The bromine is then in the number 4, or *para-*, position. The name of the group is 4-bromophenyl or *p*-bromophenyl. The compound above is therefore 3-(4-bromophenyl)-1-chlorobutane. The parentheses are used to show that the entire 4-bromophenyl group is attached to the number 3 carbon atom of the butane.

How would you name this substituted phenyl group?

A _____

1,3-dichlorophenyl group

B _____

2,4-dichlorophenyl group

C _____

2,4-dichloro-1-phenyl group

D _____

m-dichlorophenyl group

• •

A _____

Incorrect. The group represented by this formula

is not a 1,3-dichlorophenyl group. You have forgotten that the carbon with the free valence must be number 1. Keep this in mind when you go back to choose another answer.

B _____

Right. The preferred name for the aryl group (shown above) is 2,4-dichlorophenyl. Go on to the next section.

C _____

You are not quite right. If the carbon atom with the free valence is given the number 1, then the chlorine atoms are substituted on the number 2 and number 4 carbon atoms.

The custom, however, is for the carbon atom in the benzene ring with the free valence always to be number 1. For benzene rings, it doesn't have to be specified. Therefore the preferred name is just 2,4-dichlorophenyl. Go on to the next section.

D _____

Wrong. Your answer is that the group represented by the formula

Cl

Cl

is a *m*-dichlorophenyl group. This is confusing. The prefix *meta-*, or *m-,* is used to indicate two substituents on a benzene ring that are separated by one carbon atom. In this group the two chlorines are "meta" to each other, but where are they in relation to the carbon with the free valence? Your name should not leave this to the imagination. Go back and find a satisfactory answer.

Occasionally you will find compounds that have two or more identical substituents which are themselves substituted groups. Examples of this type are the insecticides DDD and DDT. The formula for DDD is

Cl

$$H-C-CHCl_2$$

Cl

The steps in naming DDD are: (1) Locate the longest carbon chain. It has two carbons and therefore the base name is ethane. (2) Name and locate the substituent groups. There are two 4-chlorophenyl groups $\left(\text{Cl}\langle\bigcirc\rangle- \right)$ on one of the carbon atoms.

You are familiar with the use of the prefixes *di-*, *tri-*, etc., used to indicate sets of identical simple substituent groups and with *bi-* and *ter-* used to describe ring assemblies. Yet another series is used to indicate sets of identical substituted, or complex, groups. The prefix for 2 is *bis-*.

Consequently, bis(4-chlorophenyl) will be part of the name of DDD. Finally, there are two chlorine atoms substituted on the other carbon of the ethane chain. Since chloro- precedes chlorophenyl- alphabetically, the latter carbon is numbered 1. The full name for DDD is 1,1-dichloro-2,2-bis(4-chlorophenyl)ethane. Compare this name carefully with the formula to be sure you understand it.

Table 2.1 summarizes the multiplying prefixes for simple groups, complex groups, and ring assemblies.

TABLE 2.1   Multiplying Prefixes

Multiplying Factor	For Simple Groups	For Complex Groups	For Ring Assemblies
2	di-	bis-	bi-
3	tri-	tris-	ter-
4	tetra-	tetrakis-	quater-
5	penta-	pentakis-	quinque-
6	hexa-	hexakis-	sexi-
7	hepta-	heptakis-	septi-
8	octa-	octakis-	octi-
9	nona-	nonakis-	novi-
10	deca-	decakis-	deci-

The formula for DDT is:

Its preferred systematic name is

A _____

dichlorophenyltrichloroethane

B _____

2,2,4-chlorophenyl-1,1,1-trichloroethane

C _____

1,1,1-trichloro-2,2-bis(4-chlorophenyl)ethane

• •

A _____

You are incorrect. By now you should know that substituent groups need to be named *and* located according to their position on the carbon chain that is used for the base name of the compound. You have correctly deduced that DDT is a substituted ethane, but have not named or located the substituents correctly. Turn back and read pages 139 and 140 again carefully.

B _____

Close but not entirely correct. You have correctly judged that DDT is a substituted ethane. You have, however, missed the point of the present discussion. Your chosen name for DDT is 2,2,4-chlorophenyl-1,1,1-trichloro-ethane. Here is the formula again:

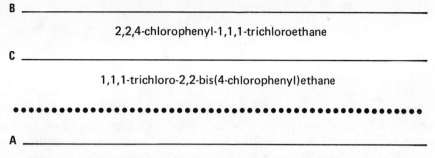

Are there two 4-chlorophenyl groups on the number 2 carbon? If so, you need to indicate this fact in some way. Go back and read pages 139 and 140 again. Then choose another answer.

C _____

Correct. The systematic name for DDT is 1,1,1-trichloro-2,2-bis(4-chloro-phenyl)ethane. Although DDT is certainly shorter, you must agree that the other name is more descriptive. *Chemical Abstracts* now names it as a substituted benzene.

2.6 NITRO COMPOUNDS

One small but synthetically useful group of substituted hydrocarbons is the *nitro* [nī′ trō] *compounds*. They are usually synthesized by treating the hydrocarbon with nitric acid. The end result is that the nitro group, $- NO_2$, replaces one or more hydrogens of the hydrocarbon. The nitroalkanes nitromethane and nitroethane are used as fuel additives in competition automobile racing.

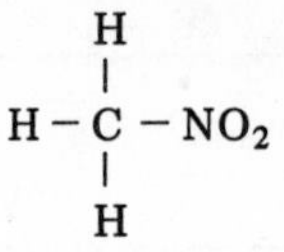

$$H-\underset{\underset{H}{|}}{\overset{\overset{H}{|}}{C}}-NO_2$$

nitromethane

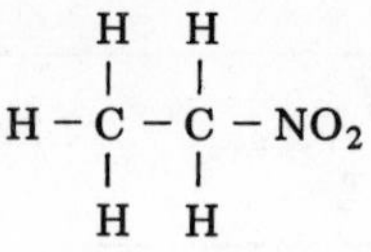

$$H-\underset{\underset{H}{|}}{\overset{\overset{H}{|}}{C}}-\underset{\underset{H}{|}}{\overset{\overset{H}{|}}{C}}-NO_2$$

nitroethane

When benzene is treated with a mixture of nitric and sulfuric acids, one of the products is nitrobenzene. It was once used as a flavoring agent because its odor resembles oil of almonds. Its high toxicity, however, has eliminated this use as well as all others in which it might come in contact with the skin.

DNP-F or FDNB, 1-fluoro-2,4-dinitrobenzene, is used to "tag" the terminal amino groups when chemists seek to establish the amino acid sequence in polypeptides and proteins (see p. 299).

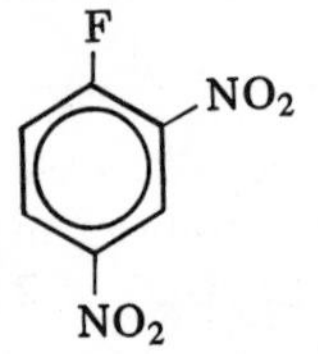

1-fluoro-2,4-dinitrobenzene

Further nitration of benzene yields 1,3,5-trinitrobenzene, a powerful explosive. Which of these is the formula for this compound?

A _____

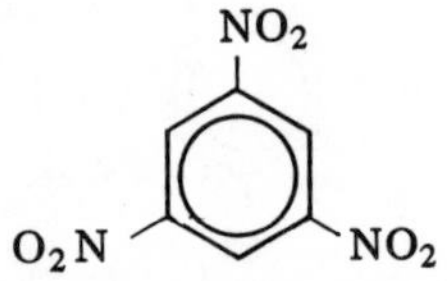

B _____

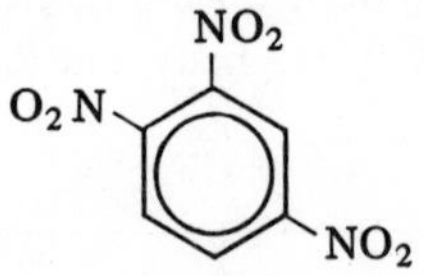

C _____

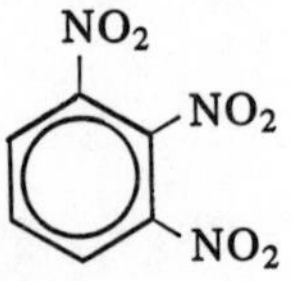

••

A _____

You are correct. The lowest set of locants is 1,3,5. Since all three of the substituent groups are alike, the carbon bearing any one of them can be numbered 1. Go on to the next section.

B _____

Wrong. The formula you chose is

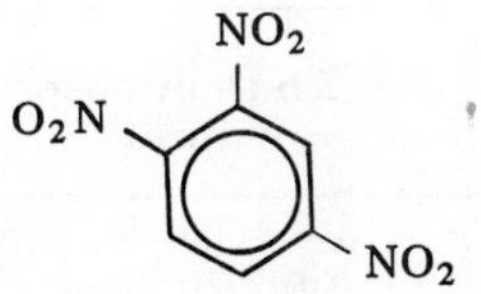

Since two of the nitro groups are on adjacent carbons, it should be apparent that the compound cannot be **1,3,5-trinitrobenzene**. How would the compound represented by this formula be named? There are several ways to number the carbons in the ring. Some are shown here.

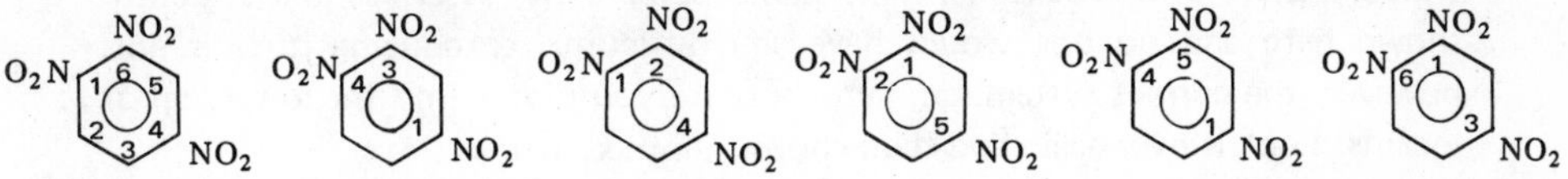

Which is correct? The one with the lowest set of numbers, or **1,2,4-trinitrobenzene**. With these principles in mind, go back and choose another answer.

C _____

You are incorrect. The formula you chose is

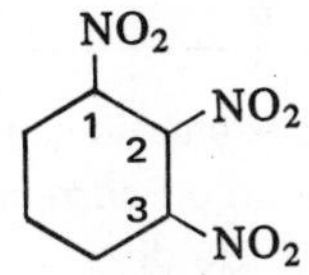

Since the three nitro- groups are on adjacent carbon atoms, the simplest (and correct) name for this compound is **1,2,3-trinitrobenzene**. Go back and choose another answer.

As it happens, 1,3,5-trinitrobenzene is difficult to prepare. A more easily made explosive, though slightly less powerful, is known as TNT. It is still an important military explosive. TNT melts at $81°C$, but does not explode until it reaches $280°C$. Consequently, it can be melted and poured into shells while liquid.

The formula for TNT is

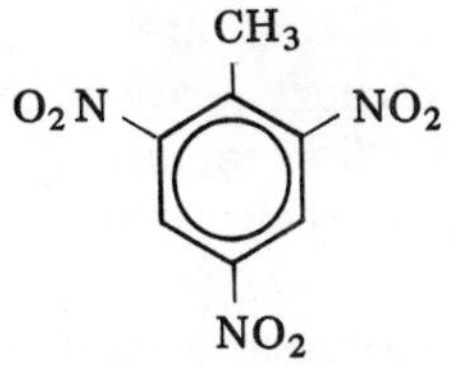

What is the systematic name for TNT?

A _____

1-methyl-2,4,6-trinitrobenzene

B _____

2-methyl-1,3,5-trinitrobenzene

C _____

2,4,6-trinitrotoluene

●●●

A _____

Almost right. The name you chose for the explosive known as TNT is 1-methyl-2,4,6-trinitrobenzene. This name completely describes the structure shown here and no one would have any difficulty reproducing it. It is not, however, the correct systematic name because you didn't find the lowest set of locants. Look it over again and then choose another answer.

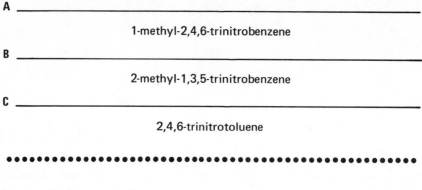

B _____

You are right. The correct systematic name for the explosive known as TNT is 2-methyl-1,3,5-trinitrobenzene. You correctly cited methyl before nitro even though the locant for methyl is 2. Go on to the next section.

C _____

Incorrect. It is easy to see that the letters TNT stand for trinitrotoluene and, in fact, 2,4,6-trinitrotoluene is a trivial name accepted by the IUPAC rules. The systematic name is based on benzene. Go back, study the formula, and choose another answer.

Summary: Nomenclature of Cyclic Hydrocarbons and Substituted Hydrocarbons

Cyclic compounds can be divided into two groups: alicyclic and aromatic. The distinction is chemical as well as structural.

Monocyclic alicyclic hydrocarbons are named according to the same principles as their aliphatic counterparts, except that the prefix *cyclo-* is placed at the beginning of the name. The first members of each homologous series are cyclopropane, cyclopropene, and cyclooctyne. *cis-trans (E,Z)* isomerism is possible in the larger cycloalkenes.

The stereochemistry of cyclohexane and its derivatives is important. Cyclohexane exists in three principal conformations: chair, boat, and twist-boat. The chair conformation is most stable. The hydrogen atoms in cyclohexane are designated axial or equatorial, depending on whether the carbon-hydrogen bond is perpendicular or roughly parallel to a plane bisecting the carbon-carbon bonds of the ring.

Univalent groups derived from cycloalkanes, cycloalkenes, and cyclo-alkynes are named according to the principles of their aliphatic analogs with the prefix *cyclo-* added. If numbers are needed, the carbon atom with the free valence is numbered 1.

$$
\begin{array}{cccc}
CH_2 & CH & CH & C \\
/ \ \backslash & / \ \backslash & \text{//} \ \backslash & / \ \text{\textbackslash\textbackslash} \\
CH_2 - CH_2 & CH_2 - CH_2 & CH - CH_2 & CH_2 - C - CH_3
\end{array}
$$

cyclopropane   *cyclopropyl group*   *cyclopropene*   *2-methyl-1-cyclopropen-1-yl group*

Spiro, fused, and bridged compounds are those in which carbon atoms are shared by more than one ring. These skeletons show the nature of the three types:

spiro   *fused*   *bridged*

There are three parts to the name of a spiro compound: (1) the prefix *spiro*; (2) brackets containing the number of carbon atoms in each ring, other than the spiro atom, arranged in ascending order and separated by a period; and (3) the name of the hydrocarbon with the same total number of carbon atoms as the spiro compound. The carbon atoms are numbered consecutively starting with an atom in the smaller ring next to the spiro atom, through the spiro atom, and around the larger ring in whichever direction gives lowest locants.

1,4-dimethylspiro[2.4]heptane
(Note: 3 chiral centers)

In bridged compounds two or more rings share non-adjacent carbon atoms. The bridges are the chain(s) of atoms or valence bond(s) connecting different parts of the molecule. The two tertiary carbon atoms at the ends of the bridges are called bridgeheads. There are three parts to the name of a bridged compound with two rings: (1) the prefix *bicyclo;* (2) brackets containing the number of carbon atoms in each of the bridges, arranged in descending order and separated by a period; and (3) the name of the hydrocarbon with the same total number of carbon atoms as the bridged compound. The carbon atoms are numbered by assigning the numeral 1 to one of the bridgehead carbon atoms, following the longest path to the other bridgehead, then the longest remaining path back, and finally the remaining path.

$$
\begin{array}{cccc}
7 & 1 & 2 \\
CH_2 - CH - CH_2 \\
| & | & | \\
 & 8\ CH_2 & CH_2\ 3 \\
 & | & | \\
CH_3 - CH - CH - CH_2 \\
6 & 5 & 4
\end{array}
$$

6-methylbicyclo[3.2.1]octane
(Note: 3 chiral centers)

Hydrocarbons containing unsaturated ring systems such as benzene and naphthalene are *aromatic* hydrocarbons. The molecular formula of benzene is C_6H_6. Its structural formula is usually written as

or

It must be remembered that each of the carbon atoms is bonded to one hydrogen atom and that all of the carbon atoms have the same chemical reactivity. The phenyl group is formed by the removal of one hydrogen atom from benzene.

Monosubstituted benzenes are named by combining the name of the substituent group with the word benzene. Disubstituted benzenes can be named by numbering the carbons in the ring consecutively, beginning with the carbon atom bearing one of the substituent groups. The position of each substituent is specified by a number.

Substituted phenyl groups are named according to the same principles as substituted benzenes. The carbon atom with the free valence is always numbered 1.

3-ethylphenyl

4-ethyl-2-methylphenyl

Ring assemblies of two or more cyclic systems joined by a single bond are named by placing the prefix *bi-* before the name of the corresponding univalent group or hydrocarbon.

CA: 1,1'-bicyclopropyl
IUPAC: 1,1'-bicyclopropane or 1,1'-bicyclopropyl

Fused aromatic ring systems generally have trivial names and specific numbering schemes. A few examples are on pages 124 and 125.

Halocarbons are named in the same way as hydrocarbons, using the prefixes *fluoro-*, *chloro-*, *bromo-*, and *iodo-* to indicate the substituent halogen atoms.

Nitro compounds contain one or more $-NO_2$ groups, the location of which is specified in a manner identical to any other substituent.

For those compounds in which there are identical substituent groups that contain substituents themselves, the prefixes *bis-*, *tris-*, *tetrakis-*, *pentakis-*, etc., are used as multiplicative prefixes. The group to be multiplied is enclosed in parentheses.

Many of these rules and conventions are illustrated by this hypothetical compound. Study its names carefully to be sure you understand them.

3,3-bis(4-bromo-2-chlorophenyl)-5-nitro-1,4-hexadiene
CA: 1,1'-(1-ethenyl-3-nitro-2-butenylidene)bis[4-bromo-2-chlorobenzene]

HALOCARBONS

General formula: $C_xH_yX_z$ Series name: *chloro-*, *bromo-*,
                                                          fluoro-, *iodo-*
C_xH_y = alkyl or aryl group
X = F, Cl, Br, I

NITRO COMPOUNDS

General formula: $C_xH_y(NO_2)_z$ Series name: *nitro-*

C_xH_y = alkyl or aryl group
$-NO_2$ = nitro group

This quiz consists of five questions. You should be able to answer all five correctly. If you make a mistake return to the page indicated for review. As usual, you should cover the answers until you have worked out your own.

1. The plastic Teflon is a polymer of tetrafluoroethene. What is the structural formula of the monomer?

2. Neoprene rubber is a polymer of the following compound. Its common name is chloroprene. What is its systematic name?

$$\begin{array}{ccc}
H & H \;\; Cl & H \\
\diagdown & | \;\;\; | & \diagup \\
& C = C - C = C & \\
\diagup & & \diagdown \\
H & & H
\end{array}$$

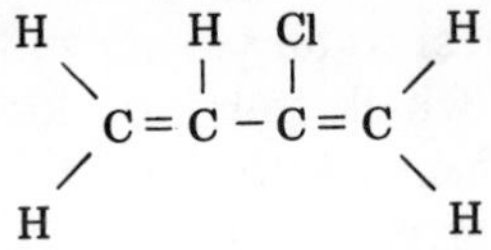

3. What is the structural formula of 2-ethylbicyclo[2.2.0] hexane?

4. Chloropicrin (tear gas) has this formula: CCl_3NO_2. What is its systematic name?

5. Neglecting stereoisomerism, what is the systematic name for the compound with this skeleton?

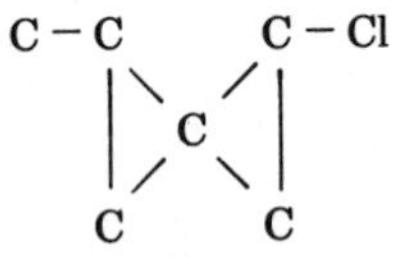

$$\begin{array}{ccc}
C - C & & C - Cl \\
| \diagdown & & \diagup \; | \\
& C & \\
| \diagup & & \diagdown \; | \\
C & & C
\end{array}$$

• •

1. $F_2C = CF_2$

Review pages 56 and 140 if you made an error.

2. 2-chloro-1,3-butadiene

Review page 58 if you were wrong.

3.
$$\begin{array}{l}
CH_2 - CH - CH - CH_2CH_3 \\
\;\; | \qquad | \qquad | \\
CH_2 - CH - CH_2
\end{array}$$

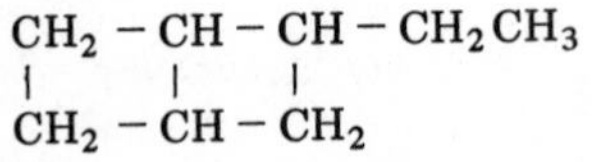

Note that the base hexane includes only the carbon atoms in the unsubstituted bridged compound. Review page 113 if you made an error.

4. trichloronitromethane

Review pages 132 and 141 if necessary.

5. 1-chloro-4-methylspiro[2.2] pentane

Notice that the base pentane includes only the carbon atoms in the unsubstituted hydrocarbon. Review page 109 if necessary.

KINDS OF SYSTEMATIC NOMENCLATURE AND CONSTRUCTION OF NAMES

In working through the first two chapters of this book, you have learned to name hydrocarbon chains and rings and have been introduced to the principles of substitutive nomenclature. Substitution means the replacement of one or more hydrogen atoms in a given compound by some other kind of atom or group of atoms, commonly called a *substituent*.

Each substituent in substitutive nomenclature is described by a prefix or suffix to the name of the compound or other substituent to which it is bonded. The latter are known as parent compounds, or parent groups. In both of these two examples, benzene is the parent compound. In addition, the ethyl group is a parent group in the second.

1,3,5-trinitrobenzene   *1,2-bis(2-chloroethyl)benzene*

Atoms or groups that lend characteristic chemical properties to the molecules in which they occur are called functional groups. Halogen atoms and nitro groups are functional groups, and the next chapter is devoted to the study of more of them. We will see that the principle of substitution makes it possible to name any functional compound.

Frequently an atom other than carbon will occur in the chain of an aliphatic or cyclic compound. Such atoms are known as hetero atoms and, for nomenclature purposes, are regarded as replacements for a carbon atom. Replacement or "a" nomenclature has been developed to provide a method for naming systems containing hetero atoms. Examples will be seen in both of the next two chapters.

Substitutive nomenclature and replacement nomenclature together are the most useful tools so far developed for the systematic naming of organic compounds. Other kinds of nomenclature have been developed over the years for special purposes, but are not favored for general use. Four of these are: radicofunctional nomenclature, conjunctive nomenclature, additive nomenclature, and subtractive nomenclature.

Radicofunctional names are frequently encountered in textbooks, trade journals, and commercial publications. In a radicofunctional name one functional group is chosen and used as the final word of a multiword name. Words to describe the remainder of the molecule precede it. For example,

CH_3OH methyl alcohol

$CH_3COC_2H_5$ methyl ethyl ketone (MEK)

Conjunctive names are used when the principal functional group occurs on a side chain of a ring system and it is undesirable to name the ring system as a substituent. For instance,

cyclohexanemethanol

Additive nomenclature involves naming atoms that have been added to the structure denoted by the rest of the name. Two examples are

CH_2Br_2

1,2,3,4-tetrahydronaphthalene   *methylene dibromide*

Subtractive names are used to express the conceptual removal of atoms or groups from a substance described by a systematic or trivial name. They occur frequently in natural products, where the removal of a hydrogen atom may be indicated by the prefix *dehydro-* and the removal of an oxygen atom by *deoxy-*. In the specialized nomenclature of carbohydrates the prefix *anhydro-* denotes the loss of the elements of water.

In constructing the systematic name to describe a given structure, the conventions for word separation, use of italics, and elision of vowels must be observed. Most systematic names will include at least one, and commonly several, punctuation and enclosing marks. Correct use of a comma or parentheses is as important as choice of the right prefix to describe a particular substituent group.

Except for substances named as acids, substitutive names are nearly always single words. Examples are acetic acid and 1-bromo-2-chloroethane. A few other multiword names will be seen in this book, but they are exceptions to the general rule.

Vowels, whether spoken or silent, are generally retained in order to avoid ambiguity. You have already seen unusual vowel combinations in cyclooctene and chloroacetic acid. There are some situations in which the vowels *a*, *e*, and *o* may be elided, that is, omitted from a name. The two most common are:

(1) preceding the suffix name of a functional group

       methanol   not methaneol

       ethanamine   not ethaneamine

(2) when both double bond(s) and triple bond(s) are present in the same chain and ring system and are named as suffixes

       1-hexen-4-yne   not 1-hexene-4-yne

As you can see from the last example, insertion of locants does not affect the elision.

Italics are used for configurational prefixes such as *cis-*, *trans-*, *E-*, and *Z-*, and for structure-defining prefixes such as *sec-* and *tert-*. However, the structure-defining prefixes cyclo-, spiro-, iso-, and neo- are treated as part of the name in which they are used and are neither separated or italicized.

When English letters are used as locants, they are italicized. You've seen *o-*, *m-*, and *p-* used to locate substituents on a benzene ring. Capital letters occur in instances of substitution on a hetero atom as, for example, *N*-hydroxybenzen-amine or 4-*O*-ethyl-D-idose. An italic capital *H* follows a locant used to express the position of a saturated atom necessary for the formation of a definable, stable ring system. The *H* is part of the locant and is not separated from the numeral. An example is 1*H*-pyrrole.

Hyphens, commas, and periods are used frequently in systematic names; the colon is used rarely and the semicolon, apostrophe, exclamation point, and quotation mark not at all.

Hyphens separate locants from the words and syllables of a name as in 2-chloropropane. When a hyphen occurs between two locants, the intention is to indicate that such locants refer to different parts of the name. For example,

N-2-naphthalenylacetamide

The *N-* places the 2-naphthalenyl substituent on the nitrogen atom of the parent compound, acetamide. Hyphens at the end of the set of substituents in the inverted name in an index signify that no space is intended when the name is uninverted for use in textual matter. Acetic acid, trichloro-, becomes trichloroacetic acid.

Commas are used to separate the members of a series of locants as in 1,3,5-trichlorobenzene. No comma is used within a locant that consists of more than one kind of symbol such as 2*H*-indene.

Periods are used within brackets to separate the numerals in names of spiro and bridged compounds. For example, bicyclo[2.2.0]hexane and spiro[3.4]-octane.

Parentheses and brackets are used as enclosing marks along with locants when the locants specify complex substituent groups. Parentheses are used first, followed by brackets, and the brackets again if necessary. Parentheses and brackets must always be used in pairs. They do not replace hyphens or otherwise affect use of punctuation marks. Brackets are, of course, used in the names of spiro and bridged compounds, as illustrated in the previous paragraph. The use of enclosing marks does not create a need for a hyphen if one is not otherwise required.

Overpunctuation and unnecessary separation of words are two common errors in substitutive nomenclature. Bromobenzene, for instance, is a single word and requires no hyphen. Amine never stands alone. It is used as a suffix, as in ethanamine, or as a parent compound as in ethylamine, but never as part of a multiword radicofunctional name.

Chapter Three

FUNCTIONAL GROUPS

Atoms or groups that change the chemical properties of hydrocarbons when they appear as substituents are called *functional groups*. Most of the homologous series of organic compounds have characteristic functional groups. Identification of the functional group of each series simplifies the naming of the compounds as well as the study of their chemistry.

Simple functional compounds contain only one kind of functional group. In a multiple functional compound the same functional group occurs more than once. Mixed functional compounds contain more than one kind of functional group. Multiple and mixed functional compounds are both called polyfunctional compounds.

If a compound is of simple or multiple function, as many of the functional group(s) as possible are cited in the suffix of the name and the remainder are cited as prefixes. There are some functional groups that do not have suffix names and are always cited as prefixes. The halogens and the nitro groups studied in the last chapter belong in this category. If a compound is one of mixed function, one of the functional groups is cited as a suffix and the remainder as prefixes.

This chapter is devoted to the nomenclature of compounds containing a number of the common functional groups. Appendix A.2 provides a complete listing of the prefixes and suffixes for functional groups.

3.1 ALIPHATIC ALCOHOLS (ALKANOLS)

These compounds have the monovalent hydroxy group, $-OH$, as their functional group. Both the name and formula of the functional group are important. The simplest member of the family is methanol. Its structural formula is

$$\begin{array}{c} H \\ | \\ H - C - OH \\ | \\ H \end{array}$$

methanol

Alkanols are given their systematic names by changing the final *-e* of the parent alkane to *-ol*. In the United States, however, the simple, unsubstituted alkanols from ethanol to dodecanol are usually called by their trivial names. For example

```
      H   H
      |   |
 H -- C - C - OH
      |   |
      H   H
```
ethyl alcohol

```
      H   H   H
      |   |   |
 H -- C - C - C - OH
      |   |   |
      H   H   H
```
propyl alcohol

```
      H   H   H   H
      |   |   |   |
 H -- C - C - C - C - OH
      |   |   |   |
      H   H   H   H
```
butyl alcohol

The contrasting name "methanol" is preferred over "methyl alcohol" in order to emphasize that the compound is poisonous.

What is the systematic name of the alkanol containing two carbon atoms?

A _____

ethanol

B _____

methylmethanol

C _____

hydroxyethane

• •

A _____

You are right. The alkanol containing two carbon atoms is ethanol. Its structural formula is

```
      H   H
      |   |
 H -- C - C - OH
      |   |
      H   H
```

To name any alkanol, merely change the final -e of the name of the hydrocarbon to -ol. Go on to the next section.

B _____

You are wrong. The systematic name of the alkanol containing two carbon atoms is not methylmethanol. Although that name does describe the compound, the correct name is derived from the name of the hydrocarbon with two carbon atoms, just as the alkanol itself is derived from that hydrocarbon. Here are the formulas:

```
      H   H
      |   |
 H -- C - C - H
      |   |
      H   H
```
ethane

```
      H   H
      |   |
 H -- C - C - OH
      |   |
      H   H
```
?

Go back and choose another answer.

C _____

Incorrect. Although the name hydroxyethane does describe the alkanol containing two carbon atoms, it is not right. The functional group of the alcohols is the hydroxy group, $-OH$, but it is not part of their names. Here are the structural formulas of ethane and the alkanol derived from it. Turn back and read page 152 carefully before choosing another answer.

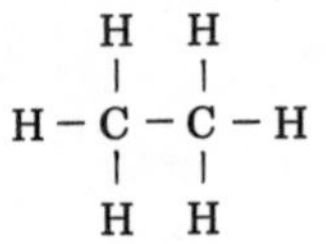

$$
\begin{array}{ccc}
\text{H} & \text{H} \\
| & | \\
\text{H} - \text{C} - \text{C} - \text{H} \\
| & | \\
\text{H} & \text{H}
\end{array}
\qquad\qquad
\begin{array}{ccc}
\text{H} & \text{H} \\
| & | \\
\text{H} - \text{C} - \text{C} - \text{OH} \\
| & | \\
\text{H} & \text{H}
\end{array}
$$

          ethane   ?

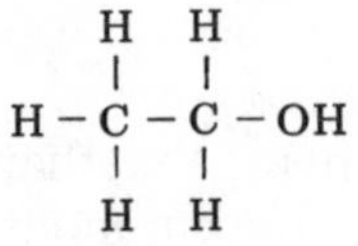

Following an almost universal custom, we shall refer to the family of compounds whose functional group is $-OH$ as alcohols rather than alkanols. You should remember, however, that the term alcohol is not a systematic name.

Constitutional isomerism becomes possible as soon as there are three carbon atoms in the molecule. The formulas of the two isomers of propanol are

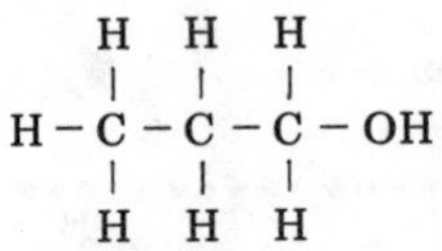

$$
\begin{array}{cccc}
\text{H} & \text{H} & \text{H} \\
| & | & | \\
\text{H} - \text{C} - \text{C} - \text{C} - \text{OH} \\
| & | & | \\
\text{H} & \text{H} & \text{H}
\end{array}
\quad\text{and}\quad
\begin{array}{cccc}
\text{H} & \text{H} & \text{H} \\
| & | & | \\
\text{H} - \text{C} - \text{C} - \text{C} - \text{H} \\
| & | & | \\
\text{H} & \text{OH} & \text{H}
\end{array}
$$

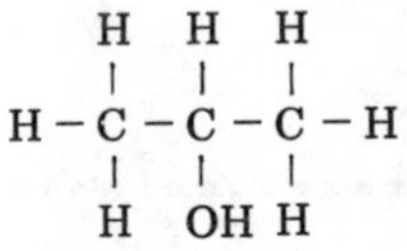

The difference in structure is more apparent if the carbon skeleton and functional group only are shown:

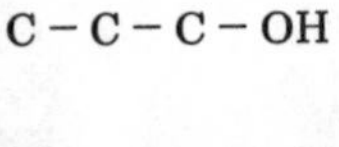

$$
\text{C} - \text{C} - \text{C} - \text{OH} \quad\text{and}\quad
\begin{array}{c}
\text{C} - \text{C} - \text{C} \\
| \\
\text{OH}
\end{array}
$$

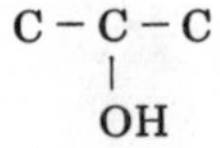

The difference between the two is in the location of the functional group. Once again, the solution to the difficulty lies in numbering the carbon atoms. After the longest continuous chain is found, it is numbered as usual beginning at the end nearer the hydroxy group. Then the location of the hydroxy group is specified. The alkanols shown above are 1-propanol and 2-propanol. The trivial names of the two compounds are propyl alcohol and isopropyl alcohol. Isopropyl alcohol is used as rubbing alcohol.

Which of these skeletons represents 3-hexanol?

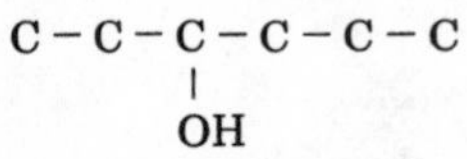

$$
\begin{array}{c}
\text{C} - \text{C} - \text{C} - \text{C} - \text{C} - \text{C} \\
| \\
\text{OH}
\end{array}
\qquad\qquad
\begin{array}{c}
\text{C} - \text{C} - \text{C} - \text{C} - \text{C} - \text{C} \\
| \\
\text{OH}
\end{array}
$$

              A 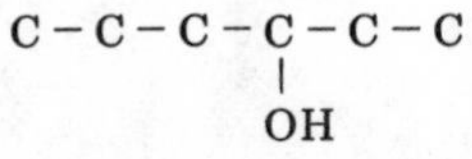   *B*

A _____

Both *A* and *B*

B _____

Only *A*

C _____

Only *B*

••

A _____

Right. Both of the skeletons shown represent 3-hexanol. Remember always to number the chain from the end nearer the functional (− OH) group. You may note that 3-hexanol is chiral. Go on to the next section.

B _____

Incorrect. Your answer is that only this skeleton

$$C - C - C - C - C - C$$
$$\overset{|}{OH}$$

represents 3-hexanol. Consider the other skeleton again for a moment. Here it is.

$$C - C - C - C - C - C$$
$$\overset{|}{OH}$$

If you number the carbon chain from the right, it's 3-hexanol, too, isn't it? Remember always to number the chain from the end nearer the functional group. Go on to the next section.

C _____

Incorrect. Your answer is that only this skeleton

$$C - C - C - C - C - C$$
$$\overset{|}{OH}$$

represents 3-hexanol. Consider the other skeleton again for a moment. Here it is.

$$C - C - C - C - C - C$$
$$\overset{|}{OH}$$

If you number the carbon chain from the left, it's 3-hexanol, too, isn't it? Remember always to number the chain from the end nearer functional group. Go on to the next section.

The series of alcohols with the hydroxy group on the terminal carbon of a straight chain alkane is often termed the *normal series*. The prefix *n-* is used with the trivial names (see p. 27). For example, 1-butanol is *n*-butyl alcohol.

When other groups, such as halogen atoms, are substituted on an alcohol, they are named and located in the same way that you have already learned for hydrocarbons. Furthermore, systematic names are always used. One illustration should make the point clear for you. Compare this skeleton and its name.

$$C - C - C - C - C$$
$$\overset{|}{C} \qquad \overset{|}{OH}$$

4-methyl-2-pentanol

It is tempting to name this compound 2-methyl-4-pentanol, but that is wrong because the lower number (2) should be used to indicate the position of the functional (− OH) group. A functional group has preference over other substituents when the sets of locants are identical.

What is the skeleton of 2-methyl-2-butanol?

A _____

$$C - C - C - C$$
$$\quad\ \ |\quad\ \ |$$
$$\quad\ \ C\quad OH$$

B _____

$$\qquad\quad C$$
$$\qquad\quad |$$
$$C - C - C - C$$
$$\qquad\quad |$$
$$\qquad\ \ OH$$

C _____

$$\qquad C$$
$$\qquad |$$
$$C - C - C$$
$$\qquad |$$
$$\quad\ OH$$

• •

A _____

Wrong. Your answer is that the formula of 2-methyl-2-butanol is

$$C - C - C - C$$
$$\quad\ \ |\quad\ \ |$$
$$\quad\ \ C\quad OH$$

This could be right only if you numbered the chain from one end to locate the methyl group and from the other to locate the hydroxy group. That's not fair. Always number the chain from the end nearer a functional group. With this in mind work out another skeleton and choose another answer.

B _____

You are right. The skeleton for 2-methyl-2-butanol is

$$\qquad\qquad\quad C$$
$$4\quad 3\quad |\ 2\ 1$$
$$C - C - C - C$$
$$\qquad\qquad |$$
$$\qquad\qquad OH$$

It has a chain of four carbons with a methyl group and an hydroxy group on the second carbon from the end. Go on to the next section.

C _____

Incorrect. The skeleton you have chosen to represent 2-methyl-2-butanol is

$$\qquad C$$
$$\qquad |$$
$$C - C - C$$
$$\qquad |$$
$$\quad\ OH$$

Although it contains four carbon atoms, it is not a butanol. The longest continuous carbon chain has three carbon atoms. It represents 2-methyl-2-propanol. (On page 158 you will learn that its trivial name is *tert*-butyl alcohol.) Work out another answer and go back to page 156.

Since the important structural feature of alcohols is the hydroxy group, the entire series of aliphatic alcohols is often represented by the general formula ROH. R− represents an alkyl group, substituted or unsubstituted. For instance, consider these formulas for ethanol:

Molecular formula: C_2H_6O

Structural formula:

$$H-\overset{\overset{\displaystyle H}{|}}{C}-\overset{\overset{\displaystyle H}{|}}{C}-OH$$
(with H below each C)

Condensed structural formula: CH_3CH_2OH

Skeleton formula: $C-C-OH$

General formula: ROH, where $R = CH_3CH_2-$

If 2-methyl-2-propanol is to be represented as ROH, what is the condensed structural formula of R−?

A _____

$$CH_3CH_2CH_2-$$

B _____

$$\overset{\overset{\displaystyle CH_3}{|}}{CH_3CHCH_2}-$$

C _____

$$CH_3-\overset{\overset{\displaystyle CH_3}{|}}{\underset{|}{C}}-CH_3$$

• •

A _____

You are wrong. This is not the condensed structural formula for the R− group of 2-methyl-2-propanol.

$$CH_3CH_2CH_2-$$

Suppose that you add the hydroxy group to form the alcohol. Isn't it ordinary propyl alcohol? Why don't you write the formula for 2-methyl-2-propanol and remove the −OH to find the R− group? Then choose another answer.

B _____

Incorrect. If an −OH group is added to the R− group you chose, the formula of the alcohol is

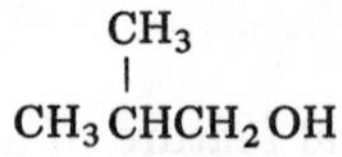

$$CH_3 \atop CH_3CHCH_2OH$$

This is the formula for 2-methyl-1-propanol, also called isobutyl alcohol. Make the change that is needed to transform it to 2-methyl-2-propanol. Then remove the − OH to leave the R − group.

Now turn back to page 157 and choose another answer.

C _____

Right. If the − OH group is added, you have the complete formula for 2-methyl-2-propanol (*tert*-butyl alcohol).

$$\begin{array}{c} CH_3 \\ | \\ CH_3CCH_3 \\ | \\ OH \end{array}$$

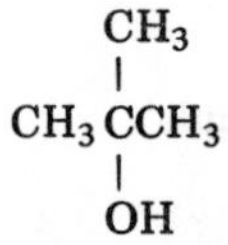

2-methyl-2-propanol

Aliphatic alcohols are often classified as primary, secondary, or tertiary. This classification is based on the number of carbon atoms joined to the carbon atom bearing the hydroxy group. This table illustrates the idea for some alcohols composed of four carbon atoms each.

TABLE 3.1   Classification of Aliphatic Alcohols

Class of Alcohol	No. of C's Joined to − OH Carbon	Carbon Skeleton	Systematic Name; Trivial Name
Primary	1	C − C − C − C − OH	1-butanol; butyl alcohol

Secondary	2	C − C − C − C
		             |
		           OH
			2-butanol; *sec*-butyl alcohol

Tertiary	3	C
		      |
		C − C − C
		     |
		    OH
			2-methyl-2-propanol; *tert*-butyl alcohol

You can see that the name for all alcohols containing four carbon atoms is butyl alcohol: *sec*-butyl and *tert*-butyl are the names for the secondary and tertiary alcohols containing four carbon atoms. The systematic name for *tert*-butyl alcohol is 2-methyl-2-propanol. The only exception to the definitions illustrated above is the simplest primary alcohol, methanol (CH_3OH).

Following are several skeleton formulas for alcohols. Classify each one as primary, secondary, or tertiary and write both the systematic name and acceptable common name if there is one. The correct answers are found below the skeletons.

1.　　　　　　　　　　　C – C – C – OH

2.　　　　　　　　　　　　　OH
　　　　　　　　　　　　　　|
　　　　　　　　　　　　C – C – C

3.　　　　　　　　　　　　　C
　　　　　　　　　　　　　　|
　　　　　　　　　　　C – C – C – OH
　　　　　　　　　　　　　　|
　　　　　　　　　　　　　　C

4.　　　　　　　　　　　　　C
　　　　　　　　　　　　　　|
　　　　　　　　　　　C – C – C
　　　　　　　　　　　　　　|
　　　　　　　　　　　　　OH

5.　　　　　　　　　C – C – C – C
　　　　　　　　　　　　　　|
　　　　　　　　　　　　　OH

• •

1. Primary; 1-propanol (propyl alcohol)

2. Secondary; 2-propanol (isopropyl alcohol)

3. Primary; 2,2-dimethyl-1-propanol

4. Tertiary; 2-methyl-2-propanol (*tert*-butyl alcohol)

5. Secondary; 2-butanol (*sec*-butyl alcohol)

There are some aliphatic hydroxy compounds that have two or three hydroxy groups. They are named by adding the suffixes *-diol* and *-triol* to the name of the parent hydrocarbon. The trivial name for a diol is glycol. The location of the hydroxy groups is specified by numbers in the usual way. Study these examples.

TABLE 3.2　Aliphatic Hydroxy Compounds

Condensed Structural Formula	Carbon Skeleton	Systematic Name; Trivial Name
CH_2OHCH_2OH or $HOCH_2CH_2OH$	HO – C – C – OH	1,2-ethanediol; ethylene glycol
$CH_3CHOHCH_2OH$	C – C – C – OH 　　　| 　　　OH	1,2-propanediol; propylene glycol
CH_2OH | CHOH | CH_2OH	C – OH | C – OH | C – OH	1,2,3-propanetriol; glycerol or glycerin(e)

The hydroxy groups in diols and triols are classified as primary, secondary, or tertiary in the same manner as the monohydroxy alcohols. As you can see by looking at its skeleton, 1,2,3-propanetriol has two primary and one secondary hydroxy groups.

The insect repellent marketed as "6-12" is named 2-ethyl-1,3-hexanediol. Which of these is the carbon skeleton of "6-12"?

A _____

$$
\begin{array}{c}
\text{OH} \\
| \\
\text{C}-\text{C}-\text{C}-\text{C}-\text{OH} \\
| \\
\text{C} \\
| \\
\text{C}
\end{array}
$$

B _____

$$
\begin{array}{c}
\text{OH} \\
| \\
\text{C}-\text{C}-\text{C}-\text{C}-\text{C}-\text{C}-\text{OH} \\
| \\
\text{C} \\
| \\
\text{C}
\end{array}
$$

C _____

$$
\begin{array}{c}
\text{OH} \\
| \\
\text{C}-\text{C}-\text{C}-\text{C}-\text{C}-\text{C}-\text{OH} \\
| \\
\text{C} \\
| \\
\text{C}
\end{array}
$$

• •

A _____

Wrong. "6-12" is 2-ethyl-1,3-hexanediol. In order for it to be a derivative of hexane, there must be a continuous carbon chain of six atoms. The skeleton you chose has only six altogether. It represents 2-ethyl-1,3-butanediol. Go back and choose another answer.

$$
\begin{array}{c}
\text{OH} \\
| \\
4 \quad 3 \quad 2 \quad 1 \\
\text{C}-\text{C}-\text{C}-\text{C}-\text{OH} \\
| \\
\text{C} \\
| \\
\text{C}
\end{array}
$$

B _____

You are correct. The skeleton of "6-12" is

$$
\begin{array}{c}
\text{OH} \\
| \\
6 \quad 5 \quad 4 \quad 3 \quad 2 \quad 1 \\
\text{C}-\text{C}-\text{C}-\text{C}-\text{C}-\text{C}-\text{OH} \\
| \\
\text{C} \\
| \\
\text{C}
\end{array}
$$

2-ethyl-1,3-hexanediol

One of the hydroxy groups is primary and the other is secondary. If you look closely, you can see that there is a continuous carbon chain that is seven carbon atoms long. Why isn't the compound named as a derivative of heptane? Simply because the seven-carbon chain does not include both of the functional groups. Go on to the next section.

C _____

You are incorrect. The skeleton you chose has all the atoms necessary to be 2-methyl-1,3-hexanediol, but they are not arranged properly. Here is the skeleton you picked.

$$
\begin{array}{ccccccc}
 & & & & & \text{OH} & \\
 & & & & & | & \\
6 & 5 & 4 & 3 & 2 & 1 & \\
\text{C} - \text{C} - \text{C} - \text{C} - \text{C} - \text{C} - \text{OH} & & & & & & \\
 & & & & & | & \\
 & & & & & \text{C} & \\
 & & & & & | & \\
 & & & & & \text{C} & \\
\end{array}
$$

Can you see that it represents 2-ethyl-1,2-hexanediol? Go back and select another answer.

3.2 AROMATIC HYDROXY COMPOUNDS

Although hydroxy groups that are united directly to a benzene or other aromatic ring system react quite differently from the hydroxy groups in aliphatic alcohols, their names are conveniently discussed at this point.

The simplest aromatic hydroxy compound might have the systematic name "benzenol," but *phenol* [fē′ nōl] has long been recognized as the accepted trivial name and is used by *Chemical Abstracts*. Aqueous solutions of phenol are used as disinfectants under the name carbolic acid. Derivatives are named as substituted phenols. A few examples are

phenol   *2-methylphenol*   *3-nitrophenol*
                  (also 3- and 4-)   (also 2- and 4-)

The trivial name for the three methylphenols is cresol.

BHT (butylated hydroxytoluene), a widely used antioxidant (see the labeled contents of many prepared breakfast foods), is a derivative of 4-methylphenol (*p*-cresol).

4-methylphenol   *2,6-bis(1,1-dimethylethyl)-4-methylphenol*
    (p-cresol)

Picric acid, a bitter-tasting compound which at times has been used both as a yellow dye and as a military explosive, has this structural formula:

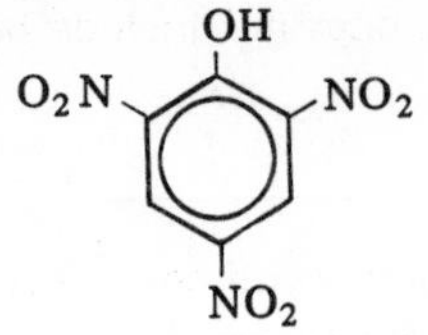

What is its systematic name?

A _____

1-hydroxy-2,4,6-trinitrobenzene

B _____

trinitrophenol

C _____

2,4,6-trinitrophenol

•••

A _____

If phenol were not an accepted trivial name, your answer would be right. Phenol, however, is a proper name and its derivatives are named as substituted phenols. It should be no trick for you to turn back and choose the correct answer.

B _____

You are wrong. Picric acid is indeed a trinitrophenol. There are, however, several separate and distinct trinitrophenols. They are distinguished by the use of numbers. Here is the formula of picric acid with numbers added.

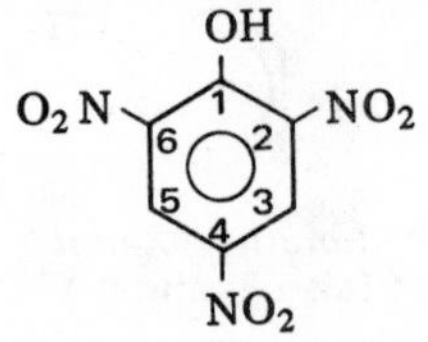

When you have decided on another answer, go back and see if it is there.

C _____

You are correct. The systematic name for picric acid is 2,4,6-trinitrophenol. Go on to the next section.

When the hydrogen is removed from the hydroxy group, alcohols and phenols can form substituent groups. Remember that these groups are not compounds, but groups with an available bond. For the first four aliphatic alcohols, the names of the groups are formed by dropping -anol and replacing it with *-oxy.* For the others, *-oxy* is added to the name of the alkyl group. Phenol forms the phenoxy group. A few examples will show this more clearly.

TABLE 3.3 Substituent Groups from Hydroxy Compounds

Hydroxy Compound		Group	
$CH_3(CH_2)_3OH$	1-butanol	$CH_3(CH_2)_3O-$	butoxy
$CH_3(CH_2)_4OH$	1-pentanol	$CH_3(CH_2)_4O-$	pentyloxy
◯— OH	phenol	◯— O —	phenoxy

As a very complex example of the use of the principles you have learned, compare the structure and systematic names of this compound.

OCH_3

$H-C-CCl_3$

OCH_3

1,1,1-trichloro-2,2-bis(4-methoxyphenyl)ethane
CA: 1,1'-(2,2,2,-trichloroethylidene)bis[4-methoxybenzene]

This substance is sold under the name *methoxychlor.* It is an insecticide reported to be as effective as DDT but less toxic. DDT is 1,1,1-trichloro-2,2-bis(4-chlorophenyl)ethane. If you want some practice, write the formula of DDT. You can check your formula against the one on page 140.

Summary: Nomenclature of Alcohols and Phenols

A functional group is a particular structure that imparts special chemical properties to a compound. An example of a functional group is the hydroxy group, $-OH$, which characterizes the alcohols and phenols. The general formula for aliphatic alcohols is ROH where R- represents an alkyl group. Alkanols are named by replacing the *-e* ending of the name of the alkane by *-ol.* These facts are illustrated by the following:

Molecular formula:	C_2H_6O	Condensed structural formula:	CH_3CH_2OH
Systematic name:	ethanol		
General formula:	ROH	Alkoxy group:	CH_3CH_2O-

When necessary, the location of the hydroxy group is indicated by a number. The carbon chain is numbered from the end nearer the hydroxy group. Other substituents are named and located as in hydrocarbons.

Primary, secondary, and tertiary alcohols have one, two, or three carbon atoms united to the carbon bearing the hydroxy group as shown in these skeletons

$$
\begin{array}{ccc}
 & \text{OH} & \text{C} \\
 & | & | \\
\text{C} - \text{C} - \text{C} - \text{OH} & \text{C} - \text{C} - \text{C} & \text{C} - \text{C} - \text{OH} \\
 & & | \\
 & & \text{C}
\end{array}
$$

primary   *secondary*   *tertiary*

Aliphatic compounds with two or more hydroxy groups are named by locating the hydroxy groups with numbers and adding the suffixes -*diol*, -*triol*, and so forth, to the name of the alkane. The common name for the series of diols is glycols.

Hydroxybenzene, ⬡OH, is named phenol.

ALIPHATIC ALCOHOLS (ALKANOLS)

Functional group: − OH   Functional group name: *hydroxy*

General formula: ROH   Series name: -*ol*

PHENOLS

Functional group: − OH   Functional group name: *hydroxy*

General formula: ⬡OH   Series name -*phenol*

Here are four questions to test your mastery of the naming of alcohols and phenols. The correct answers follow the questions. Cover the answers until you have worked out your own.

1. Write the carbon skeleton for 3-ethyl-4-methyl-2-hexanol.

2. Classify 3-methyl-2-butanol as primary, secondary, or tertiary.

3. Give the structural formula for the 4-bromophenoxy group.

4. The systematic name for ethylene glycol is 1,2-ethanediol. Write its structural formula and classify each hydroxy group as primary, secondary, or tertiary.

1.

$$
\begin{array}{c}
\quad\quad\quad\quad\quad\quad C \\
\quad\quad\quad\quad\quad\quad | \\
\quad\quad\quad\quad C \quad\; C \\
\quad\quad\quad\quad | \quad\; | \\
C - C - C - C - C - C \\
\quad\quad\quad\quad\quad\; | \\
\quad\quad\quad\quad\quad OH
\end{array}
$$

2. Secondary

$$
\begin{array}{c}
\quad\quad C \\
\quad\quad | \\
C - C - C - C \\
\quad\quad\; | \\
\quad\quad OH
\end{array}
$$

3. Br⟨◯⟩O −   (Be sure to indicate the free valence!)

4.

$$
\begin{array}{c}
\quad H \quad H \\
\quad | \quad\; | \\
HO - C - C - OH \\
\quad | \quad\; | \\
\quad H \quad H
\end{array}
$$
  Both hydroxy groups are primary.

If you answered all four questions correctly, go on to the next section. If you made errors, return to page 152 for review.

3.3 ALDEHYDES AND KETONES

The functional group common to aldehydes [ăl′ dĕ hīdz] and ketones [kē′ tōnz] is the *carbonyl* [kär′ bŏ nĭl] group, $-\overset{O}{\overset{\|}{C}}-$. When it is found at the *end* of a chain of carbon atoms, the compound is an aldehyde. If the carbonyl group appears *within* the carbon atom chain, the compound is a ketone.

Since the carbonyl group in aldehydes always occurs at the end of a carbon atom chain, the general structural formula for the series is $R - \overset{O}{\overset{\|}{C}} - H$. When written on one line as a condensed structural formula, the aldehydes are represented by RCHO. The simplest aldehyde contains only one carbon atom. Its structural formula is $H - \overset{O}{\overset{\|}{C}} - H$. (You should note that in this one instance the R− in the general formula is merely H and not an alkyl group.)

The IUPAC system names aldehydes by dropping the *-e* and adding *-al* to the name of the hydrocarbon with the longest chain containing the carbonyl group. Thus $H - \overset{O}{\overset{\|}{C}} - H$ is methanal. You are probably familiar with its trivial name formaldehyde, that is used by *Chemical Abstracts*. In aqueous solutions it is used as a fungicide.

Another aldehyde, acetaldehyde, is an important chemical in organic synthesis. Its systematic name is ethanal. What is its structural formula?

A _____

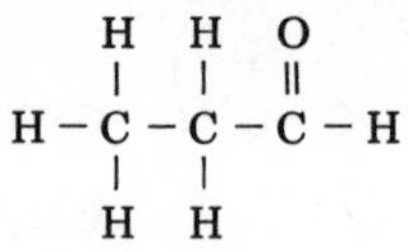

```
        H   H   O
        |   |   ||
   H -  C - C - C - H
        |   |
        H   H
```

B _____

```
        H   O
        |   ||
   H -  C - C - H
        |
        H
```

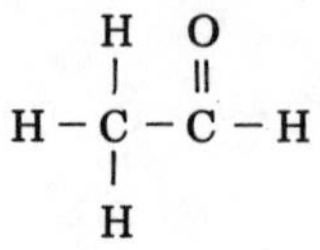

C _____

```
        H   O   H
        |   ||  |
   H -  C - C - C - H
        |       |
        H       H
```

• •

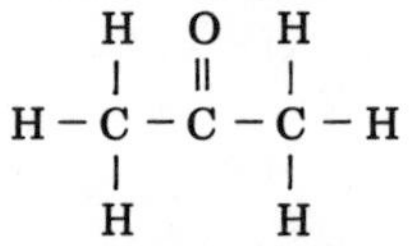

A _____

Incorrect. You were asked to write the structural formula for ethanal. You should recognize that it is a derivative of ethane and therefore should have two carbon atoms. The formula you chose is

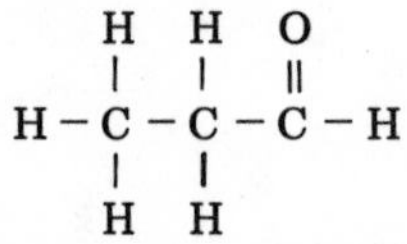

```
        H   H   O
        |   |   ||
   H -  C - C - C - H
        |   |
        H   H
```

It has three carbon atoms and is the formula for propanal. Go back and choose another answer.

B _____

You are right. The structural formula for the aldehyde ethanal is

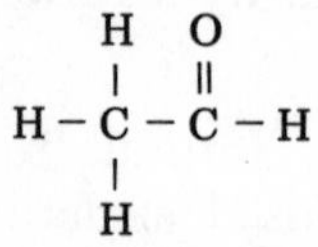

```
        H   O
        |   ||
   H -  C - C - H
        |
        H
```

The trivial name, acetaldehyde, is used by *Chemical Abstracts* for indexing. Go on to the next section.

C _____

You are wrong on two counts. The structural formula you chose for ethanal is

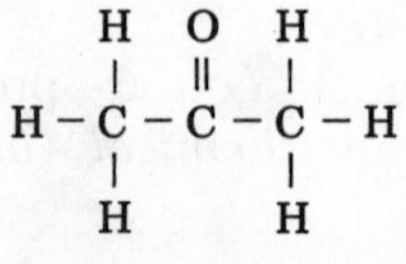

```
        H   O   H
        |   ||  |
   H -  C - C - C - H
        |       |
        H       H
```

First of all, this formula has too many carbon atoms. Ethanal, a derivative of ethane, has only two. Second, the compound shown above is not an aldehyde. It

$$\overset{O}{\overset{\|}{}}$$

is a ketone. Remember that the carbonyl group ($-\overset{O}{\overset{\|}{C}}-$) in aldehydes must be on the end of the carbon chain. Go back and choose another answer.

The trivial names of aldehydes are derived from the names of the acids that have the same number of carbon atoms. For example, the acid with one carbon atom is formic acid. The trivial name of methanal is formaldehyde. Acetaldehydehyde is the trivial name for ethanal since acetic acid has two carbon atoms. *Chemical Abstracts* uses formaldehyde and acetaldehyde as index headings.

Inasmuch as the carbonyl group of an aldehyde is always at the end of the carbon chain, it is never necessary to specify its position. For instance:

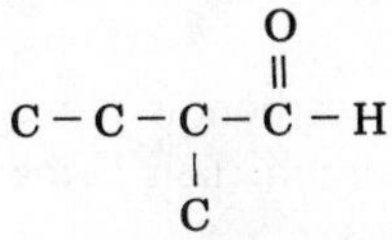

$$C - C - C - \overset{\overset{O}{\|}}{C} - H$$
$$|$$
$$C$$

2-methylbutanal

The condensed structural formula for 2-methylbutanal is $CH_3CH_2CH(CH_3)CHO$. The carbon of the carbonyl group is always the number one carbon atom.

Which of these formulas is a representation of 3-chloropentanal?

A _____

$$CH_3CH_2CHClCH_2CHO$$

B _____

$$C_5H_{10}OCl$$

C _____

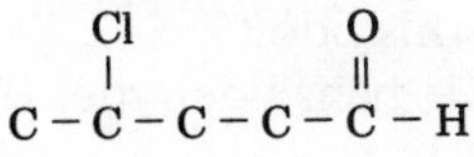

$$\overset{Cl}{\overset{|}{C}} - C - C - \overset{\overset{O}{\|}}{C} - C - H$$

••

A _____

You are right. The condensed and skeleton formulas for 3-chloropentanal are

$$CH_3CH_2CHClCH_2CHO \qquad \overset{Cl}{\overset{|}{C}} - C - C - \overset{\overset{O}{\|}}{C} - C - H$$

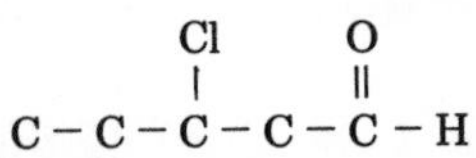

The carbon of the carbonyl group is always number 1. Go on to the next section.

B _____

You are incorrect. $C_5H_{10}OCl$ is not the molecular formula for 3-chloropentanal. Perhaps you will see your mistake if you make a structural formula by adding the requisite hydrogens to this skeleton formula.

$$\begin{array}{c} \quad\;\; Cl \quad\;\; O \\ \quad\;\; | \qquad\; || \\ C-C-C-C-C-H \end{array}$$

Now go back and pick another answer.

C _____

You are wrong. This is not the skeleton formula for 3-chloropentanal.

$$\begin{array}{c} \quad\;\; Cl \qquad\quad O \\ \quad\;\; | \qquad\quad\; || \\ C-C-C-C-C-H \\ 5 \quad 4 \quad 3 \quad 2 \quad 1 \end{array}$$

As you can see, the carbonyl carbon is number 1 and the chlorine atom is substituted on the number 4 carbon. This is the skeleton for 4-chloropentanal. Go back and choose another answer.

When the $-CHO$ group is attached directly to a carbon atom of a ring system, the suffix *-carboxaldehyde* is added to the name of the ring compound. If there are substituents on the ring, the carbon to which the $-CHO$ group is attached is numbered 1. For example:

2-cyclohexenecarboxaldehyde   *3-methylcyclobutanecarboxaldehyde*
CA: 2-cyclohexene-1-carboxaldehyde   *CA: 3-methylclyclobutane-1-carboxaldehyde*

The IUPAC rules use the suffix carbaldehyde instead of carboxaldehyde as used by *Chemical Abstracts* and in this book.

An exception to the rule is benzaldehyde, the simplest aromatic aldehyde.

CA and IUPAC: benzaldehyde

Benzaldehyde is the third of the three aldehydes that are indexed under their trivial names by *Chemical Abstracts*. The other two, as you should recall, are formaldehyde and acetaldehyde.

An important derivative of benzaldehyde is vanillin, the principal odorous constituent of vanilla beans. The name vanillin is accepted under the IUPAC rules. Its formula and systematic name are shown here.

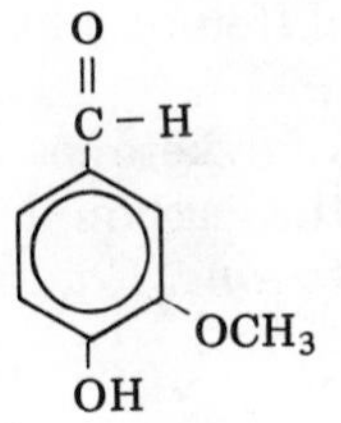

$$
\begin{array}{c}
\text{O} \\
\parallel \\
\text{C} - \text{H} \\
\end{array}
$$

CA: 4-hydroxy-3-methoxybenzaldehyde
IUPAC: *vanillin*

You may have noticed that vanillin has two functional groups: $-$ OH and $-$ CHO. The rulemakers have taken care of this by establishing an order of precedence, as given in Appendix 3 (p. A.4). In this instance the aldehyde function ranks above the alcohol. Following you will find four questions about aldehydes. After you have answered them, we'll go on to ketones.

   1. Part of the name 1-propanal is redundant. Why?

   2. Write the carbon skeleton for 3,4-dimethylpentanal.

   3. Write the general structural formula and the one-line condensed general formula for aldehydes.

   4. Write a formula for 2-methylcyclopentanecarboxaldehyde.

• •

   1. The position of the carbonyl group does not have to be specified by a number in the names of aldehydes. The carbonyl group is always the number 1 carbon atom.

   2. 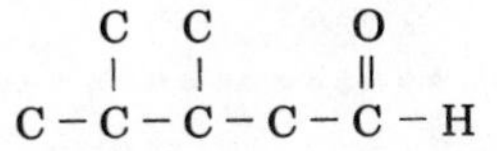
$$
\begin{array}{ccc}
\text{C} & \text{C} & \text{O} \\
| & | & \parallel \\
\text{C} - \text{C} - \text{C} - \text{C} - \text{C} - \text{H} \\
\end{array}
\quad \text{3,4-dimethylpentanal}
$$

   3. General structural formula: $\text{R} - \overset{\text{O}}{\overset{\parallel}{\text{C}}} - \text{H}$
   Condensed general formula: RCHO (*not* RCOH)

   4. 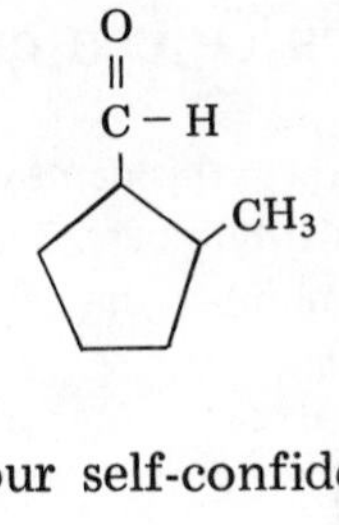
$$
\begin{array}{c}
\text{O} \\
\parallel \\
\text{C} - \text{H} \\
\end{array}
$$

Depending on the state of your self-confidence, either go on to the next section or turn back to page 165.

   Ketones have the same functional group as aldehydes: the carbonyl group, $- \overset{\text{O}}{\overset{\parallel}{\text{C}}} -$. The difference is that it is not at the end of the carbon chain in ketones.

The general formula for a ketone is $R-\overset{\overset{\displaystyle O}{\|}}{C}-R'$. The condensed structural formula is RCOR'. If $R-$ and $R'-$ are identical, the ketone is symmetrical. If $R-$ and $R'-$ represent two different alkyl groups, the compound is a mixed ketone.

The IUPAC system names for ketones are found by numbering the carbon chain from the end nearer the carbonyl group, indicating its number, and changing the -*e* on the alkane to -*one*. For example:

$$CH_3CH_2CH_2\overset{\overset{\displaystyle O}{\|}}{C}CH_3$$

2-pentanone

Substituent groups are named as before. For instance:

$$CH_3\overset{\overset{\displaystyle Cl}{|}}{C}HCH_2\overset{\overset{\displaystyle O}{\|}}{C}CH_3$$

4-chloro-2-pentanone

What is the name of this ketone?

$$CH_3CH_2\overset{\overset{\displaystyle O}{\|}}{C}CH_2CH_2Br$$

A _____

5-bromo-3-pentanone

B _____

1-bromo-3-butanone

C _____

1-bromo-3-pentanone

• •

A _____

Incorrect. You named the compound shown by this formula

$$CH_3CH_2\overset{\overset{\displaystyle O}{\|}}{C}CH_2CH_2Br$$

as 5-bromo-3-pentanone. Although this name does describe the compound, there is another name that uses smaller numbers. Bear in mind that whenever you have a choice, you should always number the carbon chain from the end nearer to any substituent groups. Try again.

B _____

Count the number of carbon atoms in this formula again.

$$CH_3CH_2\overset{\overset{\displaystyle O}{\|}}{C}CH_2CH_2Br$$

There are five, aren't there? The carbonyl group is the functional group, but it is also in the carbon chain and should be included when you count to determine the base name. The formula represents a pentanone. When you have decided on the complete name, go back and choose another answer.

C _____

Right. Go on to the next section.

The less complex ketones are often named by naming the groups attached to the carbonyl group followed by the word ketone. If the ketone is symmetrical, the prefix *di-* is used. You are probably familiar with the odor and uses of dimethyl ketone. Its trivial name, accepted by the IUPAC rules, is acetone. Its systematic name is 2-propanone.

$$\overset{\displaystyle O}{\underset{\displaystyle CH_3\,CCH_3}{\|}}$$

CA: 2-propanone
IUPAC: acetone

Ethyl methyl ketone is a widely used organic solvent with the trade designation MEK. Its formula and systematic name are

$$\overset{\displaystyle O}{\underset{\displaystyle CH_3\,CCH_2\,CH_3}{\|}}$$

CA: 2-butanone

Another widely used solvent is MIPK, isopropyl methyl ketone. Both MEK and MIPK got their acronyms when it was still permissible to list groups in order of increasing complexity instead of alphabetically.

Chemical Abstracts has abandoned all trivial names for ketones as well as the word ketone for indexing purposes. Systematic names are used exclusively. In some instances, unnecessary locants are used (e.g., 2-propanone and 2-butanone) so that the parent compounds will be found near their derivatives in the index listing.

The systematic name for MIPK is 3-methyl-2-butanone. Which of the following is its carbon skeleton?

A _____

$$
\begin{array}{ccc}
 & O & C \\
 & \| & | \\
C - & C - & C \\
 & & | \\
 & & C
\end{array}
$$

B _____

$$
\begin{array}{cccc}
 & O & & \\
 & \| & & \\
C - & C - & C - & C \\
| & & & \\
C & & &
\end{array}
$$

C _____

$$\overset{\overset{\displaystyle O}{\parallel}}{C-C-C-C-C}$$

• •

A _____

You are correct. The skeleton formula of 3-methyl-2-butanone is

$$\begin{array}{c} \overset{O}{\parallel} \quad 3 \\ C - C - C - C \\ 1 \quad 2 \quad | \\ C \\ 4 \end{array}$$

The carbon chain is numbered from the end nearer to the carbonyl group. Go on to the next section.

B _____

You are incorrect. Perhaps you were confused by the way the skeleton was written. Remember that any bends in the carbon chain are not significant. So the skeleton you chose could also be written in a straight line.

$$\begin{array}{c} \overset{O}{\parallel} \\ C - C - C - C \\ | \\ C \end{array} \qquad\qquad \overset{\overset{\displaystyle O}{\parallel}}{C - C - C - C - C}$$

Your choice   *Same skeleton*

Both represent 3-pentanone, don't they? Go back and choose another answer.

C _____

Wrong. The skeleton you chose for 3-methyl-2-butanone is

$$\overset{\overset{\displaystyle O}{\parallel}}{C-C-C-C-C}$$

Since all of the carbon atoms are contained in a single continuous chain, this is not a substituted ketone. It is plain old ordinary 2-pentanone. Read page 170 again carefully before you select another answer.

Two fairly common ketones have the trivial names acetophenone and benzophenone. They are used as intermediates in organic synthesis and to some extent in perfumery. Their formulas and systematic names are:

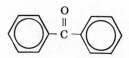

CA: 1-phenylethanone
IUPAC: acetophenone

CA: diphenylmethanone
IUPAC: benzophenone

Note that the first is a mixed alkyl-aryl ketone while the second is an aryl (aromatic) ketone.

Some carbonyl compounds undergo a spontaneous isomerization known as *tautomerism* to form an hydroxy compound as shown here:

$$\underset{\text{keto form}}{RCH_2\overset{\overset{\textstyle O}{\|}}{C}-R} \quad \rightleftharpoons \quad \underset{\text{enol form}}{RCH=\overset{\overset{\textstyle OH}{|}}{C}-R}$$

The isomers are called *tautomers*. The process is sometimes called *enolization* and the two tautomers designated as the *keto* form and *enol* form. For indexing, *Chemical Abstracts* uses the form with the highest-ranking function. In this case, it is the ketone function.

Summary: Nomenclature of Aldehydes and Ketones

Aldehydes are named by changing the -*e* ending of the parent hydrocarbon to -*al*. Since the carbonyl group is always at the end of the carbon chain, its position need not be specified by a number. When substituent groups require numbering, the carbonyl group is always numbered 1. When the −CHO group is attached to a ring system, it is denoted by the suffix -*carboxaldehyde*.

Ketones are named by changing the -*e* ending of the parent hydrocarbon to -*one*. The position of the $-\overset{\overset{\textstyle O}{\|}}{C}-$ functional group is indicated by a number. The carbon chain is numbered from the end nearer the carbonyl group. Simple ketones are often given trivial names merely by naming the substituent groups attached to the $-\overset{\overset{\textstyle O}{\|}}{C}-$ functional group.

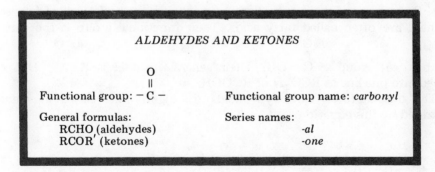

ALDEHYDES AND KETONES

Functional group: $-\overset{\overset{\textstyle O}{\|}}{C}-$ Functional group name: *carbonyl*

General formulas: Series names:
    RCHO (aldehydes)   *-al*
    RCOR′ (ketones)   *-one*

Following are three questions about aldehydes and ketones.

1. Is the name 3,4-dichloro-2-butanone correct for this compound: $CH_2ClCHClCOCH_3$?

2. Are these two compounds isomers?

        CH_3COCH_3 CH_3CH_2CHO

        2-propanone   *propanal*

3. Is this compound an aldehyde, ketone, or something else?

$$\begin{array}{ccccc} & H & O & & H \\ & | & \| & & | \\ H - & C - & C - & O - & C - H \\ & | & & & | \\ & H & & & H \end{array}$$

● ●

1. Yes. The full structural formula is

$$\begin{array}{ccccc} & H & H & O & H \\ & | & | & \| & | \\ Cl - & C - & C - & C - & C - H \\ & | & | & & | \\ & H & Cl & & H \end{array}$$

2. Yes. Both have the molecular formula C_3H_6O.

3. Something else. It is a type of compound called an ester. The general formula is

$$\begin{array}{c} O \\ \| \\ R - C - O - R' \end{array}$$

You'll see more of them later in this chapter.

Go on to the next section if you feel ready. If not, return to page 165 for review.

3.4 CARBOXYLIC ACIDS

The carboxylic [kär bŏk sĭl′ ĭk] acids are one of several classes of organic acids. Since many of them were discovered through the hydrolysis of fats, some of them are often called fatty acids. Their functional group is the carboxy [kär bŏk′ sē] group, $-\overset{\overset{\textstyle O}{\|}}{C}-OH$. Their general formula is $R-\overset{\overset{\textstyle O}{\|}}{C}-OH$, condensed into one line as RCO_2H or $RCOOH$.

The unsubstituted carboxylic acids are slightly dissociated in water as illustrated by this equation:

$$\overset{\overset{\textstyle O}{\|}}{R - C - OH} \quad \rightleftharpoons \quad \overset{\overset{\textstyle O}{\|}}{R - C - O^-} \quad + \quad H^+$$

carboxylic acid   *carboxylate ion*   *hydrogen ion*

Salts are formed when the carboxylic acids are neutralized by bases:

$$\overset{\overset{\textstyle O}{\|}}{R - C - OH} + NaOH = \overset{\overset{\textstyle O}{\|}}{R - C - O^-Na^+} + H_2O$$

The systematic names of the acids are formed by dropping the -*e* from the name of the parent alkane and adding the suffix -*oic* followed by the word *acid.* All carbon atoms in the longest chain including the carboxy group are

counted. The simplest of the series has the formula $H - \overset{\displaystyle O}{\overset{\displaystyle \|}{C}} - OH$, and takes the

name methanoic acid.

What is the formula of ethanoic acid?

A _____

$$H - \overset{\displaystyle \overset{\textstyle H}{|}}{\underset{\underset{\textstyle H}{|}}{C}} - \overset{\displaystyle \overset{\textstyle O}{\|}}{C} - OH$$

B _____

$$H - \overset{\displaystyle \overset{\textstyle O}{\|}}{C} - \overset{\displaystyle \overset{\textstyle H}{|}}{\underset{\underset{\textstyle H}{|}}{C}} - OH$$

C _____

$$H - \overset{\displaystyle \overset{\textstyle H}{|}}{\underset{\underset{\textstyle H}{|}}{C}} - \overset{\displaystyle \overset{\textstyle H}{|}}{\underset{\underset{\textstyle H}{|}}{C}} - \overset{\displaystyle \overset{\textstyle O}{\|}}{C} - OH$$

•••

A _____

Correct. The formula for ethanoic acid is

$$H - \overset{\displaystyle \overset{\textstyle H}{|}}{\underset{\underset{\textstyle H}{|}}{C}} - \overset{\displaystyle \overset{\textstyle O}{\|}}{C} - OH$$

Its trivial name, and the one used as an index heading by *Chemical Abstracts,* is acetic acid. Note again that the carbon atom in the carboxy group is counted as a member of the chain. Go on to the next section.

B _____

Incorrect. The formula you chose contains the right number of carbon atoms, but not the correct functional group. The functional group of the carboxylic acids is $- \overset{\displaystyle \overset{\textstyle O}{\|}}{C} - OH$. It is called a carboxy group and is a carbonyl group with an hydroxy group attached to it. In your choice, shown again here,

$$H - \overset{\displaystyle \overset{\textstyle O}{\|}}{C} - \overset{\displaystyle \overset{\textstyle H}{|}}{\underset{\underset{\textstyle H}{|}}{C}} - OH$$

the hydroxy group is not attached to the carbonyl group. The compound is not a carboxylic acid, but a hydroxy aldehyde. Go back and choose another answer. answer.

C _____

Wrong. The formula you chose for ethanoic acid is

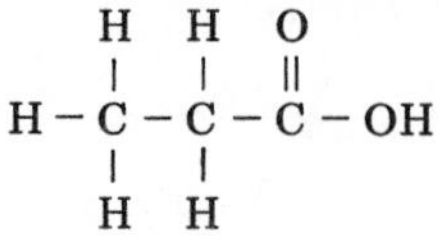

$$\begin{array}{ccc} & H & H & O \\ & | & | & \| \\ H - & C - & C - & C - OH \\ & | & | & \\ & H & H & \end{array}$$

It is an acid, all right, because it has the $-\overset{\displaystyle O}{\overset{\|}{C}} - OH$ group. Count the carbon atoms. There are three in a continuous chain; therefore, it is propanoic acid. The carboxy group in a carboxylic acid is counted as a member of the carbon chain, just as the functional group is for aldehydes and ketones. Go back and pick another answer.

The names of the carboxylate anions that result from dissociation of the acids are formed by dropping *-oic* from the name of the acid and adding the suffix *-oate*, followed by the word *ion*. The dissociation of methanoic acid is an example

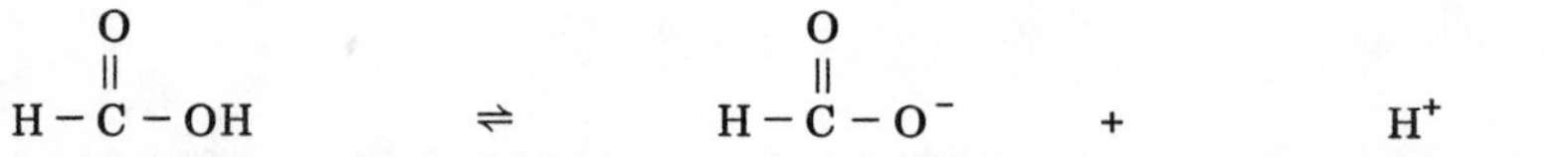

$$\begin{array}{ccccccc} & O & & & O & & \\ & \| & & & \| & & \\ H - & C - OH & \rightleftharpoons & H - & C - O^- & + & H^+ \\ \end{array}$$

methanoic acid   *methanoate ion*   *hydrogen ion*

Many of the carboxylic acids were isolated from natural products before their structural similarities were understood. Consequently, all of the lower acids are known by trivial names. The origins of the names are shown below. Under the IUPAC rules, the trivial names are preferred, but in 1972 *Chemical Abstracts* switched to the systematic names except for formic and acetic acids.

TABLE 3.4 Carboxylic Acids

Number of carbon atoms	Trivial name	Origin of trivial name	Systematic name
1	formic acid	L. *formica*, ant	methanoic acid
2	acetic acid	L. *acetum*, vinegar	ethanoic acid
3	propionic acid	Gk. *proto*, first *pion*, fat	propanoic acid
4	butyric acid	L. *butyrum*, butter	butanoic acid
5	valeric acid	valerian root	pentanoic acid

Since you'll encounter both the trivial and systematic names, you need to learn all of them.

What is the systematic name of this carboxylate ion?

$$\begin{array}{c} O \\ \| \\ CH_3\,CH_2\,C - O^- \end{array}$$

A _____

ethanoate ion

B _____

propionate ion

C _____

propanoate ion

• •

A _____

You are incorrect. This carboxylate ion is not the ethanoate ion

$$\begin{array}{c} O \\ \| \\ CH_3\,CH_2\,C - O^- \end{array}$$

The "ethan" portion of the name indicates two carbon atoms in the chain. The formula above has three. Go back and select another answer.

B _____

If you were looking for a trivial name, you would have been right. This is the formula for the propionate ion

$$\begin{array}{c} O \\ \| \\ CH_3\,CH_2\,C - O^- \end{array}$$

You were asked to give the systematic name. You should have no difficulty after looking at Table 3.4 again.

C _____

You are right. The formula represents the propanoate ion.

$$\begin{array}{c} O \\ \| \\ CH_3\,CH_2\,C - O^- \end{array}$$

It is identical to propanoic acid, less the hydrogen of the carboxy group, and bears a net negative charge.

The method of locating substituent groups and unsaturation in carboxylic acids is to number the carbon chain in the usual fashion, beginning with the carboxy group as number 1.

The skeleton of lactic acid, the acid formed when milk turns sour, is shown here.

$$
\begin{array}{c}
\overset{\displaystyle O}{\overset{\displaystyle \|}{}} \\
\text{C} - \text{C} - \text{C} - \text{OH} \\
\phantom{\text{C} - }| \\
\phantom{\text{C} - }\text{OH} \\
3 \quad 2 \quad 1
\end{array}
$$

2-hydroxypropanoic acid
(lactic acid)

Lower case Greek letters are often used with trivial names to locate substituent groups. Lactic acid may be called α-hydroxypropionic acid.

$$
\begin{array}{c}
\beta \quad \alpha \quad \overset{\displaystyle O}{\overset{\displaystyle \|}{}} \\
\text{C} - \text{C} - \text{C} - \text{OH} \\
\phantom{\text{C} - }| \\
\phantom{\text{C} - }\text{OH}
\end{array}
$$

Systematic and trivial nomenclature should not be mixed. Use numbers with systematic names and Greek letters with trivial names.

The names for carboxylic acids containing an aldehyde or ketone functional group in the principal chain or ring are derived from the names of the simple acids by adding the prefixes *oxo-*, *dioxo*, etc. For instance:

$$
\overset{\displaystyleOO}{\overset{\displaystyle\|\|}{CH_3\,CH_2\,CCH_2\,C-OH}}
$$

3-oxopentanoic acid
(β-ketovaleric acid)

$$
\overset{\displaystyleOOO}{\overset{\displaystyle\|\|\|}{HCCH_2\,CCH_2\,C-OH}}
$$

3,5-dioxopentanoic acid

What is the systematic name of the acid with this skeleton?

$$
\begin{array}{c}
\text{Br} \overset{\displaystyle O}{} \\
| \overset{\displaystyle \|}{} \\
\text{C} - \text{C} - \text{C} - \text{OH}
\end{array}
$$

A _____

2-bromopropanoic acid

B _____

3-bromopropanoic acid

C _____

β-bromopropionic acid

• •

A _____

You are incorrect. This skeleton does not represent 2-bromopropanoic acid

$$
\begin{array}{c}
\text{Br} \overset{\displaystyle O}{} \\
| \overset{\displaystyle \|}{} \\
\text{C} - \text{C} - \text{C} - \text{OH} \\
3 \quad 2 \quad 1
\end{array}
$$

Numbers have been added so you can see that the carboxy group is always number 1. Study the skeleton before you go back to choose another answer.

B _____

You are right. The skeleton represents 3-bromopropanoic acid. Go on to the next section.

C _____

If you had been asked to give a trivial name, you'd be right. β-bromopropionic acid, however, is not the systematic name. Propionic acid is the common name for propanoic acid.

$$\underset{\textit{propanoic acid}}{C - C - \overset{\displaystyle O}{\overset{\displaystyle \|}{C}} - OH}$$

β-bromopropionic acid is 3-bromopropanoic acid. Be careful to distinguish between systematic and trivial names. Go on to the next section.

Unsaturation in carboxylic acids is indicated by dropping the *-e* from the name of the corresponding alkene and adding *-oic acid*. For instance

$$CH_3 CH_2 CH = CHCH_2 CO_2 H$$

3-hexenoic acid

If an acid has two or more double bonds, the ending of the name is changed to *-dienoic*, *-trienoic*, and so on.

Among the higher carboxylic acids, i.e., those with larger numbers of carbon atoms, only the ones with an even number of carbon atoms in the chain have trivial names because they are the ones that are found in natural fats. The most widely distributed fatty acid has the trivial name oleic acid. Its systematic name is 9-octadecenoic acid and its skeleton is

$$C - C - C - C - C - C - C - C - C = C - C - C - C - C - C - C - C - \overset{\displaystyle O}{\overset{\displaystyle \|}{C}} - OH$$

9-octadecenoic acid (oleic acid)

The most important unsaturated fatty acids have 18 carbon atoms, with one of their double bonds placed in the middle of the chain as in oleic acid. If other carbon-carbon double bonds are present, they lie farther from the carboxy group.

A typical sample of butter contains 4 to 5 per cent linoleic acid. Its skeleton is

$$C - C - C - C - C - C = C - C - C = C - C - C - C - C - C - C - C - \overset{\displaystyle O}{\overset{\displaystyle \|}{C}} - OH$$

What is its systematic name?

A _____

9,12-octadecadienoic acid

B _____

9,12-octadecenoic acid

C _____

9,12-heptadecadienoic acid

• •

A _____

You are right. The systematic name of linoleic acid is 9,12-octadecadienoic acid. Just for review, analyze the parts of the name and what they indicate:

-octadeca-	18 carbon atoms
-dien-	2 double bonds
-oic acid	carboxylic acid
9,12-	locants for double bonds

Go on to the next section.

B _____

Incorrect. You chose 9,12-octadecenoic acid as the systematic name for linoleic acid. The basic name is all right and you have the correct numbers, but you left out the signal that indicates the presence of two double bonds. Read page 179 again to learn what it is. Then pick another answer.

C _____

You are incorrect. The systematic name for linoleic acid is not 9,12-heptadecadienoic acid. Your mistake was a simple one that you can avoid by remembering that all of the naturally occurring fatty acids have an *even* number of carbon atoms. Go back, count again, and choose another answer.

The most abundant fatty acid is palmitic acid. Its systematic name is hexadecanoic acid. What is its molecular formula?

A _____

$C_{16}H_{33}O$

B _____

$C_{16}H_{32}O_2$

C _____

$C_{16}H_{33}O_2$

• •

A _____

Incorrect. You chose the molecular formula $C_{16}H_{33}O$ to represent palmitic

$$\overset{O}{\underset{\|}{}}$$

or hexadecanoic acid. Since a carboxylic acid has the carboxy group, $-C-OH$, as its functional group, the molecular formula must contain at least two oxygen atoms. Perhaps it will help you to write a skeleton or structural formula before going back to choose another answer.

B _____

You are correct. The molecular formula of hexadecanoic acid is $C_{16}H_{32}O_2$. Its structural formula is

$$
\begin{array}{c}
\quad H\ \ H\ \ H\ \ H\ \ H\ \ H\ \ H\ \ H\ \ H\ \ H\ \ H\ \ H\ \ H\ \ H\ \ H\ \ O \\
\quad |\ \ \ |\ \ \ |\ \ \ |\ \ \ |\ \ \ |\ \ \ |\ \ \ |\ \ \ |\ \ \ |\ \ \ |\ \ \ |\ \ \ |\ \ \ |\ \ \ |\ \ \ \| \\
H-C-C-C-C-C-C-C-C-C-C-C-C-C-C-C-C-OH \\
\quad |\ \ \ |\ \ \ |\ \ \ |\ \ \ |\ \ \ |\ \ \ |\ \ \ |\ \ \ |\ \ \ |\ \ \ |\ \ \ |\ \ \ |\ \ \ |\ \ \ | \\
\quad H\ \ H\ \ H\ \ H\ \ H\ \ H\ \ H\ \ H\ \ H\ \ H\ \ H\ \ H\ \ H\ \ H\ \ H
\end{array}
$$

Go on to the next section.

C _____

You are incorrect. The molecular formula for hexadecanoic, or palmitic, acid is not $C_{16}H_{33}O_2$. Would you like to know an easy way to find the right formula? First, take the saturated hydrocarbon with 16 carbon atoms. Using the general formula that you have already learned (C_nH_{2n+2}), you can write its molecular formula as $C_{16}H_{34}$. Now look at the terminal carbon atom. It changes from a methyl group to a carboxy group.

$$
\begin{array}{cc}
\quad H & \quad O \\
\quad | & \quad \| \\
-C-H & -C-OH \\
\quad | & \\
\quad H &
\end{array}
$$

methyl group   *carboxy group*

Two oxygen atoms are added and the number of hydrogen atoms decreases by two. What is the net result? When you know, turn back and select another answer.

As you might imagine, the composition of a natural fat like butter varies somewhat. Here are the amounts of different fatty acids obtained by hydrolysis of a typical sample of butter.

TABLE 3.5 Fatty Acids in Butter

Systematic name	Trivial name	Per cent
butanoic	butyric	3–4
hexanoic	caproic	1–2
octanoic	caprylic	1
decanoic	capric	2–3
dodecanoic	lauric	2–3
tetradecanoic	myristic	7–9
hexadecanoic	palmitic	23–26
octadecanoic	stearic	10–13
9-hexadecenoic	palmitoleic	5
9-octadecenoic	oleic	30–40
9,12-octadecadienoic	linoleic	4–5

Following are some practice problems in naming carboxylic acids.

1. Name and give the structural formula for the functional group of the carboxylic acids.

2. What is the systematic name of this acid?

$$\begin{array}{c} \overset{\displaystyle Cl}{\underset{\displaystyle Cl}{\mid}} \quad \overset{\displaystyle O}{\parallel} \\ Cl - C - C - OH \end{array}$$

3. The insecticide 1080 is the sodium salt of fluoroacetic acid. What is its structural formula?

4. Write as many kinds of formula as you can for 4-bromo-2-butenoic acid.

● ●

1. Carboxy group: $\overset{\displaystyle O}{\overset{\displaystyle \parallel}{- C - OH}}$

2. Trichloroethanoic acid (*CA* uses trichloroacetic acid)

3.

$$F - \overset{\displaystyle H}{\underset{\displaystyle H}{\mid}}C\overset{\displaystyle O}{\overset{\displaystyle \parallel}{- C}} - O^- Na^+$$

4. Molecular: $C_4 H_5 O_2 Br$　　　Structural: $Br - \overset{\displaystyle H}{\underset{\displaystyle H}{\mid}}C - \overset{\displaystyle H}{\mid}C = \overset{\displaystyle H}{\mid}C - \overset{\displaystyle O}{\overset{\displaystyle \parallel}{C}} - OH$

Condensed: $CH_2 BrCH = CHCO_2 H$　Skeleton: $Br - C - C = C - \overset{\displaystyle O}{\overset{\displaystyle \parallel}{C}} - OH$

Go on to the next section if you feel sure of yourself. If not, go back to page 174 for review.

Several of the important carboxylic acids have two carboxy groups. They are known as dicarboxylic acids. Their trivial names are in widespread use. The lowest five dicarboxylic acids are shown in this Table. *Chemical Abstracts* uses the systematic names.

TABLE 3.6　Dicarboxylic Acids

Formula	Trivial name*	Systematic name
$HO_2 CCO_2 H$	oxalic acid	ethanedioic acid
$HO_2 CCH_2 CO_2 H$	malonic acid	propanedioic acid
$HO_2 C(CH_2)_2 CO_2 H$	succinic acid	butanedioic acid
$HO_2 C(CH_2)_3 CO_2 H$	glutaric acid	pentanedioic acid
$HO_2 C(CH_2)_4 CO_2 H$	adipic acid	hexanedioic acid

*The common names for the next four members of the series are pimelic, suberic, azelaic, and sebacic. One mnemonic for the names is "*Oh, My, Such Good Apple Pie; Sweet As Sugar.*"

Malic acid, the common name for a constituent of many fruit juices, has the formula

$$\underset{\substack{| \\ H \;\; OH}}{HO - \overset{\overset{O}{\|}}{C} - \overset{\overset{H}{|}}{C} - \overset{\overset{H}{|}}{C} - \overset{\overset{O}{\|}}{C} - OH}$$

What is its name according to the system?

A _____

hydroxysuccinic acid

B _____

2-hydroxybutanedioic acid

C _____

hydroxybutanedioic acid

D _____

Help!

● ●

A _____

You seem to be confused over the difference between systematic names and trivial names. Succinic acid is the common name for butanedioic acid. Even though malic acid is hydroxysuccinic acid, that is not its systematic name. Go back and select another answer.

B _____

Your answer is that the preferred name for malic acid is 2-hydroxybutanedioic acid. Here is its formula again.

$$\underset{\substack{| \\ H \;\; OH}}{HO - \overset{\overset{O}{\|}}{C} - \overset{\overset{H}{|}}{C} - \overset{\overset{H}{|}}{C} - \overset{\overset{O}{\|}}{C} - OH}$$

Is there any way that the hydroxy group could be on a carbon other than number 2? No, there isn't. Consequently, the name for malic acid is just hydroxybutanedioic acid. Go on to the next section.

C _____

You are right. The systematic name for malic acid is hydroxybutanedioic acid. Since the hydroxy group can only be on the number 2 carbon atom, its location does not need to be specified by a number. Go on to the next section.

D _____

So you need help. Look at the formula again.

$$\underset{\substack{| \\ H \;\; OH}}{HO - \overset{\overset{O}{\|}}{C} - \overset{\overset{H}{|}}{C} - \overset{\overset{H}{|}}{C} - \overset{\overset{O}{\|}}{C} - OH}$$

What can we see about it? It has four carbon atoms and must be a derivative of butane. There are two carboxy groups. It must be a butanedioic acid. Finally, there is an hydroxy (−OH) group on one of the other carbon atoms. Put all of these facts together and you should be able to give the correct name. Read page 182 again and select another answer.

You should remember that there are no aromatic aldehydes in which the carbonyl group is part of the ring system. The same holds true for aromatic carboxylic acids. The simplest aromatic acid is benzoic acid. Its structural formula is

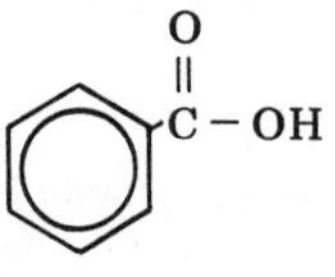

benzoic acid (benzenecarboxylic acid)

Aromatic carboxylic acids with two or more carboxy groups are named as dicarboxylic, tricarboxylic, etc., acids. The three benzenedicarboxylic acids are important in commerce. They are shown here.

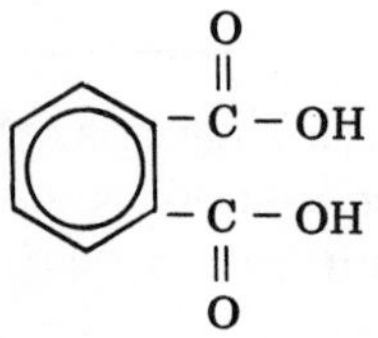

1,2-benzenedicarboxylic acid
(phthalic acid)

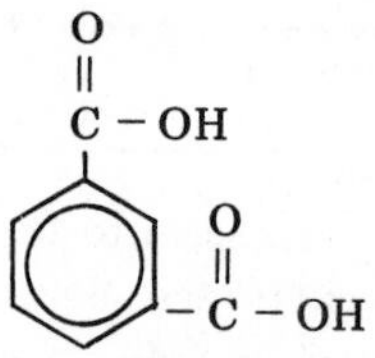

1,3-benzenedicarboxylic acid
(isophthalic acid)

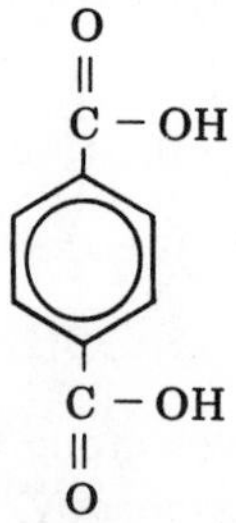

1,4-benzenedicarboxylic acid
(terephthalic acid)

Terephthalic acid is a component of the textile polymer Dacron.

Salicylic acid, an analgesic related to aspirin, is 2-hydroxybenzoic acid. Which of these is its formula?

A ——————————————————

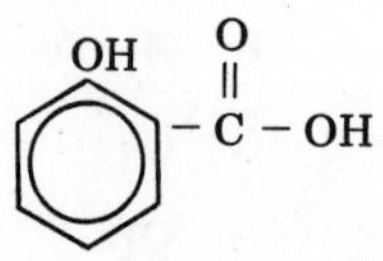

B _____

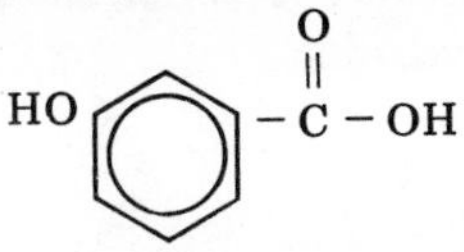

$$HO \bigcirc -\overset{\overset{\displaystyle O}{\|}}{C} - OH$$

C _____

$$\bigcirc -\overset{\overset{\displaystyle O}{\|}}{\underset{\underset{\displaystyle OH}{|}}{C}} - OH$$

• •

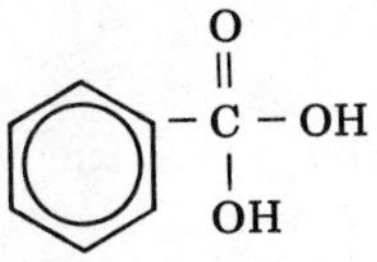

A _____

You are correct. The structural formula for salicylic acid, or 2-hydroxy-benzoic acid is

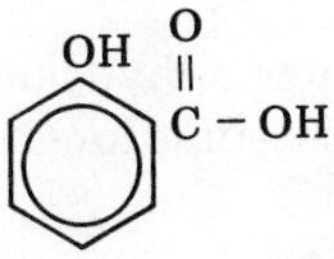

$$\bigcirc \overset{OH}{} \overset{\overset{\displaystyle O}{\|}}{C} - OH$$

The carbons to which the two substituent groups are attached are adjacent to each other. Go on to the next section.

B _____

You are incorrect. You have selected this structural formula to represent 2-hydroxybenzoic acid.

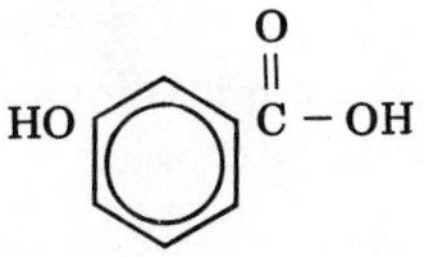

$$HO \bigcirc \overset{\overset{\displaystyle O}{\|}}{C} - OH$$

The two carbons which bear substituent groups are not adjacent, are they? This is the formula for 3-hydroxybenzoic acid. Go back and choose another answer.

C _____

You have pulled a real boner. You picked this formula for 2-hydroxybenzoic acid.

$$\bigcirc -\overset{\overset{\displaystyle O}{\|}}{\underset{\underset{\displaystyle OH}{|}}{C}} - OH$$

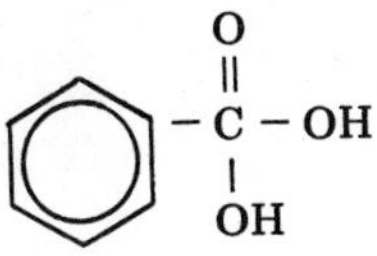

What's wrong with it? First, count the covalent bonds on the carbon atom in the carboxy group. There are five and that's one too many. Read page 184 again before you choose another answer.

Under certain conditions, two molecules of a monobasic carboxylic acid will lose a single molecule of water to form an anhydride. This equation shows the formation of ethanoic anhydride:

$$
\underset{\substack{\text{acetic acid} \\ \text{(ethanoic acid)}}}{CH_3\overset{\overset{O}{\|}}{C}-OH} + \underset{\substack{\text{acetic acid} \\ \text{(ethanoic acid)}}}{HO-\overset{\overset{O}{\|}}{C}CH_3} \rightleftharpoons \underset{\text{water}}{H_2O} + \underset{\substack{\text{acetic anhydride} \\ \text{(ethanoic anhydride)}}}{CH_3\overset{\overset{O}{\|}}{C}-O-\overset{\overset{O}{\|}}{C}CH_3}
$$

Anhydrides are named merely by using the word *anhydride* in place of the word *acid*.

Occasionally, an anhydride will be formed from molecules of two different acids. For example

$$
CH_3\,CH_2\,CH_2\,\overset{\overset{O}{\|}}{C}-O-\overset{\overset{O}{\|}}{C}CH_3
$$

acetic butanoic anhydride

Several dibasic acids form their anhydrides by losing a molecule of water from one molecule of acid. This equation shows the formation of phthalic anhydride.

$$
\underset{\substack{\text{1,2-benzenedicarboxylic acid} \\ \text{(phthalic acid)}}}{\left[\text{benzene ring with } \overset{\overset{O}{\|}}{C}-OH \text{ and } \overset{\overset{O}{\|}}{C}-OH\right]} \rightleftharpoons \underset{\text{water}}{H_2O} + \underset{\substack{\text{1,3-isobenzofurandione} \\ \text{(phthalic anhydride)}}}{\left[\text{benzene ring with } \overset{\overset{O}{\|}}{C}\text{-O-}\overset{\overset{O}{\|}}{C}\right]}
$$

Write as many formulas as you can for propanoic anhydride. When you have finished, check your answers on the next page.

Formulas for propanoic anhydride:

$$\underset{\text{\textit{propanoic acid}}}{CH_3CH_2\overset{\overset{\textstyle O}{\|}}{C}-OH} \ + \ \underset{\text{\textit{propanoic acid}}}{HO-\overset{\overset{\textstyle O}{\|}}{C}CH_2CH_3} \ \rightleftharpoons \ \underset{\text{\textit{water}}}{H_2O} \ + \ \underset{\text{\textit{propanoic anhydride}}}{CH_3CH_2\overset{\overset{\textstyle O}{\|}}{C}-O-\overset{\overset{\textstyle O}{\|}}{C}CH_2CH_3}$$

Condensed formula:　　　$CH_3CH_2\overset{\overset{\textstyle O}{\|}}{C}-O-\overset{\overset{\textstyle O}{\|}}{C}CH_2CH_3$

Line formula:　　　　　$(CH_3CH_2CO)_2O$

Empirical formula:　　　$C_6H_{10}O_3$

Skeleton:　　　　$C-C-\overset{\overset{\textstyle O}{\|}}{C}-O-\overset{\overset{\textstyle O}{\|}}{C}-C-C$

Structural formula:

$$H-\overset{\overset{\textstyle H}{|}}{\underset{\underset{\textstyle H}{|}}{C}}-\overset{\overset{\textstyle H}{|}}{\underset{\underset{\textstyle H}{|}}{C}}-\overset{\overset{\textstyle O}{\|}}{C}-O-\overset{\overset{\textstyle O}{\|}}{C}-\overset{\overset{\textstyle H}{|}}{\underset{\underset{\textstyle H}{|}}{C}}-\overset{\overset{\textstyle H}{|}}{\underset{\underset{\textstyle H}{|}}{C}}-H$$

Carboxylic acids usually do not react rapidly with other organic compounds. Chemists have found that the reactivity is enhanced if the $-OH$ group of the acid is replaced by a halogen, usually chlorine. For example:

$$\underset{\text{\textit{acetic acid}}}{CH_3\overset{\overset{\textstyle O}{\|}}{C}-OH} \qquad\qquad\qquad \underset{\text{\textit{acetyl chloride}}}{CH_3\overset{\overset{\textstyle O}{\|}}{C}-Cl}$$

As a group the compounds are known as acyl halides.

Because the compounds were in widespread use before the adoption of the IUPAC system, the names of the lowest five acyl groups are based on the trivial names of the acids. In 1972, *Chemical Abstracts* abandoned all except acetyl in favor of fully systematic names. This list shows the formulas and names of the groups.

TABLE 3.7   Acyl Groups

Acyl group	IUPAC (Trivial) name	CA name
$H-\overset{\overset{O}{\|\|}}{C}-$	formyl	formyl
$CH_3\overset{\overset{O}{\|\|}}{C}-$	acetyl	acetyl
$CH_3CH_2\overset{\overset{O}{\|\|}}{C}-$	propionyl	1-oxopropyl
$CH_3(CH_2)_2\overset{\overset{O}{\|\|}}{C}-$	butyryl	1-oxobutyl
$CH_3(CH_2)_3\overset{\overset{O}{\|\|}}{C}-$	valeryl	1-oxopentyl
$CH_3(CH_2)_4\overset{\overset{O}{\|\|}}{C}-$	hexanoyl	1-oxohexyl

Chemical Abstracts also retains the trivial name benzoyl chloride for the acyl chloride of benzoic acid. Which of these is the formula for benzoyl chloride?

A

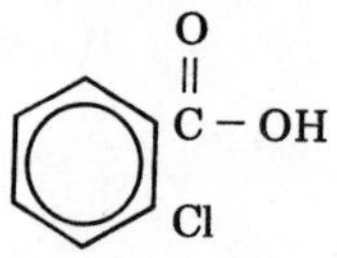

B

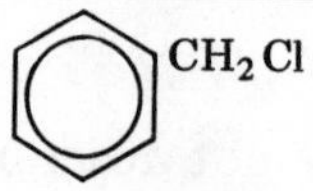

C

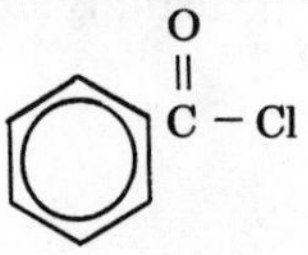

• •

A

You are wrong. You chose this formula for benzoyl chloride.

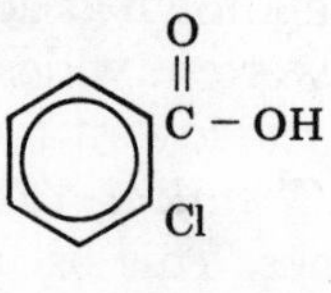

2-chlorobenzoic acid

This formula does not represent an acyl halide. Acyl halides have the general

$$R - \overset{\overset{\displaystyle O}{\|}}{C} - X$$

formula R − C − X. The formula above is for a carboxylic acid since the chlorine atom is substituted on the ring and not the carboxy group. Go back for another answer.

B _____

Incorrect. The formula you chose does not represent an acyl chloride. The

$$R - \overset{\overset{\displaystyle O}{\|}}{C} - Cl$$

acyl chlorides have the general formula R − C − Cl. A chlorine atom is substituted for the − OH group in a carboxylic acid. The correct answer to this question is the formula for the acyl chloride derived from benzoic acid.

benzoic acid

Go back and select another answer.

C _____

You are correct. Benzoyl chloride is the acyl chloride derived from benzoic acid.

benzoic acid   *benzoyl chloride*

CARBOXYLIC ACIDS

Functional group: $- \overset{\overset{\displaystyle O}{\|}}{C} - OH$ Functional group name: *carboxy*

General formula: $R - \overset{\overset{\displaystyle O}{\|}}{C} - OH$ Series name: *-oic acid*

ACYL HALIDES

Functional group: $- \overset{\overset{\displaystyle O}{\|}}{C} - X$ Functional group name: *haloformyl*

General formula: $R - \overset{\overset{\displaystyle O}{\|}}{C} - X$ Series name: *1-oxo- . . . -yl halide*

ACID ANHYDRIDES

Functional group: $- \overset{\overset{\displaystyle O}{\|}}{C} - O - \overset{\overset{\displaystyle O}{\|}}{C} -$ Functional group name: *none*

General formula: $R - \overset{\overset{\displaystyle O}{\|}}{C} - O - \overset{\overset{\displaystyle O}{\|}}{C} - R$ Series name: *-ic anhydride*

3.5 ESTERS

When hydroxy compounds (alcohols and phenols) react with acids, the product is an ester. For the present, only esters of carboxylic acids and alkanols will be considered. A typical reaction is represented by the equation

$$\underset{acid}{RC\overset{O}{\overset{\|}{-}}OH} + \underset{alcohol}{HOR'} \rightleftharpoons \underset{ester}{RC\overset{O}{\overset{\|}{}}OR'} + \underset{water}{H_2O}$$

The esters are named as if they were alkyl salts of the organic acids. This method stems from the early belief that esterification was akin to neutralization.

You must be very careful to distinguish between the portion of the ester molecule derived from the acid and the portion derived from the alcohol. It is easy if you remember that the alkyl group from the alcohol is joined to the carbonyl group through an oxygen atom. The general structural formula illustrates this point.

$$R-\overset{O}{\overset{\|}{C}}-O-R'$$

R $-$ = alkyl group from acid

R' $-$ = alkyl group from alcohol

Consider this structural formula of an ester:

$$CH_3\overset{O}{\overset{\|}{C}}-O-\underset{\underset{CH_3}{|}}{\overset{\overset{CH_3}{|}}{CH}}$$

From which of these alcohols is the ester derived?

A _____

ethanol

B _____

1-propanol

C _____

2-propanol

•••

A _____

You are wrong. Ethanol is not the alcohol which reacts to form this ester.

$$CH_3\overset{O}{\overset{\|}{C}}-O-\underset{\underset{CH_3}{|}}{\overset{\overset{CH_3}{|}}{CH}}$$

Remember that the alcohol portion of the ester is linked to the carbonyl group *through* the oxygen atom. In this example the $R' -$ group for the alcohol is

$$
\begin{array}{c}
CH_3 \\
| \\
-CH \\
| \\
CH_3
\end{array}
$$

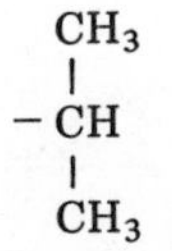

What is the alcohol with the same $R'-$ group? Go back and choose it as your answer.

B _____

Incorrect. This ester is not derived from 1-propanol.

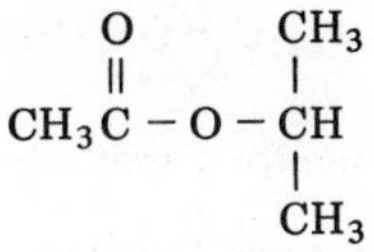

$$
\begin{array}{c}
\quad\ O \qquad CH_3 \\
\quad\ \| \qquad\ | \\
CH_3 C - O - CH \\
\qquad\qquad\ | \\
\qquad\qquad CH_3
\end{array}
$$

You know that the alcohol portion of the ester is linked to the carbonyl group through the oxygen atom. In this example the $R'-$ group for the alcohol is

$$
\begin{array}{c}
CH_3 \\
| \\
-CH \\
| \\
CH_3
\end{array}
$$

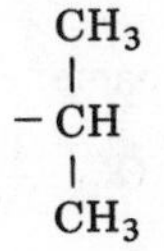

If the alcohol were 1-propanol, the $R'-$ group would be $CH_3 CH_2 CH_2-$, wouldn't it? The difference is in the position of the free valence. Go back and pick another answer.

C _____

You are right. This ester is derived from 2-propanol and acetic acid.

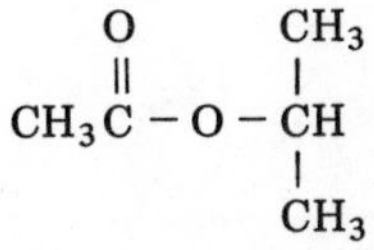

$$
\begin{array}{c}
\quad\ O \qquad CH_3 \\
\quad\ \| \qquad\ | \\
CH_3 C - O - CH \\
\qquad\qquad\ | \\
\qquad\qquad CH_3
\end{array}
$$

Esters are named by stating the alkyl radical of the alcohol as the first word. The second word is formed by dropping the *-ic* ending of the name of the acid and replacing it with *-ate*. The procedure applies to both the systematic and trivial names of the alcohols and acids.

The various combinations of atoms associated with the acids, alkanols, and esters are given special names to avoid confusion. They are shown here:

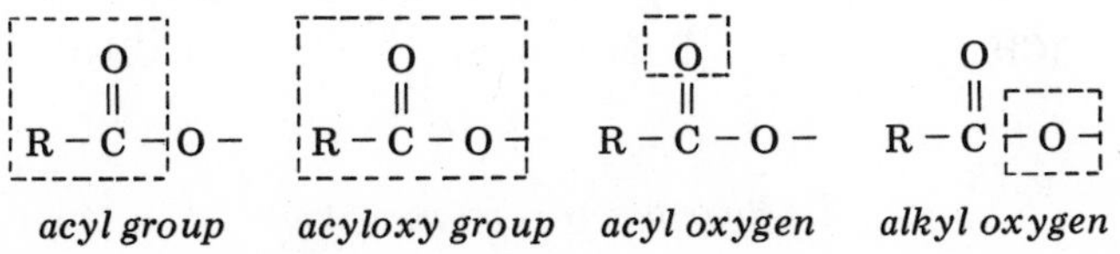

$$
\begin{array}{cccc}
O & O & O & O \\
\| & \| & \| & \| \\
R - C - O - & R - C - O - & R - C - O - & R - C - O - \\
\text{\textit{acyl group}} & \text{\textit{acyloxy group}} & \text{\textit{acyl oxygen}} & \text{\textit{alkyl oxygen}}
\end{array}
$$

Through the use of alcohols that have been isotopically enriched with ^{18}O, it has been shown that the usual esterification of a *primary* alcohol takes place by acyl-oxygen scission. The molecule of water is split out as shown:

$$CH_3C \overset{O}{\overset{\|}{}} \boxed{\,OH\,} + \boxed{H\,}{}^{18}OCH_3 \;=\; CH_3C \overset{O}{\overset{\|}{}} {}^{18}O - CH_3 \;+\; H_2O$$

    acetic acid   *methanol*   *methyl acetate*   *water*

Following are four questions concerning simple carboxylic esters. After you answer them, we will go on to more complex esters.

1. Name this ester:

$$CH_3(CH_2)_2 \overset{O}{\overset{\|}{C}} - O - CH_2CH_3$$

2. Write the structural formula for 1-methylpropyl acetate.

3. What is the parent acid of methyl pentanoate?

4. Write structural formulas for methyl benzoate and phenyl formate.

● ●

1. ethyl butanoate (Acceptable name: ethyl butyrate)

2.
$$CH_3 - \overset{O}{\overset{\|}{C}} - O - \overset{CH_3}{\overset{|}{CH}}$$
$$\overset{|}{CH_2}$$
$$\overset{|}{CH_3}$$

3. pentanoic acid

4.

     methyl benzoate   *phenyl formate*

Condensed structural formulas of esters written on one line may cause you some difficulty at first. You should have no trouble if you remember that the oxygen of the carbonyl group is always written *after* the carbon to which it is attached. For instance

$$CH_3CH_2COOC(CH_3)_3 \qquad \text{is the same as} \qquad CH_3CH_2 \overset{O}{\overset{\|}{C}} - O - \overset{CH_3}{\overset{|}{\underset{|}{C}}} - CH_3$$
$$\overset{}{\underset{CH_3}{}}$$

1,1-dimethylethyl propanoate

Consider this condensed formula of an ester:

$$CH_3 COO(CH_2)_2 CH_3$$

What is its name?

A _____

methyl butanoate

B _____

propyl acetate

C _____

I need help.

• •

A _____

Wrong. Let's write the full structural formula beside the condensed formula

$$CH_3 COO(CH_2)_2 CH_3 \qquad\qquad CH_3 - \overset{\displaystyle O}{\overset{\|}{C}} - O - CH_2 CH_2 CH_3$$

This ester is derived from

$$CH_3 \overset{\displaystyle O}{\overset{\|}{C}} - OH \qquad and \qquad CH_3 CH_2 CH_2 OH$$

Now how would you name it? Go back and choose another answer.

B _____

Correct. The condensed formula represents propyl acetate.

$$CH_3 - \overset{\displaystyle O}{\overset{\|}{C}} - O - CH_2 CH_2 CH_3$$

Go on to the next section.

C _____

Here is the condensed formula again: $CH_3 COO(CH_2)_2 CH_3$. Since the oxygen of the carbonyl group is always written after the carbon to which it is attached, the complete structural formula would be

$$CH_3 - \overset{\displaystyle O}{\overset{\|}{C}} - O - CH_2 CH_2 CH_3$$

Now divide it into acid and alcohol portions. The $R -$ group from the acid is $CH_3 -$ and the $R' -$ group from the alcohol is $CH_3 CH_2 CH_2 -$. Combine them properly into a name and you have the answer. Go back and see if it is there.

Substituent groups on substituted esters are located by numbering the carbon atoms in both directions from the oxygen atom that bridges the alcohol and acid parts of the ester. As an illustration

$$\underset{4 \quad 3 \quad 2 \quad 1 \qquad 1 \quad 2}{CH_3 CH_2 CH_2 \overset{\overset{\textstyle O}{\|}}{C} - O - CH_2 CH_3}$$

ethyl butanoate

$$\underset{4 \quad 3 \quad 2 \quad 1 \qquad 1 \quad 2}{CH_3 \overset{\overset{\textstyle OH}{|}}{C}HCH_2 \overset{\overset{\textstyle O}{\|}}{C} - O - CH_2 CH_2 Cl}$$

2-chloroethyl 3-hydroxybutanoate

What is the skeleton formula for the ester 3-bromopropyl chloroacetate?

A _____

$$Cl - C - C - \overset{\overset{\textstyle O}{\|}}{C} - O - C - C - Br$$

B _____

$$Cl - C - \overset{\overset{\textstyle O}{\|}}{C} - O - C - C - C - Br$$

C _____

$$Br - C - C - \overset{\overset{\textstyle O}{\|}}{C} - O - C - C - Cl$$

• •

A _____

You chose this skeleton to represent 3-bromopropyl chloroacetate.

$$Cl - C - C - \overset{\overset{\textstyle O}{\|}}{C} - O - C - C - Br$$

You are wrong. From the name we can tell that the compound is the ester of 3-bromopropanol and chloroacetic acid. Their formulas are

$$BrCH_2 CH_2 CH_2 OH$$

3-bromopropanol

$$Cl - \overset{\overset{\textstyle H}{|}}{\underset{\underset{\textstyle H}{|}}{C}} - \overset{\overset{\textstyle O}{\|}}{C} - OH$$

Chloroacetic acid

Write the formula for the ester and see if it appears as a choice above.

B _____

Right. This is the skeleton for 3-bromopropyl chloroacetate.

$$Cl - C - \overset{\overset{\textstyle O}{\|}}{C} - O - C - C - C - Br$$

There is no reason to designate the location of the Cl by a number because it cannot be on the other carbon atom. Continue on to the next section.

C _____

Incorrect. The skeleton you chose to represent 3-bromopropyl chloroacetate is

$$Br - C - C - \overset{\overset{\displaystyle O}{\|}}{C} - O - C - C - Cl$$

From the name you should know that the bromine atom is part of the alcohol portion of the molecule and the chlorine is substituted on the acid portion. In the skeleton above they are reversed. (Remember that the alcohol portion is

linked to the $- \overset{\overset{\displaystyle O}{\|}}{C} -$ group *through* the alkyl oxygen.) Go back, read page 194 again, and pick another answer.

Esters react with strong bases to form the parent alcohol and the salt of the parent acid. This process is called saponification. The saponification of ethyl benzoate can serve as an example.

$$\text{C}_6\text{H}_5 - \overset{\overset{\displaystyle O}{\|}}{\text{C}} - \text{O} - \text{CH}_2\text{CH}_3 + \text{NaOH} \rightleftharpoons \quad \text{C}_6\text{H}_5 - \overset{\overset{\displaystyle O}{\|}}{\text{C}} - \text{ONa} + \text{HOCH}_2\text{CH}_3$$

ethyl benzoate   *sodium benzoate*   *ethanol*

To permit information on esterified alcohols and thiols to be more readily found in the Chemical Substance Index, *Chemical Abstracts* modifies the usual rules of index name selection for some esters. The chemical functionality of certain very common acids is disregarded and the entry is made at the alcohol despite its lower functionality. A pair of examples will illustrate the procedure:

$$\text{CH}_3\overset{\overset{\displaystyle O}{\|}}{\text{C}}\text{OCH}_2\text{CH}_2 - \text{C}_6\text{H}_5$$

2-phenylethyl acetate
listed in the index under:
Acetic acid, esters
2-phenylethyl ester

$$\text{H} - \overset{\overset{\displaystyle O}{\|}}{\text{C}} - \text{OCH}_2\text{CH}_2\text{O} - \overset{\overset{\displaystyle O}{\|}}{\text{C}} - \text{H}$$

1,2-ethanediyl diformate
listed in the index under:
1,2-Ethanediol, esters
diformate

The *Chemical Abstracts* Index Guide should be consulted for additional information before a search for an ester in the Chemical Substance Index is begun.

Summary: Nomenclature of Carboxylic Acids, Anhydrides, Acyl Halides, and Esters

The functional group of the carboxylic acids is the carboxy group, $-\overset{\displaystyle O}{\overset{\|}{C}}-OH$ or $-CO_2H$. Their systematic names are formed by dropping the *-e* from the name of the parent alkane and adding *-oic* followed by the word acid. The names of the carboxylate anions are formed by dropping the *-ic* from the name of the acid and adding *-ate*. Substituent groups on carboxylic acids are located by numbering the carbon chain beginning with the carboxy carbon as number 1.

Unsaturated carboxylic acids are named in an analogous manner by dropping the *-e* from the name of the parent alkene or alkyne and adding the suffix *-oic* followed by the word *acid*.

The dicarboxylic acids are named systematically by adding the suffix *-dioic* to the name of the parent hydrocarbon. The trivial names, however, are in general use (see p. 182).

The simplest aromatic carboxylic acid is benzoic acid.

benzoic acid

Acid anhydrides are formed by the loss of a molecule of water from two carboxy groups. These groups may come from two acid molecules or a single moledule of a dibasic acid such as phthalic acid. The general formula of the acid anhydrides is $R-\overset{\displaystyle O}{\overset{\|}{C}}-O-\overset{\displaystyle O}{\overset{\|}{C}}-R$. Simple anhydrides are named by replacing the word acid with anhydride. Mixed anhydrides are named by giving the names of the two acids followed by anhydride as a third word.

Acyl halides result from the replacement of the $-OH$ group of an acid by a halogen atom (see page 187).

Esters result from the reaction of an acid and an alcohol. Their general formula is $R-\overset{\displaystyle O}{\overset{\|}{C}}-O-R'$. The alkyl group linked through the oxygen ($R'-$) is the group from the alcohol and the group attached to the carbonyl group ($R-$) is the group from the acid. Esters are named by stating the group from the alcohol as the first word and the name of the carboxylate anion as the second word. For instance:

$$CH_3CH_2\overset{\displaystyle O}{\overset{\|}{C}}-O-CH_3$$

methyl propanoate

The following terms are used to designate portions of ester molecules:

$$\begin{array}{cccc}
\underset{\textit{acyl group}}{\boxed{R-C}-O-} &
\underset{\textit{acyloxy group}}{\boxed{R-C-O-}} &
\underset{\textit{acyl oxygen}}{R-C-O-} &
\underset{\textit{alkyl oxygen}}{R-C-\boxed{O}-}
\end{array}$$

CARBOXYLIC ACID ESTERS

$$\text{Functional group:} \quad -\overset{\overset{\textstyle O}{\|}}{C}-O-R'$$ Functional group name: *alkoxycarbonyl*

$$\text{General formula:} \quad R-\overset{\overset{\textstyle O}{\|}}{C}-O-R'$$ Series name: *-yl . . . . -oate*

3.6 ETHERS

Simple ethers are organic compounds with the general formula $R-O-R'$. If $R-$ and $R'-$ represent identical alkyl or aryl groups, the ether is symmetrical. If they are different, the ether is unsymmetrical.

Chemical Abstracts has abandoned ether as a heading for indexing. Since it will be some time before the IUPAC rules and textbooks do so, this book will present both IUPAC and *Chemical Abstracts* names.

Symmetrical ethers are named by naming the $R-$ group and adding ether as a second word. Thus the simplest symmetrical ether is CH_3OCH_3, methyl ether. Ethyl ether, $CH_3CH_2OCH_2CH_3$, is used as an anesthetic and is commonly called ether. This last usage is incorrect and should be avoided.

Unsymmetrical ethers are named by giving the names of the two groups ($R-$ and $R'-$) in alphabetical order and appending ether as a third word. Thus $CH_3OCH_2CH_3$, an unsymmetrical ether, is ethyl methyl ether. As you will see and hear, many chemists are not careful to name the groups in alphabetical order.

Chemical Abstracts names symmetrical acyclic ethers as *oxy* compounds. Unsymmetrical ethers are named as alkoxy or aryloxy derivatives of the hydrocarbon with the longest chain. These examples illustrate the practice:

CH_3OCH_3 IUPAC: methyl ether
                                CA: oxybis[methane]

$CH_3CH_2OCH_2CH_3$ IUPAC: ethyl ether
                                CA: 1,1′-oxybis[ethane]

$CH_3OCH_2CH_3$ IUPAC: ethyl methyl ether
                                CA: methoxyethane

What is the IUPAC systematic name of this ether?

$$CH_3CH_2OCH_2CH_2CH_2CH_3$$

A _____

ethyl butyl ether

B _____

butyl ethyl ether

C _____

butylethyl ether

• •

A _____

The ether with the formula

$$CH_3CH_2OCH_2CH_2CH_2CH_3$$

does have an ethyl group and a butyl group. In the name, however, these groups should be given in alphabetical order. The correct name is butyl ethyl ether (ethoxybutane). Continue on to the next section.

B _____

You are right. This is the formula of butyl ethyl ether (ethoxybutane).

$$CH_3CH_2OCH_2CH_2CH_2CH_3$$

Go on to the next section.

C _____

Close, but not quite right. Unsymmetrical ethers always have three words in their names, one for each alkyl or aryl radical and then the word ether. In this example the proper name is butyl ethyl ether (ethoxybutane). Be more careful next time. Continue on to the next section.

When an organic molecule contains two or more kinds of functional groups, the principal group is expressed by the ending of the name and the others by prefixes. How is this determined? Chemists are not always consistent in this matter, but *Chemical Abstracts* uses an order of precedence in its indexing. The order does not constitute an attempt to give the relative importance of functions, but only the general usage in selection of the one to be employed for the ending of the name. The order of precedence tells one where to look in an index to find a compound of mixed function. The complete list can be found in the appendix. An abbreviated order of precedence for the functions discussed in this book is:

free radicals
onium cations
carbanions

acids
acid halides
amides
nitriles
aldehydes
ketones
alcohols and phenols (of equal rank)
amines
nitrogen compounds
oxygen compounds
sulfur compounds
carbon compounds (including hydrocarbons)

For example:

$$HOCH_2CH_2CH_2CH_2 \overset{\overset{\displaystyle O}{\|}}{C} - H$$

5-hydroxypentanal

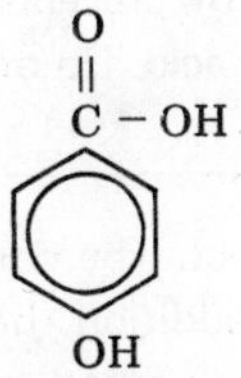

4-hydroxybenzoic acid
(p-hydroxybenzoic acid)

How would you name this compound?

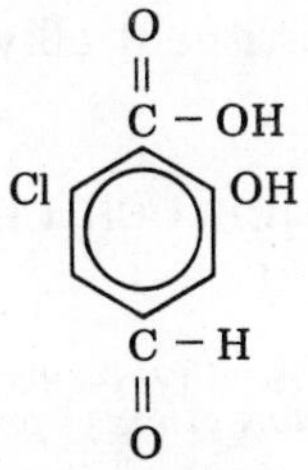

A _____

As an aldehyde

B _____

As a substituted benzene

C _____

As a carboxylic acid

D _____

As a phenol

● ●

A _____

Incorrect. The compound in question does contain the functional group of

the aldehydes, $-\overset{\overset{\displaystyle O}{\|}}{C} - H$. In addition, however, it has a carboxy group and a hydroxy group. So it might be named as an aldehyde, acid, or phenol. The correct choice depends on which function comes first in the order of precedence. Go back and choose another answer.

B _____

You are wrong. Compounds are named as substituted hydrocarbons only if they do not have a functional group. A quick look at the formula for the compound in question should show you that it has at least one functional group. Turn back and read pages 198 and 199 again before you choose another answer.

C _____

Right. There are three functional groups in the compound: carboxy, hydroxy, and carbonyl. The carboxy group and its acid function lie highest on the order of precedence. Consequently, the compound is properly named as a carboxylic acid. Go on to the next section.

D _____

Incorrect. The compound does contain the functional group of the phenols, $-OH$. In addition, however, it has a carboxy group and a carbonyl group. It might be named as an aldehyde, acid, or phenol. The proper choice depends on which function comes first in the order of precedence. Go back and choose another answer.

When necessary, the carbon atoms in the alkyl or aryl groups of ethers can be numbered. Start with 1 for both carbon atoms next to the oxygen atom and go toward the ends. Locate substituent alkyl groups or halogen atoms in the usual way.

$$ClCH_2\,CH_2\,OCH_2\,CH_2\,CH_3$$
$$2\quad 1\qquad 1\quad 2\quad 3$$

2-chloroethyl propyl ether
((2-chloroethoxy) propane)

Following are the formulas of three ethers. Name them and then check your answer with the correct names given below the formulas.

1.

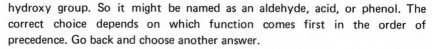

2. $CH_3\,CHCH_2\,OCH_2\,CH_2\,CH_2\,CH_2\,I$
              |
             CH_3

3. CH_3
                    |
    $CH_3\,CH_2\,OCH_2\,CCH_3$
                    |
                   CH_3

• •

1. methyl phenyl ether or anisole (*CA* name: methoxybenzene)

2. 4-iodobutyl isobutyl ether (*CA* name: 4-iodo-1-(2-methylpropoxy)-butane)

3. ethyl 2,2-dimethylpropyl ether (*CA* name: 1-ethoxy-2,2-dimethyl-propane)

Note: When unsubstituted, the $-CH_2\overset{\overset{\displaystyle CH_3}{|}}{\underset{\underset{\displaystyle CH_3}{|}}{C}}-CH_3$ group is called neopentyl.

Thus, an acceptable name is ethyl neopentyl ether.

Alkoxy groups are often present in compounds that contain other functional groups such as the hydroxy group, $-OH$, carbonyl group, $-\overset{\overset{\displaystyle O}{||}}{C}-$, or carboxy group, $-\overset{\overset{\displaystyle O}{||}}{C}-OH$. In these instances the compounds are named as alkoxy derivatives of the compound of the other function. These three examples should make the point clear.

Compound	Principal functional group		Systematic name
$CH_3OCH_2CH_2CH_2OH$	$-OH$	hydroxy	3-methoxy-1-propanol
$CH_3CH_2CH_2\overset{\overset{\displaystyle O}{||}}{C}CH\underset{\underset{\displaystyle OCH_3}{|}}{}CH_3$	$-\overset{\overset{\displaystyle O}{||}}{C}-$	carbonyl	2-methoxy-3-hexanone
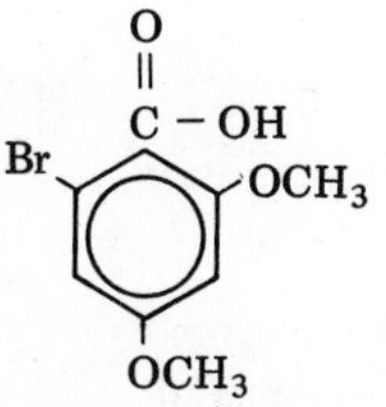 (structure)	$-\overset{\overset{\displaystyle O}{||}}{C}-OH$	carboxy	2-bromo-4,6-dimethoxybenzoic acid

What is the name of this compound?

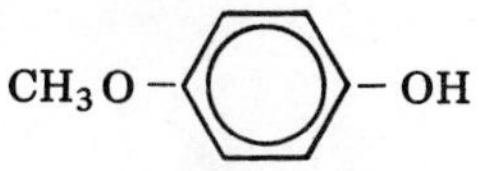

$$CH_3O - \bigcirc - OH$$

A _____

4-methoxyphenol

B _____

p-hydroxyphenyl methyl ether

C _____

1-hydroxy-4-methoxybenzene

● ●

A _____

You are right. The principal functional group in this compound is the $-$ OH group and it is properly named as an alkoxy derivative of phenol. Go on to the next section.

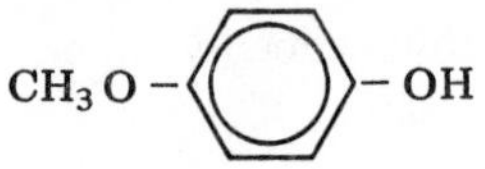

$$CH_3 O - \langle \bigcirc \rangle - OH$$

B _____

The name *p*-hydroxyphenyl methyl ether is descriptive of the compound represented by the formula

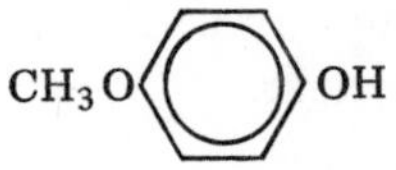

$$CH_3 O \langle \bigcirc \rangle OH$$

There is, however, an order of precedence for functional groups that is used to determine names. The hydroxy group is higher on the list than the ether function. The result is that this compound is not named as an ether. Go back and choose another answer which indicates the hydroxy group as the main functional group.

C _____

You picked 1-hydroxy-4-methoxybenzene as the name for

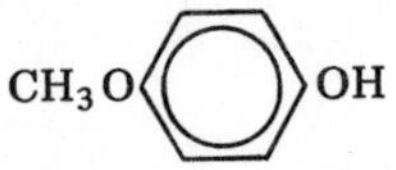

$$CH_3 O \langle \bigcirc \rangle OH$$

This name describes the compound correctly, but since there is an accepted name for hydroxybenzene, it is more properly named 4-methoxyphenol. Continue to the next section.

The herbicide 2,4-D has the systematic name (2,4-dichlorophenoxy)acetic acid. What is its structural formula?

A _____

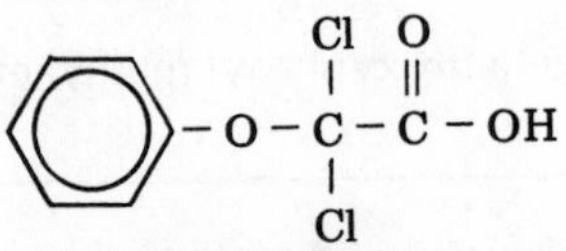

$$\langle \bigcirc \rangle - O - \underset{\underset{Cl}{|}}{\overset{\overset{Cl}{|}}{C}} - \overset{\overset{O}{\|}}{C} - OH$$

B _____

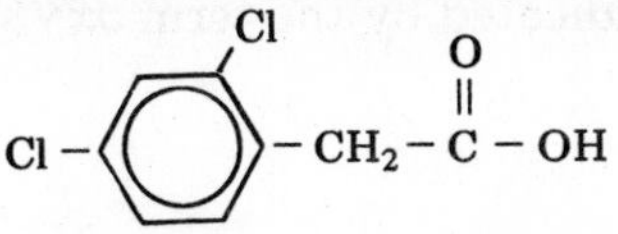

$$Cl - \underset{\underset{Cl}{}}{\bigcirc} - CH_2 - \overset{\overset{O}{\|}}{C} - OH$$

C _____

$$\underset{Cl}{\overset{Cl}{\bigcirc}} - O - CH_2 - \overset{\overset{O}{\|}}{C} - OH$$

• •

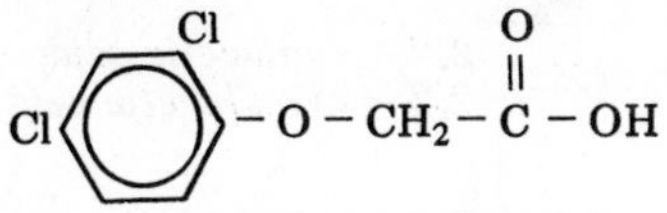

A _____

Incorrect. The formula you chose for (2,4-dichlorophenoxy)acetic acid is

$$\bigcirc - O - \overset{\overset{Cl}{|}}{\underset{\underset{Cl}{|}}{C}} - \overset{\overset{O}{\|}}{C} - OH$$

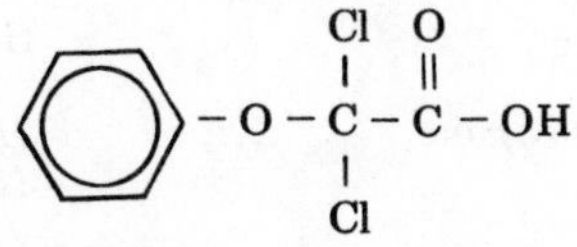

The name indicates there should be a pair of chlorine atoms substituted on the ring. Where are they? The compound shown here is 2,2-dichloro-2-phenoxy-acetic acid. Go back and choose another answer.

B _____

Your choice for the formula for (2,4-dichlorophenoxy)acetic acid is

$$Cl - \underset{\underset{Cl}{}}{\bigcirc} - CH_2 - \overset{\overset{O}{\|}}{C} - OH$$

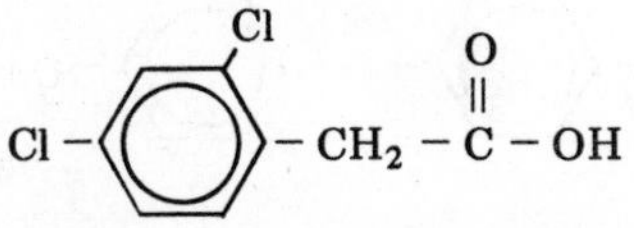

It's OK in all respects but one. Notice the *oxy* in the name. It means that the ring structure is joined to the acetic acid through an oxygen atom. The compound shown here is (2,4-dichlorophenyl)acetic acid. Go back and select another formula for 2,4-D.

C _____

You are right. The formula for (2,4-dichlorophenoxy)acetic acid is

$$\underset{Cl}{\overset{Cl}{\bigcirc}} - O - CH_2 - \overset{\overset{O}{\|}}{C} - OH$$

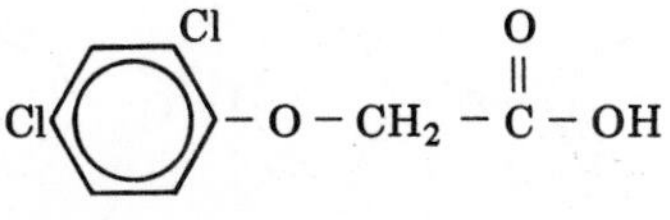

Is it any wonder it's commonly called 2,4-D?

When a functional group with higher priority is present, oxygen linking two identical groups may be indicated by the term *oxy*. For example,

$HOCH_2CH_2OCH_2CH_2OH$

IUPAC: *2,2'-oxydiethanol*
CA: *2,2'-oxybis[ethanol]*

$$HO-\overset{\overset{O}{\|}}{C}-CH_2OCH_2-\overset{\overset{O}{\|}}{C}-OH$$

2,2'-oxydiacetic acid
2,2'-oxybis[acetic acid]

3,3'-oxydiphenol
3,3'-oxybis[phenol]

Notice the use of brackets around the *Chemical Abstracts* parent heading.

The prefix *epoxy-* is used to indicate an oxygen atom directly attached to two carbon atoms already forming part of a ring or to two adjacent carbon atoms of a chain. Two examples are:

$ClCH_2CHCH_2$

IUPAC: *1-chloro-2,3-epoxypropane*
CA: *(chloromethyl)oxirane*
(see p. 247)

(E)-2,3-epoxybutane
(E)-2,3-dimethyloxirane

Compounds that have two or more $-O-$ functional groups are called polyethers. There are several ways to name them, depending upon the specific nature of the polyether.

Symmetrical linear polyethers are named in terms of the central oxygen atom when there is an odd number of ether oxygens. There are three in this example:

IUPAC: *bis(4-phenoxyphenyl) ether*
CA: *1,1'-oxybis[4-phenoxybenzene]*

If there is an even number of ether oxygen atoms, the name is based on the central hydrocarbon group. The word ether does not occur in the name. For instance:

$$CH_3CH_2CH_2OCH_2OCH_2CH_2CH_3$$

IUPAC: *dipropoxymethane*
CA: *1,1'-[methylenebis(oxy)] bis[propane]*

4-bis(4-phenoxyphenoxy)benzene
1,4-bis(4-phenoxyphenoxy)benzene

Two important solvent media for organic reactions are ethers. They have the trivial names *glyme* and *diglyme*.

$$CH_3OCH_2CH_2OCH_3 \qquad\qquad CH_3OCH_2CH_2OCH_2CH_2OCH_3$$

glyme   *diglyme*

Glyme is 1,2-dimethoxyethane. What is diglyme?

A _____

dimethoxyethyl ether

B _____

bis(2-methoxyethyl) ether

C _____

2,2′-dimethoxyethyl ether

● ●

A _____

Incorrect. Since there are three ether oxygens, the compound will be named as an ether. The term dimethoxyethyl is ambiguous, however. Does "di" refer just to methoxy or to both methoxy and ethyl? Think about it for a moment before you go back to choose another answer.

B _____

Right. Diglyme is bis(2-methoxyethyl) ether. Its *CA* name is 1,1′-oxybis[2-methoxyethane]. Go on to the next section.

C _____

Not quite. You chose the name 2,2′-dimethoxydiethyl ether for diglyme

$$CH_3OCH_2CH_2OCH_2CH_2OCH_3$$

Since the compound is symmetrical, there is a way to name the group attached to the central oxygen atom and then indicate that there are two of them. Work out the name of the group, then go back and choose another answer.

Polyethers derived from three or more molecules of aliphatic dihydroxy compounds (glycols) are named using the term *oxa* to show the substitution of oxygen, $-O-$, for a methylene group in the carbon chain. Here are the two examples cited in the IUPAC rules:

     5 4 3 2 1
$$HOCH_2OCH_2OCH_2OH$$

2,4-dioxapentane-1,5-diol

  1 2 3 4 5 6 7 8 9 10 11 12 13 14
$$HOCH_2CH_2OCH_2CH_2OCH_2OCH_2CH_2CH_2O\ CH_2CH_2OH$$

3,6,8,12-tetraoxatetradecane-1,14-diol

The following ethers with trivial names are recognized in the IUPAC rules. Their systematic names and the names used for indexing by *Chemical Abstracts* are shown in the table following the formulas.

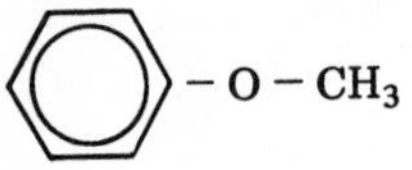

anisole

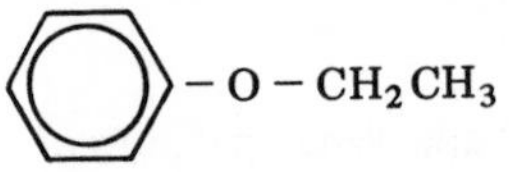

phenetole

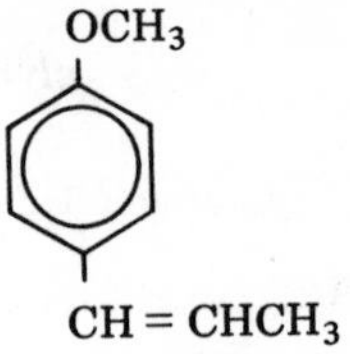

anethole

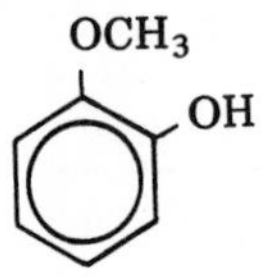

guaiacol

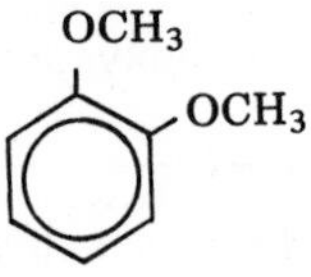

veratrole

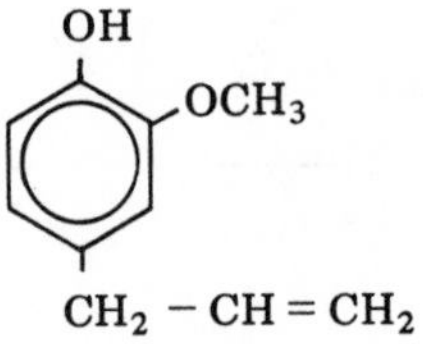

eugenol

TABLE 3.8   Ethers

IUPAC Trivial name	IUPAC Systematic name	Chemical Abstracts Index
anisole	methyl phenyl ether	methoxybenzene
phenetole	ethyl phenyl ether	ethoxybenzene
anethole	methyl 4-(1-propenyl)phenyl ether	1-methoxy-4-(1-propenyl)benzene
guaiacol	2-methoxyphenol	same as IUPAC
veratrole	1,2-dimethoxybenzene	same as IUPAC
eugenol	2-methoxy-4-(2-propenyl)phenol	same as IUPAC

Hydroperoxides, ROOH, and *peroxides*, ROOR', are named using radico-functional nomenclature. The R and R' groups are named and followed by hydroperoxide or peroxide as appropriate. For example,

CH_3CH_2OOH — ethyl hydroperoxide

$CH_3\overset{O}{\overset{\|}{C}}-O-O-\overset{O}{\overset{\|}{C}}CH_3$ — diacetyl peroxide

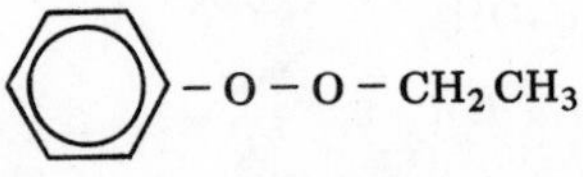

ethyl phenyl peroxide

$CH_3\overset{CH_3}{\overset{|}{C}H}-O-O-\overset{CH_3}{\overset{|}{C}}HCH_3$ — bis(1-methylethyl) peroxide

Summary: Nomenclature of Ethers

The general formula for the ethers is $R - O - R'$. Symmetrical ethers have identical alkyl or aryl groups attached to the oxygen and can be named by naming the group and adding ether as a second word. Unsymmetrical ethers can be named by giving the two alkyl or aryl groups in alphabetical order and adding ether as a third word.

The group $RO -$ is an alkoxy or aryloxy group. Alkoxy and aryloxy group names are used for compounds that have other functional groups besides the ether linkage.

The prefix *oxy-* may be used to indicate oxygen linking two identical parent compounds when a functional group of higher priority is present. The prefix *epoxy-* designates an oxygen atom attached to two adjacent carbon atoms in a ring or chain.

Polyethers with odd numbers of ether oxygens are named as ethers, and those with even numbers as derivatives of the central hydrocarbon.

Chemical Abstracts has abandoned the use of ether as an indexing parent. Ethers are indexed as substituted hydrocarbons or other parent compounds.

ETHERS

Functional group: $- O -$ Functional group name: *oxy*

General formula: $R - O - R'$ Series name: *-yl . . . . -yl ether*

3.7 AMINES AND IMINES

Amines [ă-mĕnz′] are organic bases that are derived from ammonia, NH_3. One, two, or all three hydrogen atoms of the ammonia may be replaced by alkyl or aryl groups. These are known as primary, secondary, and tertiary amines, respectively. Table 3.9 will show you the general formulas and functional groups.

TABLE 3.9   Types of Amines

Class of amine	General formula	Functional group
Primary	$R - NH_2$	$- NH_2$
Secondary	$\begin{matrix} R \\ R' \end{matrix} \!\! > NH$	$> NH$
Tertiary	$\begin{matrix} R \\ R' - N \\ R'' \end{matrix}$	$\geq N$

Primary amines are named by adding the suffix -*amine* to the name of the hydrocarbon group. If it is necessary to number the carbon atoms in order to designate substituents, start numbering with the carbon atom next to the $-NH_2$ group.

To insure unique names and to avoid any possibility of ambiguity, *Chemical Abstracts* indexes all amines as primary or *N*-substituted primary amines. The names are formed by eliding the -*e* from the name of the corresponding hydrocarbon and adding the suffix -*amine*. These indexing names will be given along with the IUPAC names throughout this section. This table illustrates the two systems.

TABLE 3.10   Naming of Amines

Formula	*IUPAC name*	*CA indexing name*
CH_3NH_2	methylamine	methanamine
$(CH_3)_2NH$	dimethylamine	*N*-methylmethanamine
$(CH_3)_3N$	trimethylamine	*N,N*-dimethylmethanamine

What is the IUPAC name of the amine represented by this formula:

$$ClCH_2CH_2NH_2$$

A _____

chloromethylamine

B _____

chloroethylamine

C _____

2-chloroethylamine

• •

A _____

Incorrect. You named this compound chloromethylamine.

$$ClCH_2CH_2NH_2$$

As the name implies, chloromethylamine is a derivative of methylamine, CH_3NH_2. The formula above has two carbon atoms and must, therefore, be a derivative of ethylamine. Go back and choose another answer.

B _____

Your answer is that this formula represents chloroethylamine.

$$ClCH_2CH_2NH_2$$

You are right, but doesn't this formula also represent chloroethylamine?

$$CH_3 CHClNH_2$$

You can distinguish between the two by means of numbers. Work out a name for both of these amines. Then go back and pick another answer.

C _____

Correct. The name for the compound is 2-chloroethylamine.

$$ClCH_2 CH_2 NH_2$$

The carbon chain is numbered from the $-NH_2$ functional group. The *CA* name is 2-chloroethanamine. Continue to the next section.

Secondary and tertiary amines that have the same unsubstituted alkyl groups may be named by using the prefixes *di-* and *tri-* before the name of the group. When numbers are needed, primes and double primes are used to indicate the second and third groups. As an example consider this formula for the tertiary amine, tripropylamine (*CA* name: *N,N*-dipropyl-1-propanamine).

$$\begin{array}{cccccc} 3 & 2 & 1 & 1' & 2' & 3' \end{array}$$
$$CH_3 CH_2 CH_2 - N - CH_2 CH_2 CH_3$$
$$\underset{\begin{array}{ccc}1'' & 2'' & 3''\end{array}}{CH_2 CH_2 CH_3}$$

Symmetrically substituted derivatives of symmetrical secondary or tertiary amines can be named either by locating each substituent or by placing the complete name of the substituent group in parentheses after *bis-* or *tris-* as appropriate. For example:

$(ClCH_2 CH_2)_2 NH$

IUPAC: *2,2'-dichlorodiethylamine*
or
bis(2-chloroethyl)amine

CA: 2-chloro-N-(2-chloroethyl)ethanamine

$(ClCH_2 CH_2)_3 N$

2,2',2''-trichlorotriethylamine
or
tris(2-chloroethyl)amine

2-chloro-N,N-bis(2-chloroethyl)ethanamine

What is the formula for 1,1'-dichlorotriethylamine? (*Chemical Abstracts* would call it 1-chloro-*N*-(1-chloroethyl)-*N*-ethylethenamine.)

A _____

$$\underset{\begin{array}{c}|\\Cl\end{array}}{CH_3 CH} - N - \underset{\begin{array}{c}|\\Cl\end{array}}{CHCH_3}$$
$$\overset{CH_3 CHCl}{\underset{}{|}}$$

B _____

$$ClCH_2 CH_2 - N - \underset{\begin{array}{c}|\\CH_2 CH_2 Cl\end{array}}{CH_2 CH_2 Cl}$$
$$\overset{CH_2 CH_2 Cl}{|}$$

C _____

$$CH_3CH - \underset{\underset{Cl}{|}}{\overset{\overset{CH_2CH_3}{|}}{N}} - \underset{\underset{Cl}{|}}{CHCH_3}$$

• •

A _____

You chose this formula for 1,1′-dichlorotriethylamine.

$$CH_3CH - \underset{\underset{Cl}{|}}{\overset{\overset{CH_3CHCl}{|}}{N}} - \underset{\underset{Cl}{|}}{CHCH_3}$$

It appears that all three groups attached to the nitrogen are identical. According to the name, only two of them are to have chlorine atoms substituted on the ethyl groups. Think this over and select another answer.

B _____

You picked this formula to represent 1,1′-dichlorotriethylamine.

$$ClCH_2CH_2 - \underset{\overset{|}{CH_2CH_2Cl}}{N} - CH_2CH_2Cl$$

All three groups attached to the nitrogen are the same, aren't they? They are $ClCH_2CH_2-$. These are 2-chloroethyl groups. From the name you can see that the compound you seek is a derivative of triethylamine.

$$CH_3CH_2 - \underset{\overset{|}{CH_2CH_3}}{N} - CH_2CH_3$$

Chlorine atoms are to be substituted on the number 1 and 1′ carbons. When you have done this, go back and choose another answer.

C _____

You are correct. This formula depicts 1,1′-dichlorotriethylamine.

$$CH_3CH - \underset{\underset{Cl}{|}}{\overset{\overset{CH_2CH_3}{|}}{N}} - \underset{\underset{Cl}{|}}{CHCH_3}$$

The *CA* name is 1-chloro-*N*-(1-chloroethyl)-*N*-ethylethenamine.

Mixed amines (i.e., those in which the alkyl or aryl groups are not identical) are named as nitrogen-substituted derivatives of the primary amine with the largest group. Substituents on the nitrogen atom are indicated by an italic *N-*. Thus

$$\underset{\overset{|}{N}CH_3}{CH_3\,CH_2\,CH_2}$$

H

$$CH_3\,CH_2\,CH_2\,NCH_3$$

N-methylpropylamine
CA name: *N-methyl-1-propanamine*

Tertiary amines that have two identical groups can be named as *N,N-*derivatives of primary amines or as *N-* derivatives of secondary amines. For instance

$$CH_3\,CH_2\,CH_2\,CH_2 - N \overset{\displaystyle CH_2\,CH_3}{\underset{\displaystyle CH_2\,CH_3}{}}$$

N,N-diethylbutylamine
or
N-butyldiethylamine
CA name: *N,N-diethyl-1-butanamine*

Aromatic amines may also be primary, secondary, or tertiary. The latter two may be mixed aliphatic-aromatic. The primary amines are properly named as amino derivatives of the aromatic hydrocarbon, but many of them have trivial names accepted under the IUPAC rules. Here they are:

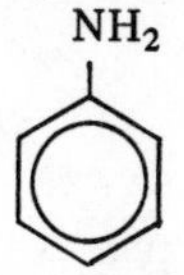

IUPAC name: *aniline*
CA name: *benzenamine*

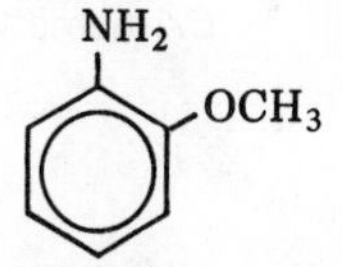

o-anisidine (also *m-, p-*)
2-methoxybenzenamine (also *3-, 4-*)

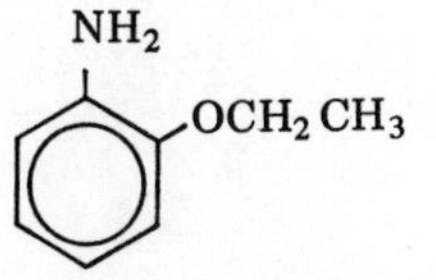

IUPAC name: *o-phenetidine* (also *m-, p-*)
CA name: *2-ethoxybenzenamine* (also *3-, 4-*)

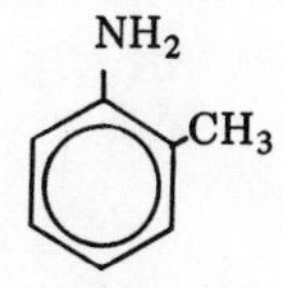

o-toluidine (also *m-, p-*)
2-methylbenzenamine (also *3-, 4-*)

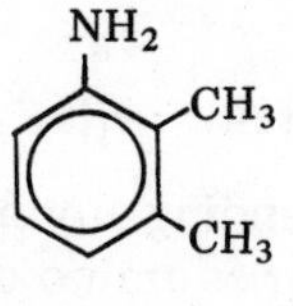

IUPAC name: *2,3-xylidine* (also *2,4-,2,5-,2,6-,3,4-* and *3,5-*)
CA name: *2,3-dimethylbenzenamine* (also *2,4-, 2,5-,2,6-,3,4-,* and *3,5-*)

Following are the common and systematic names of a few more aromatic amines. Write their structural formulas. The correct formulas are below the names.

1. *m*-diaminobenzene (*m*-phenylenediamine)

2. 2-methylaniline (*o*-toluidine)

3. 2,4,6-trinitroaniline (picramide)

4. 2,4,6-trimethylaniline (mesidine)

5. *N,N*-dimethylaniline

••

1.

CA name: 1,3-benzenediamine

2.

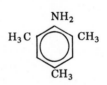

CA name: 2-methylbenzenamine

3.

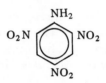

CA name: 2,4,6-trinitrobenzenamine

4.

CA name: 2,4,6-trimethylbenzenamine

5.

CH_3NCH_3

CA name: *N,N*-dimethylbenzenamine

Many compounds that contain $-NH_2$, $>NH$, and $\geqslant N$ groups also contain other functional groups. Depending upon the order of precedence adopted, they may be named as amines or as amino derivatives of the other function.

Amino alcohols are an example. If the IUPAC and *Chemical Abstracts* order of precedence is followed, they are named as alcohols.

$H_2NCH_2CH_2OH$ $(HOCH_2CH_2)_2NH$

2-aminoethanol   *bis(2-hydroxyethyl)amine*
                                    CA: 2,2′-iminobis[ethanol]

The names monoethanolamine and diethanolamine are improperly formed and should be avoided.

2-Phenylethanamine is the parent of a large number of medicinally important compounds known as sympathomimetic amines (so called because their effect mimics the action of the sympathetic nervous system).

$$\text{C}_6\text{H}_5\text{—CH}_2\,\text{CH}_2\,\text{NH}_2$$

2-phenylethanamine

One of its derivatives is adrenaline.

$$\text{HO—C}_6\text{H}_3(\text{HO})\text{—CHCH}_2\,\text{NH—CH}_3,\; \text{OH}$$

adrenaline
CA: *(−)-4 [1-hydroxy-2-(methylamino)ethyl] -1,2-benzenediol*

Adrenaline is named as a phenol because the phenol function is higher in the order of precedence than the amine function.

Amino acids are important physiological compounds. The α-amino acids result from the hydrolysis of proteins. Two examples are:

$$\underset{\text{CH}_2\text{C—OH}}{\overset{\text{H}_2\text{N}\quad\text{O}}{|\qquad\;\|}} \qquad\qquad \underset{\text{CH}_3\text{CHC—OH}}{\overset{\text{H}_2\text{N}\;\;\text{O}}{|\;\;\|}}$$

aminoacetic acid (glycine)   *2-aminopropanoic acid (alanine)*

Further discussion of the naming of amino acids will be found later in this chapter.

What is the formula of 1-amino-2-propanol?

A _____

$$\underset{\text{CH}_3\text{CHCH}_2\,\text{OH}}{\overset{\text{NH}_2}{|}}$$

B _____

$$\underset{\text{CH}_3\text{CHCH}_2\,\text{NH}_2}{\overset{\text{OH}}{|}}$$

C _____

$$\underset{\text{CH}_3\text{CCH}_2\,\text{NH}_2}{\overset{\text{O}}{\|}}$$

•••

A _____

Incorrect. The formula you chose for 1-amino-2-propanol is

$$\underset{3\quad 2\quad\; 1}{CH_3\overset{\displaystyle NH_2}{\overset{|}{C}}HCH_2\,OH}$$

Since the − OH group is attached to the number 1 carbon atom, this formula must represent a derivative of 1-propanol. What is it? 2-amino-1-propanol, of course. Go back and pick another answer.

B _____

Right. The formula for 1-amino-2-propanol is

$$CH_3\overset{\displaystyle OH}{\overset{|}{C}}HCH_2\,NH_2$$

Go on to the next section.

C _____

You picked this formula to represent 1-amino-2-propanol.

$$CH_3\overset{\displaystyle O}{\overset{||}{C}}CH_2\,NH_2$$

Does it have the right functional groups? According to the name, the compound has an amino group, − NH_2, and is an alcohol. The functional group of the alcohols is − OH. Your formula doesn't have it. Read page 212 again carefully before you select another answer.

Compounds containing two or more amino groups are named by adding *-diamine, -triamine*, etc., to the name of the parent compound or multivalent group. Many have trivial names as well. Two diamines with eminently descriptive trivial names occur among the decomposition products of proteins.

$$H_2N(CH_2)_4NH_2 \qquad\qquad\qquad H_2N(CH_2)_5NH_2$$

1,4-butanediamine (putrescine)   *1,5-pentanediamine (cadaverine)*

Complex linear polyamines are best named with the use of the term *aza* to designate the replacement of a methylene group by a nitrogen atom in the carbon chain. Here is an example:

$$HO-\overset{\displaystyle O}{\overset{||}{C}}-CH_2-\overset{\displaystyle CH_3}{\overset{|}{N}}-CH_2-CH_2-\overset{\displaystyle H}{\overset{|}{N}}-CH_2-\overset{\displaystyle H}{\overset{|}{N}}CH_2-\overset{\displaystyle CH_3}{\overset{|}{N}}-CH_2-CH_2-CH_2-\overset{\displaystyle O}{\overset{||}{C}}-OH$$

3,10-dimethyl-3,6,8,10-tetraazatetradecanedioic acid

An important intermediate in the synthesis of nylon 66 is 1,6-hexanedi-
amine. What is its condensed structural formula?

A _____

$$H_2N(CH_2)_6NH_2$$

B _____

$$\overset{\displaystyle H}{\underset{\displaystyle |}{}}$$
$$H_2N(CH_2)_3N(CH_2)_2CH_3$$

C _____

$$\overset{\displaystyle NH_2}{\underset{\displaystyle |}{}}$$
$$CH_3(CH_2)_4CHNH_2$$

• •

A _____

You are correct. The condensed structural formula for 1,6-hexanediamine is

$$H_2N(CH_2)_6NH_2$$

It is also known as hexamethylenediamine, but this name should be avoided. Go
on to the next section.

B _____

You are incorrect. You picked this formula for 1,6-hexanediamine.

$$\overset{\displaystyle H}{\underset{\displaystyle |}{}}$$
$$H_2N(CH_2)_3N(CH_2)_2CH_3$$

There are two amino groups in this formula; one primary and one secondary.
But there is no continuous chain to form the basis for the name. The compound
here would be named as a secondary amine: N-propyl-1,3-propanediamine.
Study this name. When you understand it fully, go back and choose another
answer.

C _____

Incorrect. This is the formula you selected to represent 1,6-hexanediamine.

$$\overset{\displaystyle NH_2}{\underset{\displaystyle |}{}}$$
$$CH_3(CH_2)_4CHNH_2$$

It's a diamine and there is a continuous chain of six carbon atoms. The amino
groups, however, are both on the same carbon atom. The name for the
compound shown here is 1,1-hexanediamine. Go back and choose another
formula to represent 1,6-hexanediamine.

Imines are compounds containing the imino group, $=NH$. They are named
by eliding the *-e* of the corresponding hydrocarbon and adding the suffix
-imine. For example,

$$CH_3CH=NH$$

ethanimine

Because the imine function is the lowest compound class named by use of a functional suffix, polyfunctional imines are generally described by the prefix *imino-*, as shown in this example:

$$CH_3 - \overset{\overset{\displaystyle O}{\|}}{C} - CH_2 - CH = N - \bigcirc$$

4-(phenylimino)-2-butanone

3.8 AMIDES AND IMIDES

When ammonia or a primary amine or a secondary amine reacts with an acid, the product is an amide [ăm′ ĭd]. An ammonium or substituted ammonium salt is formed first, followed by loss of water, as shown in these overall equations:

$$R - \overset{\overset{\displaystyle O}{\|}}{C} - OH \quad + \quad NH_3 \quad \rightarrow \quad R - \overset{\overset{\displaystyle O}{\|}}{C} - NH_2 \quad + H_2O$$

acid   *ammonia*   *amide*

$$R - \overset{\overset{\displaystyle O}{\|}}{C} - OH \quad + \quad R' - NH_2 \quad \rightarrow \quad R - \overset{\overset{\displaystyle O}{\|}}{C} - \overset{\overset{\displaystyle H}{|}}{N} - R' + H_2O$$

acid   *primary amine*   *amide*

$$R - \overset{\overset{\displaystyle O}{\|}}{C} - OH \quad + \quad R' - \overset{\overset{\displaystyle H}{|}}{N} - R'' \quad \rightarrow \quad R - \overset{\overset{\displaystyle O}{\|}}{C} - N \overset{\nearrow R'}{\searrow R''} + H_2O$$

acid   *secondary amine*   *amide*

Tertiary amines do not react to produce amides.

There is a similarity between esters and amides in that the imino group, $-\overset{\overset{\displaystyle H}{|}}{N}-$, replaces oxygen as the link between the carbonyl group, $-\overset{\overset{\displaystyle O}{\|}}{C}-$, and an alkyl group.

$$R - \overset{\overset{\displaystyle O}{\|}}{C} - O - R'$$

ester

$$R - \overset{\overset{\displaystyle O}{\|}}{C} - \overset{\overset{\displaystyle H}{|}}{N} - R'$$

amide

Simple amides are named by replacing the *-oic acid* in the name of the acid by *amide*. Carbon atoms are numbered in the same way as esters.

$$CH_3\overset{\overset{\displaystyle O}{\|}}{C} - NH_2$$

acetamide or *ethanamide*

$$CH_3\overset{\overset{\displaystyle Cl}{|}}{C}H\overset{\overset{\displaystyle O}{\|}}{C} - NH_2$$

α-chloropropioanamide
or *2-chloropropanamide*

What is the name of this amide?

$$BrCH_2CH_2CH_2\overset{\overset{\displaystyle O}{\|}}{C}-NH_2$$

A _____

bromobutanamide

B _____

4-bromobutylamine

C _____

4-bromobutanamide

• •

A _____

Incorrect. You named the compound shown by this formula as bromobutanamide.

$$BrCH_2CH_2CH_2\overset{\overset{\displaystyle O}{\|}}{C}-NH_2$$

There are two other carbon atoms that can bear the bromine atom. Wouldn't each compound so formed also be bromobutanamide? As usual, the solution is to number the carbon atoms in the chain. When you've added a number to the name, go back and choose another answer.

B _____

You are wrong. Look at this formula carefully. What is the functional group?

$$BrCH_2CH_2CH_2\overset{\overset{\displaystyle O}{\|}}{C}-NH_2$$

Is it a primary amine with the general formula $R-NH_2$, or is it an amide with the general formula $R-\overset{\overset{\displaystyle O}{\|}}{C}-NH_2$? It's an amide. You had better turn back and read page 216 again before selecting another answer.

C _____

Right you are. The formula shows 4-bromobutanamide.

$$BrCH_2CH_2CH_2\overset{\overset{\displaystyle O}{\|}}{C}-NH_2$$

It may also be named γ-bromobutyramide. Go on to the next section.

Chemical Abstracts uses trivial names for these three amides:

$$
\begin{array}{ccc}
\quad O & \quad O & \quad O \\
\quad \| & \quad \| & \quad \| \\
H - C - NH_2 & CH_3C - NH_2 & C - NH_2
\end{array}
$$

formamide   *acetamide*   *benzamide*

Amides that are derived from amines rather than ammonia are named as *N*-substituted products of the simple amide. For instance:

$$
\begin{array}{ccc}
O \; H & O \; H \;\; CH_3 & O \quad\quad CH_2CH_3 \\
\| \;\; | & \| \;\; | \;\;\; | & \| \quad\quad / \\
CH_3C - NCH_3 & CH_3 - C - N - CH & H - C - N \\
 & \quad\quad\quad | & \quad\quad\quad \backslash \\
 & \quad\quad\quad CH_3 & \quad\quad\quad CH_2CH_3
\end{array}
$$

N-methylacetamide   *N-ethyl-N-methylacetamide*   *N-N-diethylformamide*

The first is the amide formed from methanamine and acetic acid. The second is formed from 2-propanamine and acetic acid. The third is from methanoic (formic) acid and *N*-ethylethanamine.

The compound acetanilide, used as a pain reliever and an intermediate in organic synthesis, is *N*-phenylacetamide.

Which of these represents its structural formula?

A _____

$$
\begin{array}{c}
\quad\quad\quad O \\
\quad\quad\quad \| \\
- CH_2 CNH_2
\end{array}
$$

B _____

$$
\begin{array}{c}
O \; H \\
\| \;\; | \\
CH_3C - N -
\end{array}
$$

C _____

$$
\begin{array}{c}
O \\
\| \\
CH_3C - N -
\end{array}
$$

•••

A _____

Incorrect. This is not the formula for *N*-phenylacetamide.

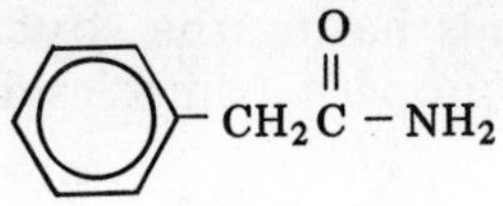

$$\text{⬡} - CH_2\overset{\overset{\displaystyle O}{\|}}{C} - NH_2$$

From the name you can see that acetanilide is a relative of acetamide,

$$CH_3\overset{\overset{\displaystyle O}{\|}}{C} - NH_2$$ It is the product of the reaction between acetic acid and aniline.

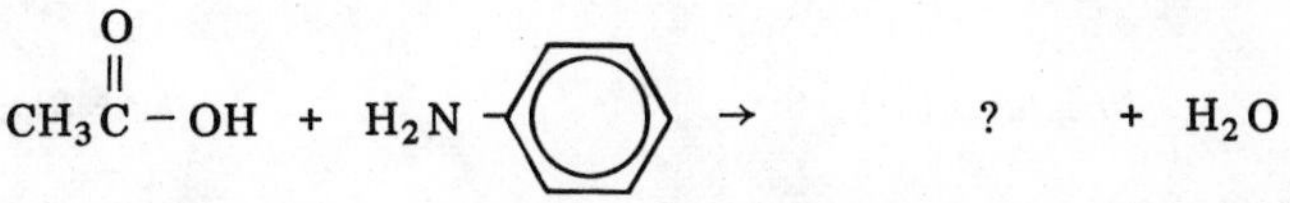

$$CH_3\overset{\overset{\displaystyle O}{\|}}{C} - OH \ + \ H_2N - \text{⬡} \ \rightarrow \ ? \ + \ H_2O$$

acetic acid   benzenamine   N-phenylacetamide

Go back and choose another answer.

B _____

You are correct. N-phenylacetamide is the product of the reaction between ethanoic acid and benzenamine (aniline).

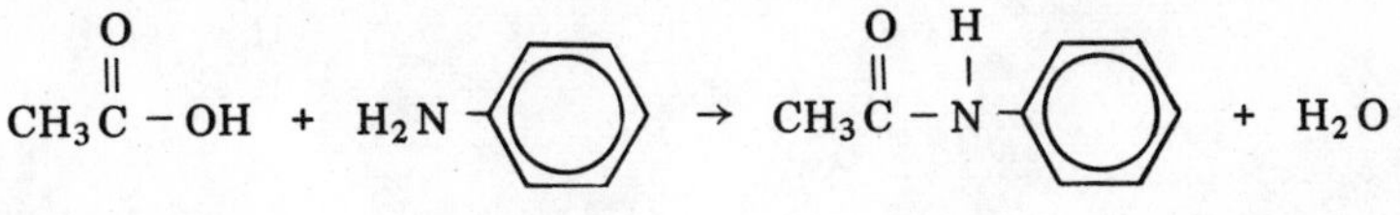

$$CH_3\overset{\overset{\displaystyle O}{\|}}{C} - OH \ + \ H_2N - \text{⬡} \ \rightarrow \ CH_3\overset{\overset{\displaystyle O}{\|}}{C} - \overset{\overset{\displaystyle H}{|}}{N} - \text{⬡} \ + \ H_2O$$

acetic acid   benzenamine   N-phenylacetamide
(aniline)

Go on to the next section.

C _____

No. The formula for N-phenylacetamide is not

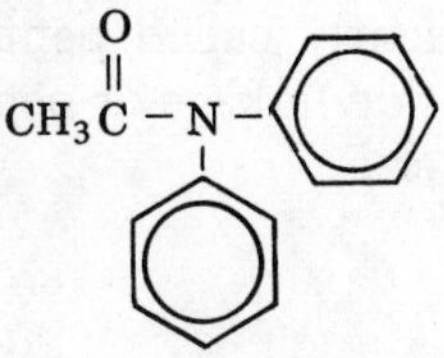

$$CH_3\overset{\overset{\displaystyle O}{\|}}{C} - N - \text{⬡}$$
$$\text{⬡}$$

This formula shows two phenyl groups attached to the nitrogen. Is there anything in the name to make you think there should be two? If not, the formula must be wrong. Go back and pick another answer.

Imides are cyclic secondary amides. For example:

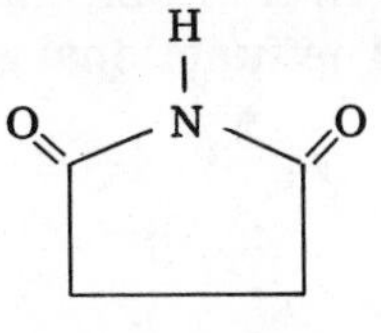

$$O = \underset{\underset{\displaystyle }{}}{C} \quad \overset{\overset{\displaystyle H}{|}}{N} \quad C = O$$

butanimide

In the past, acyclic compounds having the structure RCONHCOR have been named as imides, but they are properly named as diacylamides. For instance:

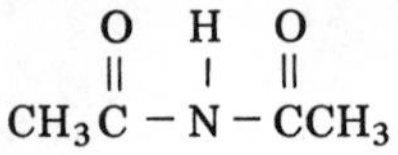

$$CH_3\overset{\overset{\textstyle O}{\|}}{C}-\overset{\overset{\textstyle H}{|}}{N}-\overset{\overset{\textstyle O}{\|}}{C}CH_3$$

diacetamide 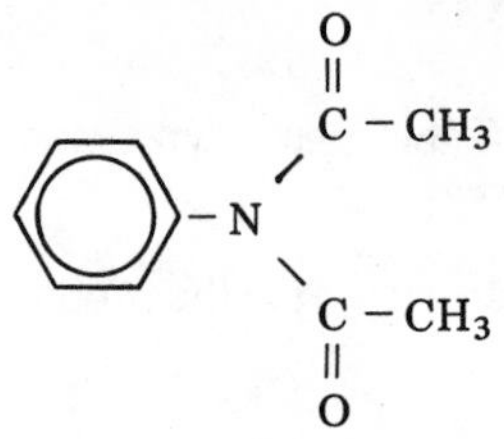   *N-phenyldiacetamide*

Amide linkages are important in both natural and synthetic polymers. The synthetic fiber nylon 66 is formed by the reaction between hexanedioic acid and 1,6-hexanediamine. Water is formed as well. The reaction is shown here for only one active functional group on each molecule.

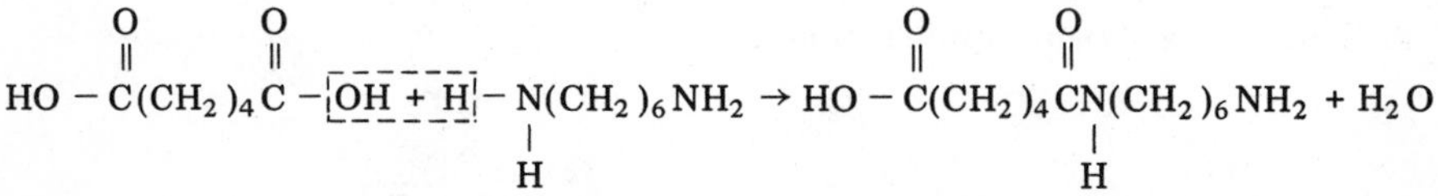

$$HO-\overset{\overset{\textstyle O}{\|}}{C}(CH_2)_4\overset{\overset{\textstyle O}{\|}}{C}-\underline{[OH+H]}-\underset{\underset{\textstyle H}{|}}{N}(CH_2)_6NH_2 \rightarrow HO-\overset{\overset{\textstyle O}{\|}}{C}(CH_2)_4\overset{\overset{\textstyle O}{\|}}{C}\underset{\underset{\textstyle H}{|}}{N}(CH_2)_6NH_2 + H_2O$$

hexanedioic acid   1,6-hexanediamine
(adipic acid)   (hexamethylenediamine)

Notice that both ends of the product molecule have reactive functional groups. Nylon 66, a polymer, is formed when these ends continue to react and form long molecules. The acid and amine themselves are not used in actual practice, but the principle is as shown above.

Chains of amino acids that are linked through carboxy and amino groups in similar amide linkages are frequently called peptides. This particular type of amide linkage is known as a peptide linkage or peptide bond. Further examples of peptides will be seen in Chapter 5.

Summary: Nomenclature of Amines and Amides

Amines are organic bases derived from ammonia and can be classified as primary, secondary, or tertiary. The functional groups may be summarized:

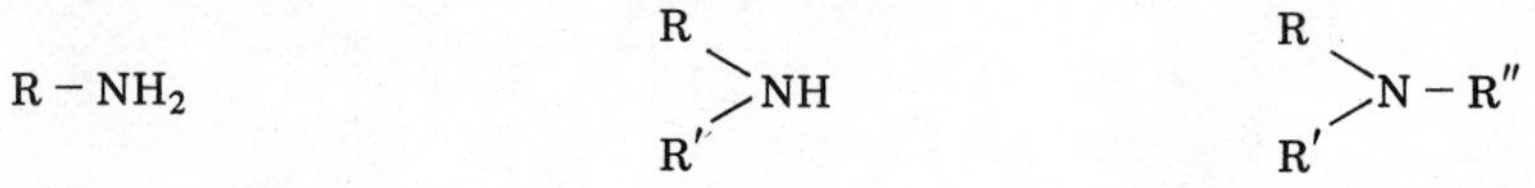

$R-NH_2$ $\overset{\textstyle R}{\underset{\textstyle R'}{>}}NH$ $\overset{\textstyle R}{\underset{\textstyle R'}{>}}N-R''$

primary   *secondary*   *tertiary*

Primary amines are named by adding the suffix *-amine* to the name of the hydrocarbon group. Secondary and tertiary amines may be named by enumerating the groups in alphabetical order or as *N*-substituted derivatives of a primary amine. When numbers are needed to locate substituents, the carbon attached to the nitrogen is always number 1.

Aromatic amines may be primary, secondary, or tertiary. Most common is benzenamine, or aniline.

benzenamine
(aniline)

A number of other aromatic amines have trivial names accepted under the IUPAC rules. They are the anisidines, phenetidines, toluidines, and xylidines.

Chemical Abstracts indexes all secondary and tertiary monoamines as substituted primary amines.

Imines are compounds containing the imino group, $= NH$, and are named by eliding the *-e* of the corresponding hydrocarbon and adding the suffix *-imine*.

Amino acids contain both an amino group and a carboxy group. Aminoacetic acid (glycine) is an example. These acids will be discussed in the next section.

Compounds which contain more than one amino group are named by adding *-diamine*, *-triamine*, etc., to the name of the parent compound or multivalent group.

$$H_2N(CH_2)_4NH_2$$

1,4-butanediamine

Amides result from the reaction between amines or ammonia and carboxylic acids. Ethanamide is an example: $CH_3\overset{\text{O}}{\overset{\|}{C}} - NH_2$. The functional group of the amides is $-\overset{\text{O}}{\overset{\|}{C}} - N\diagup$ and the general formula for amides is $R - \overset{\text{O}}{\overset{\|}{C}} - N\diagup^{R'}_{\diagdown R''}$. $R' -$ and $R'' -$ may be hydrogen or aryl or alkyl groups. Amide linkages are important in biochemistry since they join amino acids to form proteins.

Imides are cyclic secondary amides and are named by dropping *-edioic acid* from the name of the corresponding acid and adding the suffix *-imide*.

AMINES

Functional groups: $-NH_2$ Functional group names: *amino*

$$
\begin{array}{c}
H \\
| \\
-N- \\
| \\
\end{array}
$$
                                                            imino

$$
\begin{array}{c}
| \\
-N- \\
| \\
\end{array}
$$
                                                            nitrilo

General formulas: RNH_2 (primary)   Series name: *-amine*

$$
\begin{array}{c}
H \\
| \\
R-N-R' \text{ (secondary)}
\end{array}
$$

$$
\begin{array}{c}
R'' \\
| \\
R-N-R' \text{ (tertiary)}
\end{array}
$$

AMIDES

Functional group: $-\overset{\overset{\textstyle O}{\|}}{C}-NH_2$ Functional group name: *carbamoyl*

General formula: $R-\overset{\overset{\textstyle O}{\|}}{C}-NH_2$ Series name: *-amide*

3.9 AMINO ACIDS

The importance of amino acids rests largely on their occurrence in animal proteins, the chief components of muscle fiber, skin, nerves, and blood. As you already know, amino acids contain both the amino group and the carboxy group. Two examples are glycine [glī-sēn′] and alanine [ăl′ à-nēn] :

$$
H_2NCH_2\overset{\overset{\textstyle O}{\|}}{C}-OH
$$

glycine
(aminoacetic acid)

$$
CH_3\overset{\overset{\textstyle H_2N}{|}}{C}H\overset{\overset{\textstyle O}{\|}}{C}-OH
$$

alanine
(2-aminopropanoic acid)

Hydrolysis of proteins yields mixtures of these and other 2-amino acids. They are generally called α-amino acids. The condensed formula of the α-amino acids is $RCH(NH_2)CO_2H$.

As will be seen on the pages that follow, all of the naturally occurring amino acids except glycine have at least one chiral center. Every amino acid obtained from the hydrolysis of proteins, except glycine, has been found to be optically active. Investigation of the configuration of the α-carbon atoms has shown every one to have the (S)- configuration. A correct Fischer projection and a perspective formula for these acids are shown here.

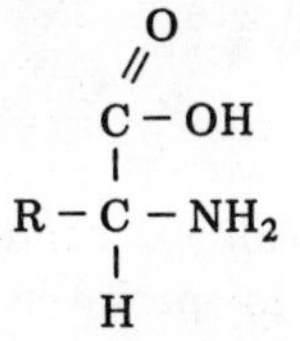

$$
\begin{array}{c}
O \\
\parallel \\
C-OH \\
| \\
R-C-NH_2 \\
| \\
H
\end{array}
\qquad\qquad
\begin{array}{c}
O \\
\parallel \\
C-OH \\
| \\
R\blacktriangleright C\blacktriangleleft NH_2 \\
| \\
H
\end{array}
$$

(S)-amino acid 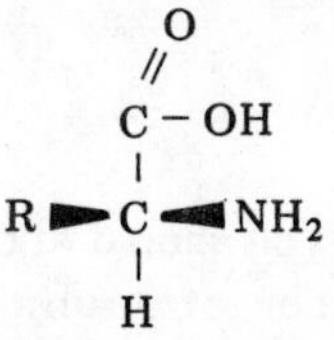   (S)-amino acid

You may see the naturally occurring amino acids termed as L-amino acids. The L- refers to the configuration of the α-carbon atom.

In the specialized nomenclature of carbohydrates (see p. 267) the D- or L-prefix is used to designate the configuration of the highest-numbered chiral center. Because these prefixes are used to designate the configuration of the α-carbon atom of amino acids, ambiguity may arise. It may be forestalled by amplifying the prefixes to D_s and L_s. The subscript s indicates that serine is taken as the standard.

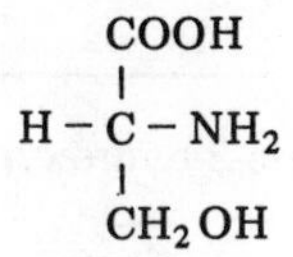

$$
\begin{array}{c}
COOH \\
| \\
H-C-NH_2 \\
| \\
CH_2OH
\end{array}
\qquad\qquad
\begin{array}{c}
COOH \\
| \\
H-C-NH_2 \\
| \\
HO-C-H \\
| \\
CH_3
\end{array}
$$

D-*serine* 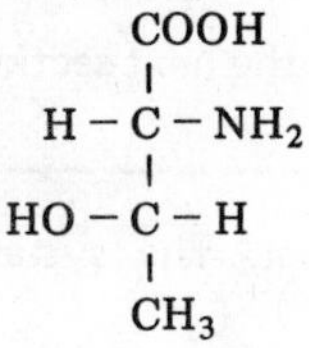 D_s-*threonine*

Carbohydrate nomenclature may be applied to amino acids, with D- and L-prefixes designating the configuration of the highest-numbered chiral center and using the prefixes D_g- and L_g- to indicate that glyceraldehyde is the standard. The D_s-threonine and L_g-threonine are names for the same structure.

Unfortunately, the amplified prefixes are not widely used. In their absence, you should assume that configurational prefixes for amino acids are based on the α-carbon atom.

What is the systematic name for serine?

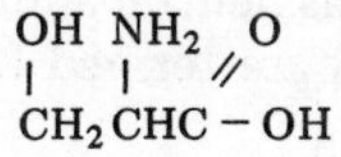

$$
\begin{array}{c}
OH\ NH_2\ \ O \\
|\ \ \ \ |\ \ \ \ \parallel \\
CH_2\,CHC-OH
\end{array}
$$

A _____

aminohydroxypropanoic acid

B _____

2-amino-3-hydroxypropionic acid

C _____

2-amino-3-hydroxypropanoic acid

● ●

A _____

Incorrect. Although the name aminohydroxypropanoic acid does describe the acid represented by this formula, it is not a unique and completely unambiguous name.

$$\overset{OH}{\underset{|}{}}\ \overset{NH_2}{\underset{|}{}}\ \overset{O}{\overset{/\!/}{}}$$
$$CH_2\,CHC-OH$$

By now you should know that whenever there is a possibility of confusion over the location of a substituent group, a number is used to specify its location. Go back and choose another answer.

B _____

Not quite right. The trivial name for the carboxylic acid with three carbon atoms in a chain is propionic acid. The systematic name is propanoic acid. Since you were asked for the systematic name of serine, it must be 2-amino-3-hydroxypropanoic acid.

$$\overset{OH}{\underset{|}{}}\ \overset{NH_2}{\underset{|}{}}\ \overset{O}{\overset{/\!/}{}}$$
$$CH_2\,CHC-OH$$

Go on to the next section.

C _____

Correct. The systematic name for serine is 2-amino-3-hydroxypropanoic acid.

$$\overset{OH}{\underset{|}{}}\ \overset{NH_2}{\underset{|}{}}\ \overset{O}{\overset{/\!/}{}}$$
$$CH_2\,CHC-OH$$

Amino acids are named as acids because of the presence of the carboxy group. Their aqueous solutions may be acidic or basic, however, depending on the relative strength and number of acidic carboxy groups and basic amino groups in the molecule. Most amino acids have equal numbers of amino and carboxy groups and are known as neutral amino acids. A few have more amino groups than carboxy groups and are termed basic amino acids. An example is hydroxylysine.

$$H_2N\ \ OH\ \ \ \ H_2N\ O$$
$$CH_2\,CHCH_2\,CH_2\,CHC-OH$$

2,6-diamino-5-hydroxyhexanoic acid
(hydroxylysine)

A few others have more carboxy than amino groups and are known, not surprisingly, as acidic amino acids. Glutamic acid, whose sodium salt is used as a flavor intensifier in foods, is one of them. Glutamic acid is 2-aminopentane-dioic acid.

$$\underset{\substack{\text{O} \quad \text{NH}_2 \qquad \text{O} \\ \parallel \quad | \qquad \parallel}}{\text{HO} - \text{CCHCH}_2\,\text{CH}_2\,\text{C} - \text{OH}}$$

2-aminopentanedioic acid
(glutamic acid)

The formula for aspartic acid is

$$\underset{\substack{\text{O} \quad \text{NH}_2 \quad \text{O} \\ \parallel \quad | \quad \parallel}}{\text{HO} - \text{C} - \text{CHCH}_2\,\text{C} - \text{OH}}$$

Is aspartic acid a basic, acidic, or neutral amino acid?

A _____

acidic

B _____

basic

C _____

neutral

• •

A _____

You are right. Since aspartic acid has two carboxy groups and one amino group, it is classed as an acidic amino acid.

$$\underset{\substack{\text{O} \quad \text{NH}_2 \quad \text{O} \\ \parallel \quad | \quad \parallel}}{\text{HO} - \text{C} - \text{CHCH}_2\,\text{C} - \text{OH}}$$

aminobutanedioic acid
(aspartic acid)

Go on to the next section.

B _____

Incorrect. This classification of amino acids is based entirely on the relative number of amino and carboxy groups that are present. If they are equal, the acid is termed neutral. If one group or the other predominates, the acid is either acidic or basic. Look at the formula for aspartic acid again carefully before you go back to choose another answer.

$$\underset{\substack{\text{O} \quad \text{NH}_2 \quad \text{O} \\ \parallel \quad | \quad \parallel}}{\text{HO} - \text{C} - \text{CHCH}_2\,\text{C} - \text{OH}}$$

aspartic acid

C _____

You are incorrect. Aspartic acid is not classed as a neutral amino acid. This classification of amino acids is based on the relative number of acidic carboxy

$$\overset{\displaystyle O}{\overset{\displaystyle \|}{}}$$

groups, $-C-OH$, and basic amino groups, $-NH_2$. Look at the formula again and then go back for another answer.

$$HO - \overset{\displaystyle O}{\overset{\displaystyle \|}{C}} - \overset{\displaystyle NH_2}{\overset{\displaystyle |}{C}}HCH_2\overset{\displaystyle O}{\overset{\displaystyle \|}{C}} - OH$$

aspartic acid

There are eight amino acids known as essential amino acids. This means that to sustain life they must be present in the diet of human beings. All other amino acids needed in metabolism can be synthesized in the body. The eight essential amino acids are isoleucine [ī sṓ lōō′ sĕn], leucine [lōō′ sĕn], lysine [lī′ sĕn], methionine [mĕ thī′ ṓ nĕn], phenylalanine [fĕn ĭl ăl′ ă nĕn], threonine [thrĕ′ ṓ nĕn], tryptophan [trĭp′ tṓ făn], and valine [văl′ ĕn].

Methionine contains sulfur and will be mentioned in the next section. Its systematic name is 2-amino-4-(methylthio)butanoic acid.

$$CH_3SCH_2CH_2\overset{\displaystyle NH_2}{\overset{\displaystyle |}{C}}H\overset{\displaystyle O}{\overset{\displaystyle /\!/}{C}} - OH$$

methionine

Tryptophan is (S)-α-amino-1H-indole-3-propanoic acid.

tryptophan

The systematic names for the other six essential amino acids, however, are well within your capability. They are given below, followed by their structural formulas. Write the formulas before looking at them.

1. isoleucine: 2-amino-3-methylpentanoic acid

2. leucine: 2-amino-4-methylpentanoic acid

3. lysine: 2,6-diaminohexanoic acid

4. phenylalanine: 2-amino-3-phenylpropanoic acid

5. threonine: 2-amino-3-hydroxybutanoic acid

6. valine: 2-amino-3-methylbutanoic acid

1.

$$\underset{\underset{CH_3CH_2CHCHC-OH}{|\quad/\quad//}}{CH_3\;NH_2\;O}$$

2.

$$\underset{\underset{CH_3CHCH_2CHC-OH}{|\qquad|\quad//}}{CH_3\qquad NH_2\;O}$$

3.

$$\underset{\underset{CH_2CH_2CH_2CH_2CHC-OH}{|\qquad\qquad\qquad|\quad//}}{NH_2\qquad\qquad\quad NH_2\;O}$$

4.

$$\underset{\underset{CH_2CHC-OH}{|\quad//}}{NH_2\;O}$$

5.

$$\underset{\underset{CH_3CHCHC-OH}{|\;\;/\;\;//}}{OH\;NH_2\;O}$$

6.

$$\underset{\underset{CH_3CHCHC-OH}{|\;\;/\;\;//}}{CH_3\;NH_2\;O}$$

3.10 NITRILES

Nitriles are compounds containing the structure $-C \equiv N$. Acyclic compounds RCN in which the *nitrilo* group, $\equiv N$, replaces $-H_3$ at the end of the main chain are named by adding *-nitrile* or *-dinitrile* to the name of the appropriate hydrocarbon. These suffixes represent only the triple-bonded nitrogen atom and not the carbon to which it is attached. If numbering is required, number one is assigned to that carbon. For example:

butanedinitrile $\qquad\qquad\qquad$ $NCCH_2CH_2CN$

2-chloropropanenitrile $\qquad\qquad$ $\underset{\underset{Cl}{|}}{CH_3CHCN}$

Since *Chemical Abstracts* retains acetic acid, the trivial name acetonitrile is used for CH_3CN, as well as benzonitrile for C_6H_5CN.

Compounds RCN that can be considered as derivatives of acids via the dehydration of the corresponding amide are named by changing the *-oic acid* or *-ic acid* to *-onitrile.* For instance:

benzoic acid $\qquad\qquad\qquad\qquad\qquad\qquad\qquad\qquad\qquad$ *benzonitrile*

Nitriles that can be considered as derivatives of acids whose systematic names end in *-carboxylic acid* are named by changing the ending to *-carbonitrile*.

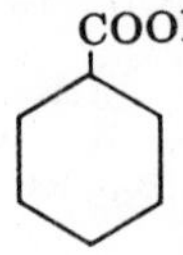

COOH

cyclohexanecarboxylic acid

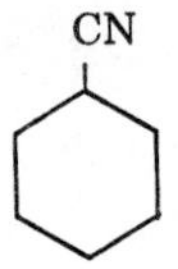

CN

cyclohexanecarbonitrile

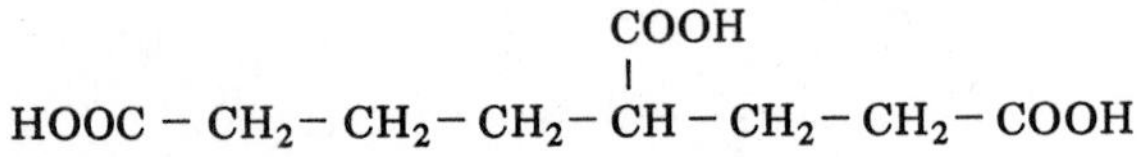

$$\text{COOH}$$
$$|$$
$$\text{HOOC}-\text{CH}_2-\text{CH}_2-\text{CH}_2-\text{CH}-\text{CH}_2-\text{CH}_2-\text{COOH}$$

1,3,6-hexanetricarboxylic acid

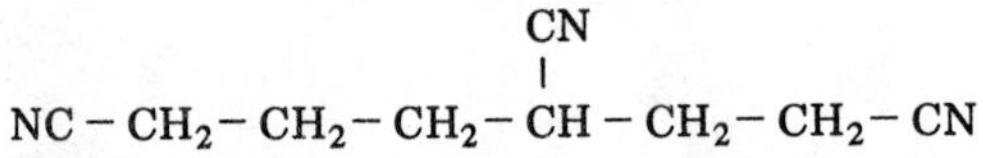

$$\text{CN}$$
$$|$$
$$\text{NC}-\text{CH}_2-\text{CH}_2-\text{CH}_2-\text{CH}-\text{CH}_2-\text{CH}_2-\text{CN}$$

1,3,6-hexanetricarbonitrile

Carbonitrile denotes the structure $-\text{CN}$. Consequently, the carbon atom in the group is excluded from the numbering in the chain shown above.

When $-\text{CN}$ occurs in a compound with a group that has priority over $-\text{CN}$ for forming the name, the $-\text{CN}$ group is designated by the prefix *cyano-*, as shown in these compounds:

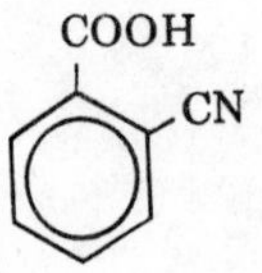

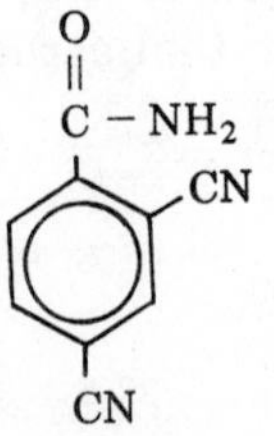

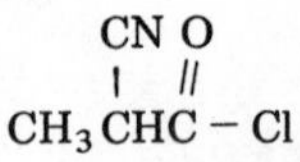

COOH   CN

$$\begin{array}{c}\text{O}\\||\\\text{C}-\text{NH}_2\end{array}$$

CN

CN

$$\begin{array}{cc}\text{CN} & \text{O}\\|&||\\\text{CH}_3\,\text{CHC}-\text{Cl}\end{array}$$

2-cyanobenzoic acid   *2,4-dicyanobenzamide*   *2-cyanopropanoyl chloride*

After referring to the order of precedence on pages 198 and 199, choose the systematic name for this compound.

$$\text{HOCH}_2\,\text{CH}_2\,\text{CN}$$

A _____

2-cyanoethanol

B _____

2-hydroxyethanenitrile

C —————————————————————————————————

3-hydroxypropanenitrile

• •

A —————————————————————————————————

Incorrect. While the name 2-cyanoethanol does describe the structure $HOCH_2CH_2CN$, it mistakenly gives precedence to the alcohol function. Look carefully at the order of precedence before choosing another answer.

B —————————————————————————————————

You are wrong even though you did correctly give precedence to the nitrile function over the alcohol function. You seem to have forgotten that the suffix nitrile refers to the ≡ N entity only. Go back and choose another answer.

C —————————————————————————————————

Right. The systematic name for $HOCH_2CH_2CN$ is 3-hydroxypropanenitrile. The nitrile function has priority over the alcohol function and all three carbon atoms are counted in determining the length of the carbon chain.

3.11 MISCELLANEOUS FUNCTIONAL GROUPS CONTAINING NITROGEN

There are many classes of compounds with functional groups containing nitrogen atoms. Among those with two nitrogen atoms are azo and azoxy compounds and hydrazines, as well as diazonium compounds (see p. 129). Compounds with three or more contiguous nitrogen atoms include amidines, amide oximes, amidrazones, hydrazidines, formazans, carbodiimides, and many others.

The nomenclature of these compounds is continually moving toward greater systemization. The practice of *Chemical Abstracts* is more advanced than the IUPAC rules. Most textbooks published through the early 1970's use the older names. For this reason you need to be familiar with both.

What are commonly called azo compounds have the general formula $RN = NR$, and the $- N = N -$ group is called the azo group. According to the IUPAC rules, if the azo group joins parent molecules that are identical when unsubstituted, the resulting compound is named by adding the prefix *azo-* to the name of the unsubstituted parent molecule. Such names are objectionable, however, because they do not clearly indicate the presence of *two* parent molecules. The present practice of *Chemical Abstracts* is to name the compound as a derivative of diazene, $HN = NH$, and the group $- N = N -$ as the diazenediyl group. Diazene is not to be confused with diazine (see p. 249). For example:

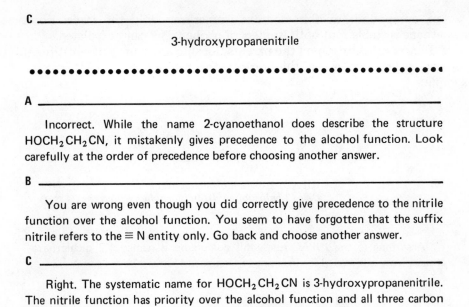

$$CH_3N = NCH_3$$

IUPAC: *azomethane*
CA: *dimethyldiazene*

azobenzene
diphenyldiazene

Azo compounds exhibit *cis-trans* isomerism. Only the *trans-* isomer of azomethane is stable, but both isomers of azobenzene can be isolated.

If there is substitution on the parents, locants and names of substituents are prefixed, using primes on the locants for the second parent group. For example:

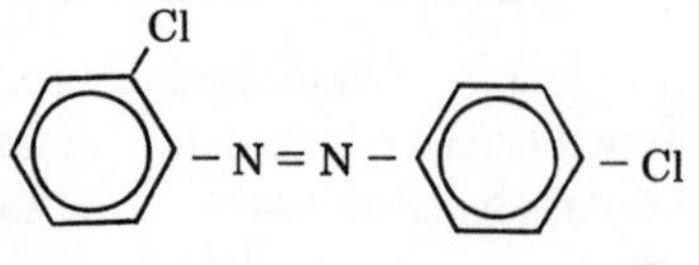

IUPAC: *2,4'-dichloroazobenzene*
CA: *(2-chlorophenyl)(4-chlorophenyl)diazene*

According to the IUPAC rules, this compound is named 1,2'-dichloroazo-ethane. How is it named for indexing in *Chemical Abstracts*?

$$\overset{\displaystyle Cl}{\underset{\displaystyle |}{}}$$
$$CH_3 CH - N = N - CH_2 CH_2 Cl$$

A _____

1,2-bis(2-chloroethyl)diazene

B _____

(1-chloroethyl)(2-chloroethyl)diazene

C _____

bis(chloroethyl)diazene

• •

A _____

Incorrect. You chose the name bis(2-chloroethyl)diazine for the structure

$$\overset{\displaystyle Cl}{\underset{\displaystyle |}{}}$$
$$CH_3 CH - N = N - CH_2 CH_2 Cl$$

For the prefix *bis-* to be correct, the substituents must be identical. Here they are different. Go back and select another name.

B _____

You are right. *Chemical Abstracts* names azo compounds as derivatives of diazene. The name for this structure is shown here:

$$\overset{\displaystyle Cl}{\underset{\displaystyle |}{}}$$
$$CH_3 CH - N = N - CH_2 CH_2 Cl$$

(1-chloroethyl)(2-chloroethyl)diazene

Go on to the next section.

C _____

You are wrong. You selected bis(chloroethyl)diazene. The prefix *bis-* implies that the substituent groups are identical. Go back and you'll see they aren't.

If the $-N=N-$ group links structures derived from parent molecules that are different when unsubstituted, the IUPAC rules form the name by placing azo between the complete names of the substituted parent molecules. The more complex parent is cited first, and any locants needed to indicate the point of attachment of the $-N=N-$ group are placed between azo and the name of the parent to which each locant refers. The present practice of *Chemical Abstracts* is simpler: name the compound as a substituted diazene. For example,

IUPAC: *naphthalene-2-azobenzene*
CA: *(2-naphthalenyl)phenyldiazene*

Although it is longer and appears more complicated, the second name is easier to interpret and definitely is unambiguous.

Azoxy compounds have the general formula $R-N=N(O)-R$ or

$$R-N \overset{\overset{\textstyle O}{\uparrow}}{=} N - R.$$

If the position of the azoxy-oxygen atom is unknown or unimportant, the compound is named in the same manner as an azo compound with *azo* replaced by *azoxy*. *Chemical Abstracts* names such compounds as oxides of substituted diazenes. For instance:

IUPAC: *azoxybenzene*
CA: *diphenyldiazene 1-oxide*

When it is desired to express the position of the azoxy-oxygen atom in an unsymmetrical compound, a prefix *ONN-* or *NNO-* is used by the IUPAC rules, depending upon whether the first or second cited parent group is attached directly to the $-NO-$ grouping. Here is an example:

IUPAC: *(ethyl-ONN-azoxy)benzene*
CA: *ethylphenyldiazene 1-oxide*

Another class of compounds containing the $-N=N-$ group is commonly called *diazo compounds*. Their general formula is $RN=NX$ and they differ from azo compounds in that the group X is not joined by a carbon-nitrogen bond except for X = − CN. Formed by the IUPAC rules, their names consist of three parts: (1) the name of the parent compound, RH; (2) diazo; and (3) the name of the group, X. *Chemical Abstracts* names them under diazene headings. An example is:

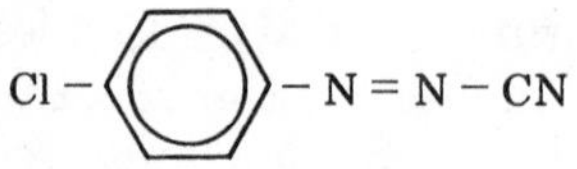

$$Cl - \bigcirc - N = N - CN$$

IUPAC: *p-chlorobenzenediazocyanide*
CA: *(4-chlorophenyl)diazenecarbonitrile*

Use of diazene as a parent compound simplifies the naming of azo, azoxy, and diazo compounds considerably.

Because nitrogen compounds have low precedence among functional groups, the presence of another functional group usually relegates the − N = N − group to the status of a substituent. For instance:

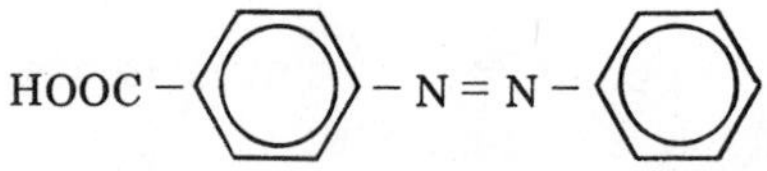

$$HOOC - \bigcirc - N = N - \bigcirc$$

IUPAC: *carboxyphenyl-4-azobenzene*
CA: *4-(phenylazo)benzoic acid*

$$CH_3 N = NCH_2 CH_2 OH$$

IUPAC: *hydroxyethyl-2-azomethane*
CA: *2-(methylazo)ethanol*

Which of the following structures represents 4-(methylazo)-2-butanone?

A _____

$$CH_3 N = NCH_2 \overset{\overset{\displaystyle O}{\|}}{C} CH_2 CH_3$$

B _____

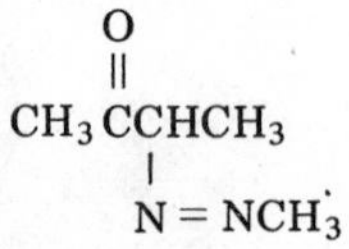

$$CH_3 \overset{\overset{\displaystyle O}{\|}}{C} \underset{\underset{\displaystyle N = NCH_3}{|}}{C} HCH_3$$

C _____

$$CH_3 N = NCH_2 CH_2 \overset{\overset{\displaystyle O}{\|}}{C} CH_3$$

A _____

Incorrect. You may have forgotten that numbering of the carbon chain in ketones starts at the end nearer the carbonyl group. With this in mind, go back and pick another answer.

B _____

Not quite. The main chain of the structure you selected is indeed 2-butanone, but the methylazo group is not a substituent on the number 4 carbon atom. Go back and carefully choose another answer.

C _____

Right. This structure represents 4-(methylazo)-2-butanone.

From the foregoing it would seem reasonable to name H_2NNH_2 as diazane. It is, in fact, a recommendation. The IUPAC approved name that is used by *Chemical Abstracts* is hydrazine. Its derivatives have some importance. A common one is phenylhydrazine (phenyldiazane):

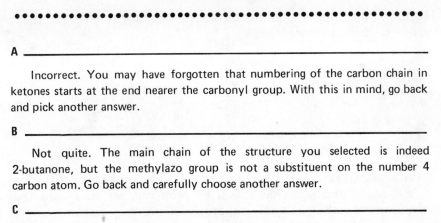

When necessary, the nitrogen atoms may be numbered 1 and 2.

IUPAC and *CA: 1,1-dimethyl-2-phenylhydrazine*
Recommended: *1,1-dimethyl-2-phenyldiazane*

The IUPAC rules allow an alternative procedure for compounds of the type $R-NH-NH-R'$ if the groups R and R' are derived from molecules that are identical when unsubstituted. They are named in the same manner as azo compounds using *hydrazo* in place of azo. The recommendation is to name them as substituted diazanes. Here are two examples:

IUPAC: *2,2'-dibromohydrazobenzene*
CA: 1,2-bis(2-bromophenyl)hydrazine
Recommended: *1,2-bis(2-bromophenyl)diazene*

IUPAC: *1,2'-hydrazonaphthalene*
CA: 1-(1-naphthalenyl)-2-(2-naphthalenyl)hydrazine
Recommended: *1-(1-naphthalenyl)-2-(2-naphthalenyl)diazane*

When a functional group of higher precedence is present, the prefix *hydrazino-* is used, as shown here:

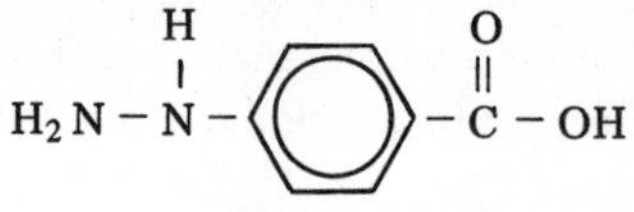

$$H_2N - \overset{\overset{\displaystyle H}{|}}{N} - \langle\bigcirc\rangle - \overset{\overset{\displaystyle O}{||}}{C} - OH$$

IUPAC and *CA: 4-hydrazinobenzoic acid*
Recommended: *4-diazanylbenzoic acid*

A further extension might be to name NH_3 as azane rather than ammonia, but it is unlikely that this change will be proposed, let alone adopted. It is possible, however, that NH_2OH, hydroxylamine, may become azanol (see methanol, p. 152).

Here are some exercises for you. Fill in the empty spaces for each compound. The answers follow.

Structural Formula	IUPAC Name	CA Name
		diethenyldiazene
	1-methyleneazoethane	
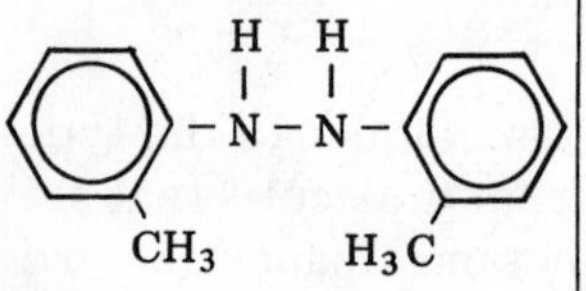 (o-tolyl)–N(H)–N(H)–(o-tolyl), CH₃ and H₃C substituents		

Answers:

$CH_2 = CHN = NCH = CH_2$	azoethene	
$CH_3CH_2N = C = CH_2$ (with CH_3 on C)		ethyl(1-methylethenyl)diazene
	hydrazo-*o*-toluene	1,2-bis(2-methylphenyl)hydrazine

AZO COMPOUNDS

Functional group: $- N = N -$

Functional group name:
    azo (IUPAC and *CA*)

General formula: $RN = NR$

Series name:
    azo-.... (IUPAC)
    ....*-diazene* (*CA*)

DIAZO COMPOUNDS

Functional group: $- N = N -$

Functional group name:
    diazo (IUPAC)
    diazenediyl (*CA*)

General formula: $RN = NX$
    (X = halogen, cyanide, hydroxide)

Series name:
    diazo-... (IUPAC)
    ...*diazene* (*CA*)

HYDRAZO COMPOUNDS

Functional group: $- NHNH -$

Functional group name:
    hydrazo (IUPAC and *CA*)
    1,2-diazanediyl (Recommended)

General formula: $RNHNHR'$

Series name:
    hydrazo-.... (IUPAC)
    ....*-hydrazine* (*CA*)
    ....*-diazane* (Recommended)

3.12 ORGANIC SULFUR COMPOUNDS

Many classes of organic compounds contain sulfur. Since sulfur has the same number of valence electrons as oxygen, it forms a number of families of compounds in which it replaces oxygen wholly or partially. A few of them will be discussed in this section.

The term *thio* has always played a key role in the naming of sulfur-containing organic compounds. It nearly always denotes replacement of oxygen by sulfur. Thio is generally placed in front of the term that denotes an oxygen-containing group or an oxygen atom. For example, *-ol* denotes $- OH$; *-thiol* denotes $- SH$. Just as *-one* denotes $= O$ attached to carbon, *-thione* denotes $= S$ attached to carbon.

Thio is distinct from *thia*, which usually denotes replacement of a methylene group by sulfur or the presence of sulfur in a heterocyclic system.

Thiols are the sulfur analogs of alcohols and phenols. They are named by adding the suffix *-thiol* to the name of the parent hydrocarbon or ring system. Thiol should not be confused with thiole, the name of a heterocyclic compound (see p. 249).

$CH_3 CH_2 SH$ 　　　　　 ⬡ $- SH$ 　　　　　 (S ring structure) 　　　　　 $HSCH_2 CH_2 CH_2 SH$

ethanethiol 　　　　 *benzenethiol* 　　　　 *thiole* 　　　　 *1,3-propanedithiol*
　　　　　　　　　　 (thiophenol) 　　　　 *(thiophene)*

Formerly, trivial names were formed by naming the alkyl or aryl group and following it with the word mercaptan. This practice has been abandoned by both the IUPAC and *Chemical Abstracts*. The root is still used as the prefix *mercapto-* for an unsubstituted $HS -$ group.

Most thiols have an unpleasant odor. Ethanethiol can be detected by smell when mixed with air in a ratio of 1 to 50,000,000 volumes of air.

Substituted mercapto- groups, $RS -$, are *alkylthio-* groups. When they occur along with a functional group of higher precedence such as carboxy, amino, or hydroxy, the appropriate prefix is used.

$HSCH_2 CH_2 OH$ 　　　　 $CH_3 CH_2 SCH_2 CH_2 OH$ 　　　　 $HS -$ ⬡ $- COOH$

2-mercaptoethanol 　　　 *2-(ethylthio)ethanol* 　　　 *4-mercaptobenzoic acid*

$$HS - CH - COOH$$
$$|$$
$$HS - CH - COOH$$

2,3-dimercaptobutanedioic acid

The groups $- SH$ and $RS -$ occur in many proteins, especially enzymes.

Methionine, one of the essential amino acids, is 2-amino-4-(methylthio)-butanoic acid. What is its structural formula?

A _____

```
        H   H   NH2  O
        |   |   |    ||
   H -  C - C - C  —  C - OH
        |   |   |
        SH  H   H
```

B _____

```
        H   H   NH2  O
        |   |   |    ||
   H -  C - C - C  —  C - OH
        |   |   |
        S   H   H
        |
   H -  C - H
        |
        H
```

C _____

```
        H   H   H   NH2  O
        |   |   |   |     ||
   H -  C - C - C - C   —  C - OH
        |   |   |   |
        H   SH  H   H
```

• •

A _____

Incorrect. This formula does not represent 2-amino-4-(methylthio)butanoic acid.

$$
\begin{array}{c}
\quad\;\; H \;\; H \;\; NH_2 \; O \\
\quad\;\; | \quad\; | \quad\; | \quad\; || \\
H - C - C - C \!\!-\!\! C - OH \\
\quad\;\; | \quad\; | \quad\; | \\
\quad\;\; SH \;\; H \;\; H
\end{array}
$$

Your error is one of omission rather than commission. Compare the name with the formula item by item. You will find something in the name that is not in the formula. It is a methyl group. Read page 236 again to learn where to place it in the formula.

B _____

Right. The structural formula for methionine is

$$
\begin{array}{c}
\quad\;\; H \;\; H \;\; NH_2 \; O \\
\quad\;\; | \quad\; | \quad\; | \quad\; || \\
H - C - C - C \!\!-\!\! C - OH \\
\quad\;\; | \quad\; | \quad\; | \\
\quad\;\; S \;\;\; H \;\; H \\
\quad\;\; | \\
H - C - H \\
\quad\;\; | \\
\quad\;\; H
\end{array}
$$

In the L form it is one of the eight essential amino acids. Go on to the next section.

C _____

You are incorrect. Methionine is 2-amino-4-(methylthio)butanoic acid. Look carefully at the formula you picked.

$$
\begin{array}{c}
\quad\;\; H \;\; H \;\; H \;\; NH_2 \; O \\
\quad\;\; | \quad\; | \quad\; | \quad\; | \quad\; || \\
H - C - C - C - C \!\!-\!\! C - OH \\
\quad\;\; | \quad\; | \quad\; | \quad\; | \\
\quad\;\; H \;\; SH \;\; H \;\; H
\end{array}
$$

Notice that is has five carbon atoms in the continuous chain. That would make it a derivative of pentanoic acid, wouldn't it? Your formula shows a mercapto-group whereas the name calls for a methylthio- group. Make both of these changes and then choose another answer.

Sulfides are the sulfur analogs of ethers and have the general formula $R - S - R$. Their nomenclature is analogous to that of ethers. They are named as sulfides under the IUPAC rules, but *Chemical Abstracts* has abandoned the term. Here are some examples:

$CH_3 SCH_3$ ⬡ $- S -$ ⬡ $CH_3 CH_2 S -$ ⬡

IUPAC: *methyl sulfide*   *phenyl sulfide*   *ethyl phenyl sulfide*
CA: *thiobis[methane]*   *1,1'-thiobis[benzene]*   *(ethylthio)benzene*

Mustard gas, one of the powerful vesicants used in chemical warfare, is not really a gas but an oily liquid that boils at $217^{\circ}C$. Its condensed formula is

$$ClCH_2 CH_2 SCH_2 CH_2 Cl$$

What is its IUPAC systematic name?

A _____

2-chloroethyldisulfide

B _____

bis(2-chloroethyl) sulfide

C _____

2-chloroethyl sulfide

● ●

A _____

Incorrect. Organic sulfides have the general formula $R-S-R$. The family called the disulfides has the general formula $R-S-S-R$. Which family does mustard gas belong to?

$$ClCH_2 CH_2 SCH_2 CH_2 Cl$$

Go back and choose another answer.

B _____

Right. The multiplicative prefixes *bis-, tris-,* etc., are used to prevent confusion when a substituted group appears more than once in an organic molecule. In mustard gas, the 2-chloroethyl group needs to be given the *bis-* prefix. Mustard gas is bis(2-chloroethyl) sulfide. *Chemical Abstracts* indexes it as 1,1'-thiobis[2-chloroethane] . Go on to the next section.

C _____

Incorrect. The preferred systematic name for mustard gas is not 2-chloroethyl sulfide. When the group is complex, as the 2-chloroethyl group is, the multiplicative prefixes *bis-, tris-,* etc., must be used. Mustard gas is properly named bis(2-chloroethyl) sulfide. *Chemical Abstracts* indexes it as 1,1'-thiobis[2-chloroethane] .

When sulfide names become complex because of the presence of several sulfur atoms in a carbon chain, replacement nomenclature can be used. The term *thia*, with multiplying prefixes if needed, denotes the replacement. This usage parallels the use of *oxa* for replacement by oxygen (see p. 205). For example,

$$CH_3 - S - \overset{\overset{\displaystyle CH_3}{|}}{CH} - S - CH_3$$

3-methyl-2,4-dithiapentane

Chemical Abstracts uses acyclic thia-, oxa-, etc., names only when there are four or more such atoms or groups present.

Cyclic sulfides are named either by replacement nomenclature or by the Hantzsch-Widman method, to be discussed in Chapter 4.

thiacycloheptane

1-thiaspiro[4.4] nonane

Sulfoxides and sulfones are compounds which contain both sulfur and oxygen. Their general formulas are

$$\begin{array}{c} O \\ \uparrow \\ R - S - R \end{array} \qquad\qquad \begin{array}{c} O \\ \uparrow \\ R - S - R \\ \downarrow \\ O \end{array}$$

sulfoxide   *sulfone*

Methyl sulfoxide has been used for many years as an organic solvent.

$$\begin{array}{c} O \\ \uparrow \\ CH_3 - S - CH_3 \end{array}$$

It recently came into prominence under the name dimethyl sulfoxide (DMSO).

Chemical Abstracts has dropped sulfoxide and sulfone as indexing names. They are named as sulfinyl or sulfonyl compounds, respectively. The *sulfinyl* group is $-\underset{\downarrow}{\overset{}{S}}-$ and the *sulfonyl* group is $-\underset{\downarrow O}{\overset{\uparrow O}{S}}-$. DMSO is indexed as sulfinylbismethane.

Sulfonyl halides are often used as intermediates. One of them, with the trivial name dansyl chloride, is used to convert end amino groups in proteins to sulfonamides that are more resistant to hydrolysis.

SO_2Cl

H_3CNCH_3

5-(dimethylamino)-1-naphthalenesulfonyl chloride
(dansyl chloride)

Sulfonic acids contain the functional group $-SO_3H$. Although they are not prepared in this way, they may be viewed as the product obtained by replacing one hydroxy group in sulfuric acid with an alkyl or aryl group

$$\begin{array}{c} O \\ \uparrow \\ HO - S - OH \\ \downarrow \\ O \end{array} \qquad\qquad \begin{array}{c} O \\ \uparrow \\ CH_3CH_2 S - OH \\ \downarrow \\ O \end{array}$$

sulfuric acid   *ethanesulfonic acid*

The names are formed by adding the functional ending *sulfonic acid* to the name of the corresponding hydrocarbon. The general formula is RSO_3H.

Benzenesulfonic acid is an important intermediate in organic synthesis. What is its formula?

A _____

B _____

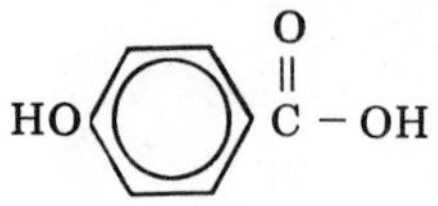

C _____

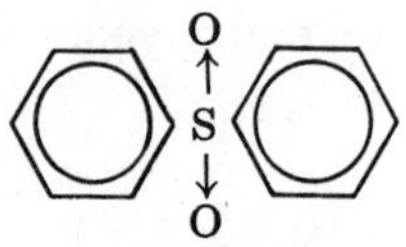

• •

A _____

You are correct. The formula for benzenesulfonic acid is

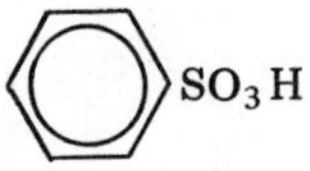

Continue on to the next section.

B _____

Incorrect. This formula does not have the proper functional group to be a sulfonic acid. The functional group of the sulfonic acids is $-SO_3H$. The formula shown here

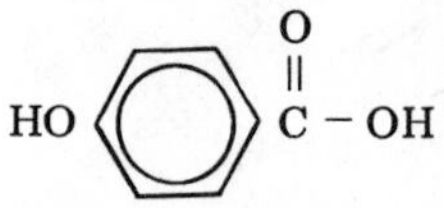

is a derivative of benzoic acid, a carboxylic acid. Go back and choose another answer.

C _____

You are wrong. This formula does not represent a sulfonic acid. It does not have the right functional group.

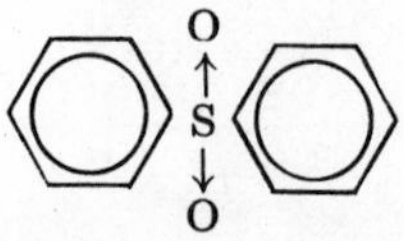

Sulfonic acids have the $-SO_3H$ group. When you read page 239 again, you will see that this formula represents phenyl sulfone, indexed by *Chemical Abstracts* as 1,1'-sulfonylbis[benzene].

Sulfonic acids form both esters and amides. The amide of benzenesulfonic acid has the formula

$$SO_2NH_2$$

benzenesulfonamide

The amide can be viewed formally as the result of the reaction between the hydroxy group and ammonia as shown here

$$\text{benzene} - \overset{\overset{O}{\uparrow}}{\underset{\underset{O}{\downarrow}}{S}} - OH \;+\; NH_3 \;\longrightarrow\; \text{benzene} - \overset{\overset{O}{\uparrow}}{\underset{\underset{O}{\downarrow}}{S}} - NH_2 \;+\; H_2O$$

  benzenesulfonic acid ammonia   benzenesulfonamide   water

Sulfanilamide, one of the earliest of the sulfa drugs, is the amide of another aromatic sulfonic acid.

$$SO_2NH_2$$

$$NH_2$$

sulfanilamide

The acid, sulfanilic acid, is

$$SO_3H$$

$$NH_2$$

sulfanilic acid

What is the systematic name of sulfanilic acid?

A _____

4-aminobenzenesulfonic acid

B _____

aminobenzenesulfonic acid

C _____

4-aminophenylsulfonic acid

• •

A _____

Right you are. Sulfanilic acid is 4-aminobenzenesulfonic acid.

$$SO_3H$$

$$NH_2$$

4-aminobenzenesulfonic acid
(sulfanilic acid)

Go on to the next section.

B _____

You named sulfanilic acid as aminobenzenesulfonic acid.

$$SO_3H$$

$$NH_2'$$

You are right as far as you went, but you did not go far enough. Aren't there other places where the amino group could be attached to the ring? If there are, the location must be specified. Go back and choose another answer.

C _____

Incorrect. Sulfanilic acid is not 4-aminophenylsulfonic acid. When a functional group such as $-SO_3H$ or $-COOH$ is substituted on a benzene ring, the name benzene is maintained. For example, phthalic acid is 1,2-benzenedicarboxylic acid:

Benzenesulfonic acid is

$$SO_3H$$

Go back and choose another answer.

Alkylated benzenesulfonic acid salts are used in detergents. They are present as mixtures rather than individual pure compounds. Typical formulas for the parent acids are:

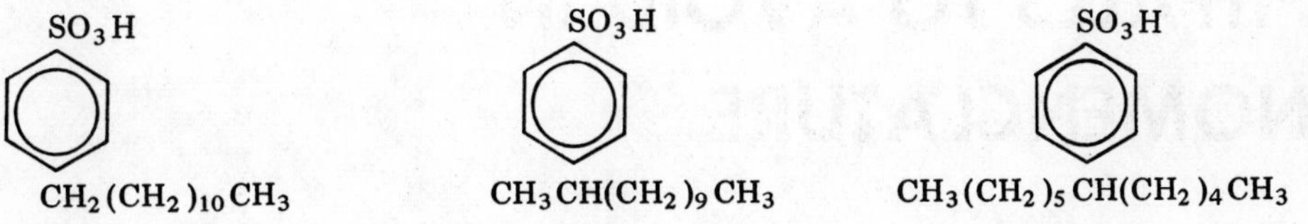

SO_3H

$CH_2(CH_2)_{10}CH_3$

SO_3H

$CH_3CH(CH_2)_9CH_3$

SO_3H

$CH_3(CH_2)_5CH(CH_2)_4CH_3$

The aromatic ring may be substituted on any carbon of the 12-carbon chain. Detergents having the ring at or near the end of the chain are claimed to be most efficient.

Guidepost:

PITFALLS TO AVOID IN NOMENCLATURE

Organic chemical nomenclature is a dynamic subject that undergoes continuing change and improvement. We have seen how the practice of *Chemical Abstracts* often differs from the IUPAC Rules. It is worthwhile to note some of the common errors that should be avoided regardless of the rules being followed.

A widespread error is the use of names derived partly from one nomenclature system and partly from another. While isopropyl alcohol is a correct trivial name for 2-propanol, the name isopropanol is an objectionable combination of the systematic and trivial names.

Selection of the principal function for naming polyfunctional compounds should follow the order of preference given in Appendix 3 (p. A.4). For example, $HOCH_2CH_2COOH$ is named 3-hydroxypropanoic acid rather than 2-carboxy-1-propanol because acids are higher in the order of precedence than alcohols.

Care must be exercised to avoid adding spaces to names that should not have them, and omitting spaces when they are needed. In general, a compound derived by the replacement of hydrogen with some other atom or group is named by prefixing the name of the substituent group to the name of the unsubstituted compound without a space. Methylbenzene is correctly written as a single word because benzene is the name of a compound into which methyl is substituted. Acetyl chloride, on the contrary, is two words because "chloride" is not the name of a compound.

The radicofunctional names of ketones, ethers, sulfides, anhydrides, etc., are written as two words. Diethyl ketone and acetic anhydride are examples. The term amine is an apparent exception. Because it is a shortened form of ammonia, a compound, it is not used as a separate word. Consequently, methylamine is correct, not methyl amine.

Another pitfall to avoid is the contraction of two-word class names to a single word. Examples are aminoacids rather than amino acids, ethylesters rather than ethyl esters, etc.

Structural prefixes that are attached to a name by a hyphen and frequently appear in italic type should not be capitalized at the beginning of a sentence and do not serve as the initial capital for a sentence. A few of them are *cis-*, *trans-*, *ortho-*, *meta-*, and *para-*. The same holds true for abbreviations such as *o-*, *m-*, *p-*, *sec-*, *tert-*, and *n-*. Proper capitalization at the beginning of a sentence is:

> *cis-*Butene can be . . .   not   *Cis-*butene can be . . .
> *sec-*Butyl alcohol boils . . .   not   *Sec-*butyl alcohol boils . . .

Likewise, a prefix that is a capital letter is not used as the initial capital: *N*-Methylacetamide is . . . not *N*-methylacetamide is . . . . When used as adjectives, however, all of these terms may be capitalized to begin sentences. For example: *Cis* isomers are . . . .

Organic nomenclature in the chemical literature of today is greatly improved over that used only a few years ago. Nonetheless, occasional errors still occur, and the student of nomenclature should strive to recognize them as well as to avoid them in his own work.

Chapter Four

HETEROCYCLIC COMPOUNDS

Any atom or atom group capable of forming two or more covalent bonds may be a member of a heterocyclic ring system, but the ones most often encountered along with carbon are nitrogen, oxygen, and sulfur. In the past few years rings containing phosphorus, silicon, or boron have received increased attention.

The *Ring Index* lists about 15,000 carbocyclic and heterocyclic ring systems. It would be impossible to present the systematic nomenclature for all of them. We will discuss the basic rules and look at enough examples so at least you will be able to recognize the names of the common heterocyclic systems.

By now you should be accustomed to the idea that trivial names are irreplaceably entrenched in the literature and in everyday usage by chemists. The situation is no different with heterocyclic compounds. When there is an accepted trivial name, it will be given in parentheses after the systematic name. *Chemical Abstracts*, moreover, has made a few refinements of the IUPAC rules for its own purposes. Some of these will be pointed out.

This presentation of systematic names will be limited to monocyclic systems. The names of those with three to ten members are based on the extended Hantzsch-Widman stems. Each name has two parts: one or more prefixes to describe the hetero-atom(s), followed by a stem that tells the number of atoms in the ring, whether or not nitrogen is one of the hetero-atoms, and whether or not the ring is saturated.

Before considering the details, let's look at an example. Dioxane, a common solvent and paint remover, is really 1,4-dioxane. What does this name tell us?

1,4-	hetero-atoms in 1 and 4 positions
diox(a)	both hetero-atoms are oxygen
-ane	6-membered ring, no nitrogen, saturated

1,4-dioxane

Notice that the *a* in oxa is elided before the stem beginning with an *a* (or other vowel).

The two basic tools for building systematic names are the lists of prefixes and stems given in Tables 4.1 and 4.2.

TABLE 4.1   Replacement Prefix Order of Precedence

Element	Group	Oxidation State	Prefix
oxygen	VIA	II	oxa
sulfur	VIA	II	thia
selenium	VIA	II	selena
tellurium	VIA	II	tellura
nitrogen	VA	III	aza
phosphorus	VA	III	phospha[a]
arsenic	VA	III	arsa[a]
antimony	VA	III	stiba[a]
bismuth	VA	III	bismutha[b]
silicon	IVA	IV	sila
germanium	IVA	IV	germa[c]
tin	IVA	IV	stanna
lead	IVA	IV	plumba
boron	IIIA	III	bora
mercury	IIB	II	mercura

[a]When immediately followed by *-in* or *-ine*, these replacements are made:

    phospha becomes *phosphor*
    arsa becomes *arsen*
    stiba becomes *antimon*

[b]*Chemical Abstracts* now uses *bisma*, previously used in the IUPAC rules.

[c]*Chemical Abstracts* adds this to note [a]:

    germa becomes *german*

TABLE 4.2   Hantzsch-Widman Stems

Number of atoms in ring	Rings containing nitrogen unsaturated[a]	saturated	Rings containing no nitrogen unsaturated[a]	saturated
3	-irine	-iridine	-irene	-irane
4	-ete	-etidine	-ete	-etane
5	-ole	-olidine	-ole	-olane
6	-ine[b]	c	-in[b]	-ane[d,e]
7	-epine	c	-epin	-epane
8	-ocine	c	-ocin	-ocane
9	-onine	c	-nin	-nane
10	-ecine	c	-ecin	-ecane

[a]Stem corresponds to the maximum number of non-cumulative double bonds when the heteroatoms have valences shown in Table 4.1.

[b]See special provisions for phosphorus, arsenic, antimony, and germanium in footnotes [a] and [c] in Table 4.1.

[c]Expressed by prefixing *perhydro-* to the name of the unsaturated compound. *Chemical Abstracts* uses *tetrahydro-, hexahydro-*, etc.

[d]Not applicable to silicon, germanium, tin, and lead. In these instances *perhydro-* is prefixed to the name of the unsaturated compound. *Chemical Abstracts* uses *tetrahydro-, hexahydro-*, etc.

[e]*Chemical Abstracts* denotes saturation of six-membered hetero systems based on boron or phosphorus by the stem *-inane*.

The order of precedence in Table 4.1 is by descending group number in the Periodic Table and increasing atomic number within a group. As you can see, all of the group VI elements precede those in group V, etc. The order is from top to bottom within the groups.

Now for the Hantzsch-Widman stems. They apply to monocyclic systems of three through ten members.

Believe it or not, there's some logic behind the syllables denoting the ring size for 3, 4, and 7 to 10 members. They are derived as follows:

No. members	Syllable	Root
3	ir	tri
4	et	tetra
7	ep	hepta
8	oc	octa
9	on	nona
10	ec	deca

The presence of an *e* at the end of a stem lengthens the preceding vowel. For example, *-in* [ĭn] changes to *ine* [ēn].

As used in Table 4.2, unsaturated means the maximum number of non-cumulative double bonds with the hetero-atoms having the normal oxidation states (valences) shown in Table 4.1. We'd better examine this statement closely.

Cumulative double bonds are those in which at least three contiguous atoms are joined by double bonds. Non-cumulative double bonds are every other arrangement of two or more double bonds in a single structure.

$$CH_2 = C = C = C = CH_2 \qquad\qquad CH_3 - CH = CH - CH = CH_2$$

cumulative   *non-cumulative*

It is necessary to find the maximum number of non-cumulative double bonds in order to use the stem for unsaturated compounds. Benzene, for example, has the maximum number; 1,3-cyclohexadiene does not.

1,3-cyclohexadiene   *benzene*

Here are some other examples with the maximum number of non-cumulative double bonds:

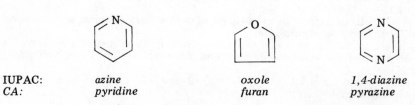

| IUPAC: | *azine* | *oxole* | *1,4-diazine* |
| CA: | *pyridine* | *furan* | *pyrazine* |

The *diazines* should not be confused with *diazene*, HN = NH.

Chemical Abstracts indexes this compound by its trivial name, thiophene. What is its Hantzsch-Widman name? (See Table 4.2.)

A _____

thiolane

B _____

thiole

C _____

thiete

• •

A _____

Incorrect. You chose thiolane as the Hantzsch-Widman name for this compound.

If you'll look at Table 4.2 again, you'll see that *-olane* is the stem for a saturated compound. Try another answer.

B _____

Right. The Hantzsch-Widman name for thiophene is thiole. Go on to the next section.

C _____

You are wrong. You chose thiete as the Hantzsch-Widman name for this compound.

Look again at Table 4.2. The stem *-ete* is for a four-membered ring. Thiophene has a five-membered ring. Work out the correct name and go back to see if it's one of the answers given.

In a few instances there will be more than one possible structure with the maximum number of non-cumulative double bonds. Both of these structures for oxin have two double bonds:

IUPAC: *2H-oxin*   *4H-oxin*

CA: *2H-pyran*   *4H-pyran*

This difficulty is overcome by means of indicated hydrogen, a designation consisting of a locant followed by an italic capital *H* placed before the name of the ring system. It expresses the position of each saturated atom necessary to form a definable, stable ring system.

> Indicated hydrogen is needed to specify the structures of azole (*CA:* pyrrole). 1*H*-Azole is shown here.

Which other isomer(s) exists?

A _____

2*H*-

B _____

2*H*- and 3*H*-

C _____

Help!

••

A _____

You are partially right. Your answer is that the isomers of azole are 1*H*-azole and 2*H*-azole as shown here.

The other possible isomer is 3*H*-azole. Draw a formula for it to see if it's different from the two above. Now read answer B below.

B _____

You are right. The three isomers of azole are shown here.

	1H-azole	*2H-azole*	*3H-azole*
CA:	*1H-pyrrole*	*2H-pyrrole*	*3H-pyrrole*

Each of the three represents a definable, stable ring system. Go on to the next section.

C _____

So you need help. Let's take a look at the ring structure of azole, with no double bonds shown.

Because it is a five-membered ring, the maximum number of non-cumulative double bonds is two. The question is: How many different combinations of two non-cumulative double bonds can there be? And how can they be differentiated? The answer to the second question is *indicated hydrogen* to specify a ring member that will be saturated — that is, not involved in a double bond. Work out an answer to the first question, then go back and choose another answer.

By now you have guessed that a single hetero-atom is given the locant 1. When the same hetero-atom occurs more than once in a ring, the appropriate prefix is added and the numbering is chosen to give the lowest set to the hetero-atoms. For instance,

	1,4-diazine	*1,2,4-triazine*	*1,3-dioxane*
CA:	*pyrazine*		

When two different hetero-atoms are present, the order of citation is by descending periodic group number and increasing atomic number within a group (i.e., the order in Table 4.1). Two examples are:

1,2-oxathiolane   *1,3-thiazole*
                                     CA: thiazole

The trivial name for this compound, used by *Chemical Abstracts,* is furazan. What is its systematic (Hantzsch-Widman) name?

A ————————————————————————————————————

1,2,5-oxazole

B ————————————————————————————————————

2,5-oxadiazole

C ————————————————————————————————————

1,2,5-oxadiazole

● ●

A ————————————————————————————————————

You are wrong. There are three hetero-atoms: one oxygen and two nitrogens. The name fragment oxazole specifies only two, one oxygen and one nitrogen. Choose another answer.

B ————————————————————————————————————

Incorrect. There are three hetero-atoms in the compound, but your chosen name, 2,5-oxadiazole, has only two locants. There must be a locant for each hetero-atom. Go back and choose another answer.

C ————————————————————————————————————

Right. The systematic name for furazan is 1,2,5-oxadiazole. The three hetero-atoms are identified and the position of each is specified by a locant. Go on to the next section.

If the number of double bonds is less than the maximum, prefixes are used to indicate the number of hydrogen atoms that have been added. The number is always even, since two hydrogens are needed to saturate each double bond. A few examples will illustrate the point:

maximum unsaturation	partial saturation	complete saturation
oxole furan	2,3-dihydrooxole 2,3-dihydrofuran	oxolane tetrahydrofuran

CA:

Which of the following is the structure for 2,3-dihydroazine? (The *CA* name is 2,3-dihydropyridine.)

A _____

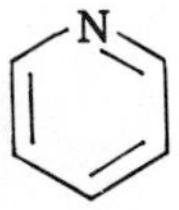

B _____

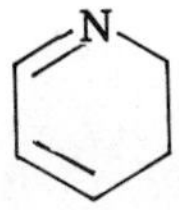

C _____

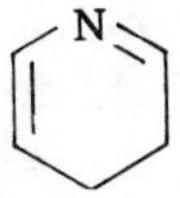

• •

A _____

Wrong. The structure you have chosen has three double bonds, the maximum number of non-cumulative double bonds for a six-membered ring. It is the structure for azine. What structure results when you saturate the double bond joining the number 2 and number 3 carbon atoms? Go back and select another answer.

B _____

Right you are. Go on to the next section.

C _____

Incorrect. You chose this structure for 2,3-dihydroazine.

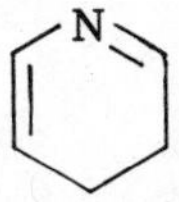

Perhaps you have forgotten that the hetero-atom is numbered 1. This is the structure for 3,4-dihydroazine. Go back and choose another answer.

As you can see from Table 4.2, fully saturated rings with six to ten members that contain nitrogen are named under the IUPAC rules by adding the prefix *perhydro-* to the name of the unsaturated compound. As shown here, different numbers of hydrogen atoms may be required, depending on the number of double bonds in the unsaturated ring. *Chemical Abstracts* tends to avoid terms that may have more than one meaning and does not use perhydro-, choosing instead to specify the exact number of hydrogens by the prefix.

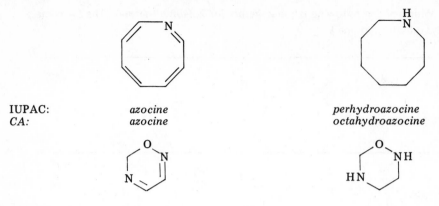

| IUPAC: | *azocine* | *perhydroazocine* |
| CA: | *azocine* | *octahydroazocine* |

| IUPAC: | *6H-1,2,5-oxadiazine* | *perhydro-1,2,5-oxadiazine* |
| CA: | *6H-1,2,5-oxadiazine* | *tetrahydro-2H-1,2,5-oxadiazine* |

Under the IUPAC rules there are special terminating stems for four- and five-membered rings with one double bond. They are given in Table 4.3.

TABLE 4.3   **Hantzsch-Widman Stems for Partly Saturated Rings**

Number of members in ring	Ring contains nitrogen	No nitrogen in ring
4	-etine	-etene
5	-oline	-olene

The position of the double bond is specified by a locant as usual.

Chemical Abstracts has abandoned these special endings and names the compounds as dihydro- derivatives of the unsaturated compounds. Here are two illustrations:

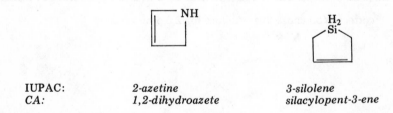

| IUPAC: | *2-azetine* | *3-silolene* |
| CA: | *1,2-dihydroazete* | *silacylopent-3-ene* |

Now it's time for you to practice. Here are six monocyclic heterocyclic structures that *Chemical Abstracts* indexes according to the trivial name given. Work out the systematic name for each and check it against the one given in the list below.

1.

piperidine

2.

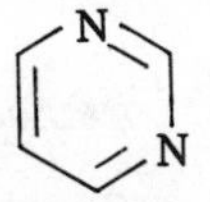

pyrimidine

3.

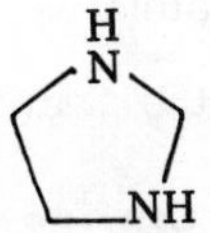

imidazolidine

4.

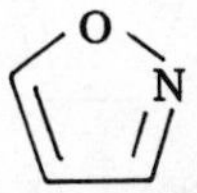

isoxazole

5.

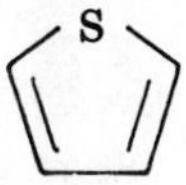

thiophene

6.

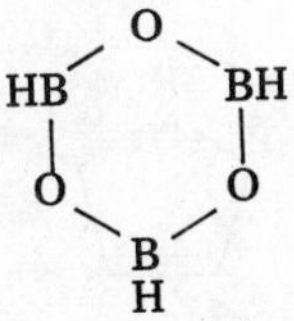

boroxin

• •

1. IUPAC: perhydroazine

2. 1,3-diazine

3. 1,3-diazine

4. 1,2-oxazole

5. thiole

6. 1,3,5,2,4,6-trioxotriborinane (note use of *-inane* to show saturation)

Here are another half-dozen compounds indexed at trivial names by *Chemical Abstracts.* Work out the structures and check them against those shown below.

1. IUPAC: 1,2-diazolidine; *CA:* pyrazolidine

2. IUPAC: perhydro-1,4-oxazine; *CA:* morpholine

3. IUPAC: 1,2-selenazole; *CA:* isoselenazole

4. IUPAC: 3-azoline; *CA:* 2,5-dihydro-1*H*-pyrrole

5. IUPAC: 1,2-thiazole; *CA:* isothiazole

6. IUPAC: perhydro-1,3,5,2,4,6-triazatriborine; *CA:* borazine

• •

1.

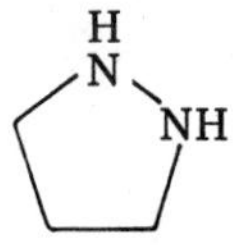

2.

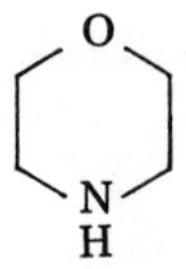

3.

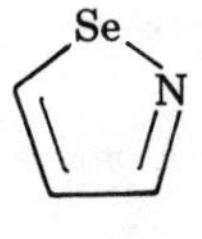

4.

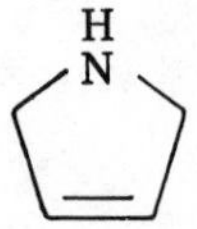

5.

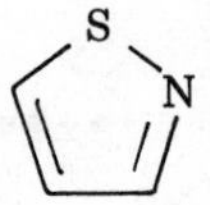

6.

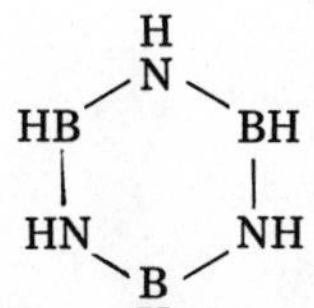

Heterocyclic rings larger than ten members not containing silicon are named as replacement derivatives of the carbocyclic compound. The highest-ranking hetero-atom is given the locant 1. Double and triple bonds are located as usual. Here are two illustrative examples:

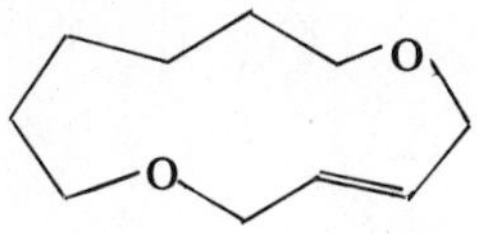

1,6-dioxacyclododec-3-ene

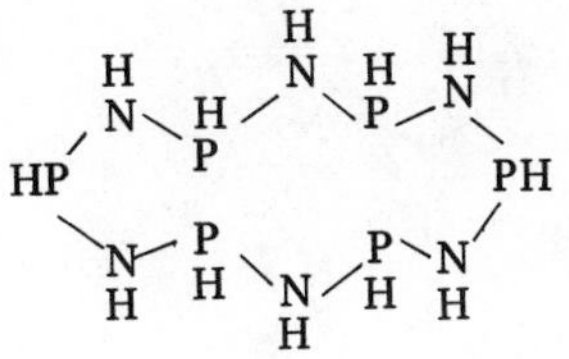

*1,3,5,7,9,11-hexaaza-2,4,6,8,10,12-
hexaphosphacyclododecane*

Many fused heterocyclic ring systems have trivial names. Here are a few of the ones you're apt to encounter.

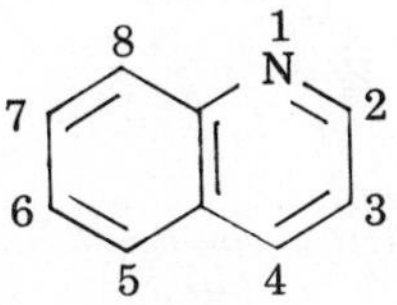

quinoline

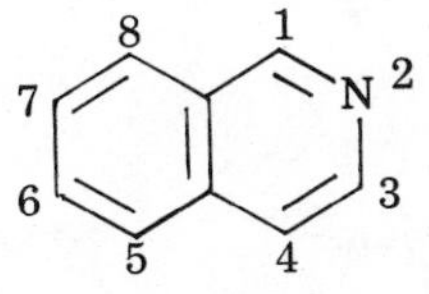

isoquinoline

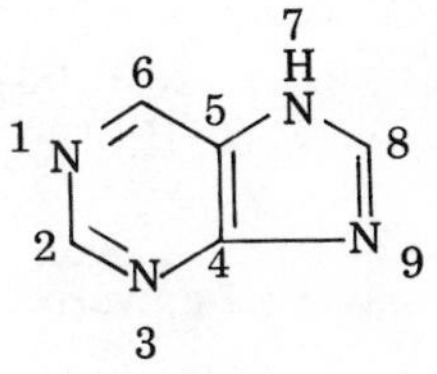

purine
(exception to systematic numbering)

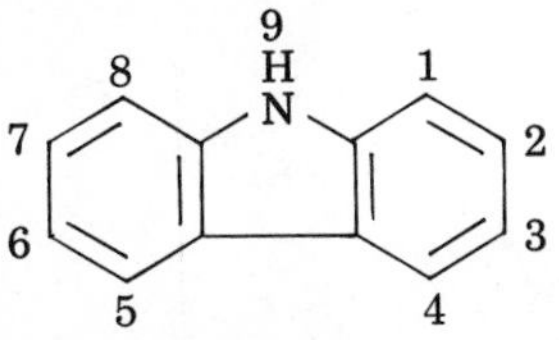

carbazole

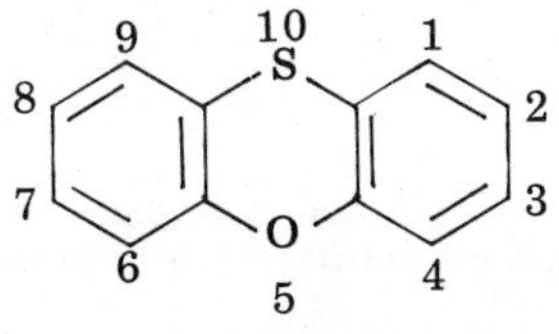

phenoxathiin

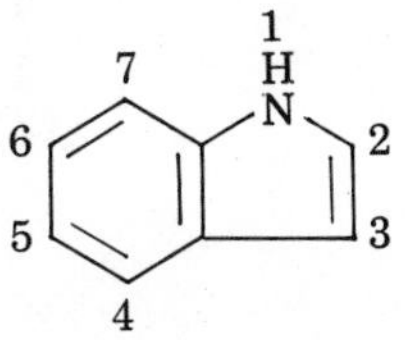

indole

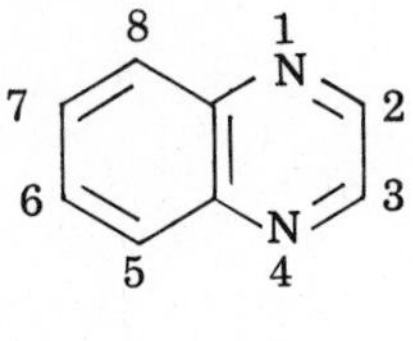

quinoxaline

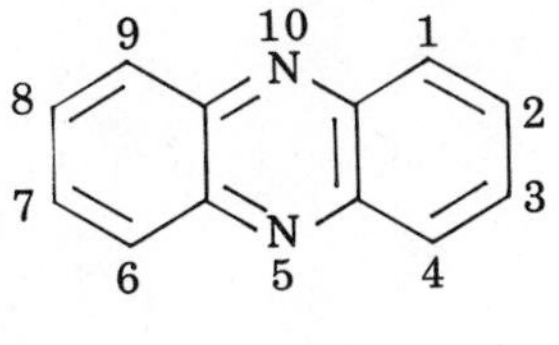

phenazine

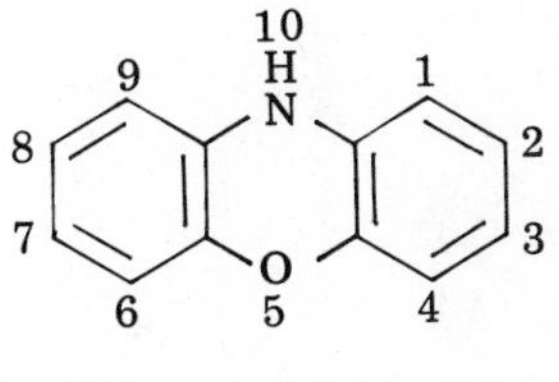

phenoxazine

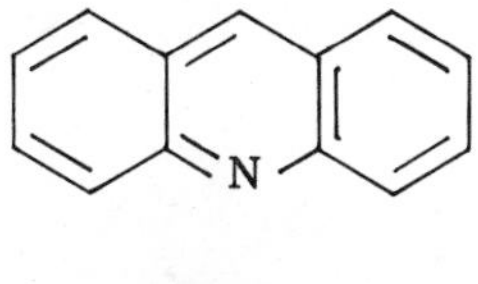

acridine

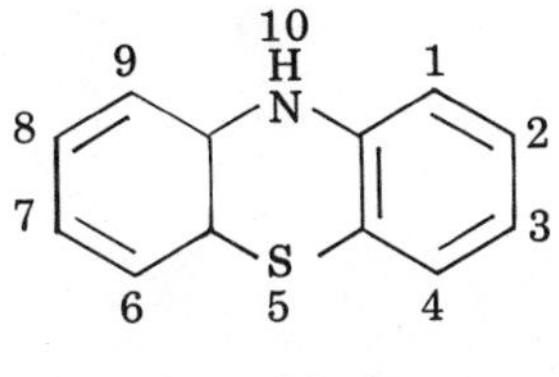

phenothiazine

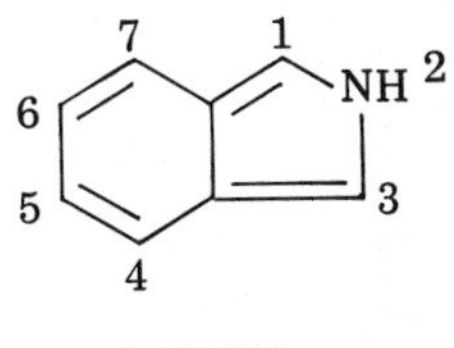

isoindole

Six heterocyclic bases occur in DNA and RNA. They are shown here. You may find formulas in other books which differ slightly in the placement of double bonds and of hydrogen on the nitrogen atoms.

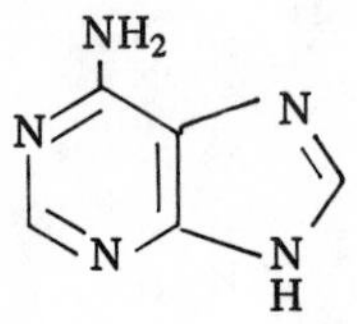

adenine
1H-purin-6-amine

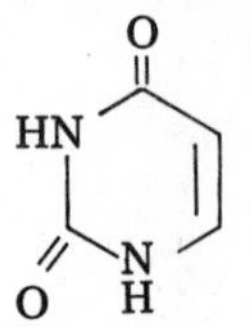

uracil
2,4(1H,3H)-pyrimidinedione

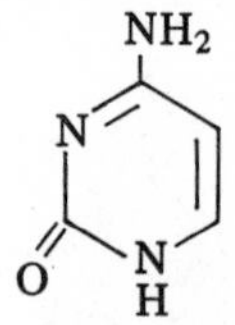

cytosine
4-amino-2(1H)-pyrimidinone

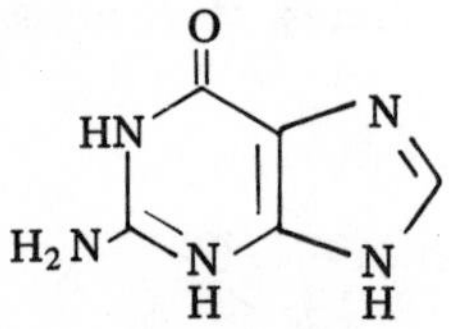

guanine
2-amino-1,7-dihydro-6H-purin-6-one

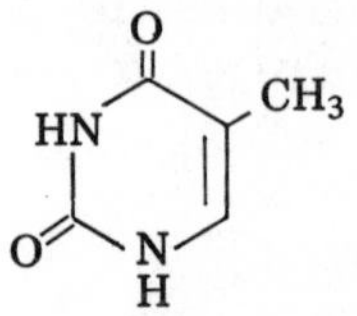

thymine
5-methyl-2,4(1H,3H)-pyrimidinedione

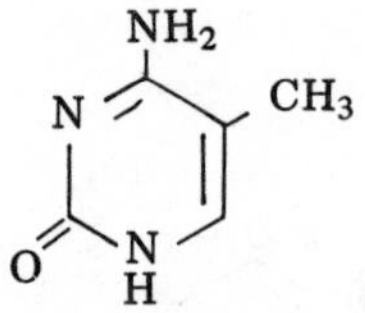

5-methylcytosine
4-amino-5-methyl-2(1H)-pyrimidinone

CHEMICAL ABSTRACTS

Sooner or later every serious student of chemistry wants to know more than his elementary textbooks can tell him about a particular compound or class of compounds. Sometimes advanced textbooks and compendia available in a library will serve the purpose, but often the comprehensive coverage of *Chemical Abstracts* is required.

Chemical Abstracts (CA) is known as the "Key to the World's Chemical Literature" and has published English-language abstracts and indexes of the literature of chemistry and chemical engineering from all over the world since 1907. Journal articles, patents, review articles, technical reports, monographs, conference proceedings, symposia, Ph.D. dissertations, and new books are covered. A glance at the bound volumes on library shelves dramatically illustrates the rapid growth of the literature in recent years. A total of 356,549 documents was cited in 1973! Because the annual subscription rate is high ($2900 for 1974–75), few chemists own individual copies but rely instead on the holdings of libraries.

A volume of *CA* now covers a six-month period and includes 26 weekly issues and a volume index. The abstracts are presently arranged into 80 subject sections. The sections covering biochemistry (1–20) and organic chemistry (21–34) appear biweekly in the odd-numbered issues. The sections on macromolecular chemistry (35–46), applied chemistry and chemical engineering (47–64), and physical and analytical chemistry (65–80) are published in the even-numbered issues.

The arrangement of abstracts within each of the 80 sections places abstracts of journal articles and proceedings first, followed by monographs and the abstracts of patents. Cross-references at the end of each section indicate abstracts appearing in other sections that may also be relevant.

Each *CA* abstract provides a brief summary of the novel content of the original document and gives the reader its citation, but is not intended to replace the document. The primary purpose is to give the reader enough information for him to decide whether to refer to the original publication. The initial sentence of the abstract often focuses on the primary findings and conclusions reported in the original document.

Each issue of *CA* contains a four-part index related to that issue: the Keyword Subject Index, the Author Index, the Numerical Patent Index, and the Patent Concordance. Each is explained in the first issue of a volume. These indexes do not substitute for the in-depth treatment of Volume Indexes.

Since Volume 66, for 1967, the abstracts have been numbered continuously through an entire six-month volume. The letter following each number is a computer-generated check, allowing computer validation of each reference to the abstract. Before Volume 66, references were to column numbers with an appended letter designating the position in the column.

A student's first use of *Chemical Abstracts* is usually a search for information in a Volume Index or a Collective Index covering a period of five or ten years.

Absolutely indispensable to successful use of *CA* is the Index Guide. Time spent reading the Index Guide for the Ninth Collective Index Period (1972–1976) will be repaid many times over in time saved in use of the work. The Index Guide is divided into five sections: Introduction, Indexes to Chemical Abstracts, General Subjects, Selection of Names for Chemical Substances, and the Alphabetic Sequence of Cross-References. The section on selection of names is a treasury of additional background information on chemical nomenclature for the student.

The Alphabetic Sequence of Cross-References is the longest and most important section and is commonly what is intended by reference to the Index Guide.

Before consultation of the indexes, the *CA* user should compare his proposed list of search subjects with the cross-reference to insure he has selected appropriate index headings.

In a complex printed index the name that best locates a substance adjacent to substances of similar structure is chosen for indexing and, once chosen, must be used consistently. Because many substances have widely used trivial names as well as one or more systematic names, cross-references are needed to lead the index user to the one chosen for indexing.

For example, the preferred index name for the structure $CH_3COCH_2CH_3$ is **2-butanone**. It is frequently referred to as ethyl methyl ketone, methyl ethyl ketone, MEK, or methylacetone, and might well be described as acetylethane or 2-oxobutane. The number of possible alternatives increases rapidly as the complexity of substances increases. If a user cannot find a particular name, he must consult the indexing policy notes in the Guide. The Index Guide note at **Ketones** explains the principle of adding the suffix "-one" to the name of the molecular skeleton (butane in the example). Although the indexing name assigned for even a simple compound can vary over the nearly 70 years of publication of *Chemical Abstracts*, it is held the same within a Collective Period.

Seven types of volume indexes are now published for *CA*: Author Index, Chemical Substance Index, Index of Ring Systems, General Subject Index, Formula Index, Patent Index, and Patent Concordance. Each is described fully at its beginning. The Chemical Substance Index, General Subject Index, and the Index of Ring Systems are of most value to the ordinary user.

The Chemical Substance Index includes entries for all completely defined substances such as: elements, compounds, alloys, minerals, mixtures and polymers of compounds, antibiotics, enzymes, hormones, polysaccharides, elementary particles, alphanumeric, and trade-name designations. All consistently defined substances in this index are now processed through the Chemical Abstracts Service (*CAS*) Registry System and each index entry includes the assigned *CAS* Registry Number (see p. 261).

Substances are listed alphabetically by index heading parent. For example, **Benzoic acid** is an index heading parent. In recent years, subdivisions are used to organize references of related interest. There may be subdivisions for **analysis, biological studies, preparation,** etc. Specific derivatives are indexed with inverted names. For instance, **4-acetylbenzoic acid** is written as **Benzoic acid, 4-acetyl** so that it will appear "under" benzoic acid in the index.

The General Subject Index includes all index headings that do not refer to specific chemical substances. Included are classes of chemical substances, incompletely defined materials, applications, properties, reactions, apparatus and processes, and the common and scientific names of animals and plants. While references to benzoic acid appear in the Chemical Substance Index as explained above, references to the class Carboxylic Acids would be found in the General Subject Index.

The Formula Index often provides the quickest and most direct access to the *CA* index names and relevant abstracts for a single compound when the complete molecular formula is known. In this index, molecular formulas are the headings and the arrangement is by the Hill system:

a. For carbon-containing compounds − C first, followed directly by **H** (if present), then the remaining element symbols alphabetically.

b. For compounds that do not contain carbon, strictly alphabetically by symbols.

The complete molecular formulas, thus arranged, are listed in alphanumeric order with each chemical element symbol and its numerical subscript considered as a separate unit. Hence H_4 precedes H_5 and all formulas beginning with CH appear before all beginning with CH_2. The following formulas are arranged correctly and illustrate the Hill system:

$Al_6 Ca_5 O_{14}$	$CH_3 Cl$	$C_2 H_5 DO$
CCl_4	CO	$C_2 H_7 NO_3 S$
$CHBrCl_2$	$C_2 Ca$	$CaO_3 Ti$
$CHCl_3$	$C_2 H_2 Cl_2 O_2$	$H_4 Sn$
$CHNO$	$C_2 H_4 O_2$	
$CH_2 O$	$C_2 H_5 Cl_3 Si$	

Corresponding index heading parent names are given for each molecular formula, now together with the *CAS* Registry Number and often with specific *CA* references.

The object of the *CA* Index of Ring Systems is to lead the user to the assigned *CA* index names for unsubstituted cyclic compounds. Once the correct index parent name is found, the structural formula with its accompanying numbering system is found in the Index Guide. Specific references to the compound and its derivatives are found at the same index name in the *CA* Chemical Substance Index. The introduction to the Index of Ring Systems should be consulted for an explanation of its organization.

The Index of Ring Systems and structural formulas will be brought together in time when a Parent Compound Handbook is published to replace their separate publication. The handbook will be devoted to ring systems and cyclic stereoparents with emphasis on the relations between the two.

The *CAS* Chemical Registry is a computer-based system that identifies entered chemical substances on the basis of structure and assigns to each definable chemical entity a unique Registry Number that is computer verifiable. For example, D-Aspartic acid [1783-96-6], L-Aspartic acid [56-84-8], DL-Aspartic acid [617-45-8], and unspecified Aspartic acid [6899-03-2] all have different Registry Numbers. The Registry began operation in 1965 and at the end of 1973 contained just under 2.7 million entries. Some 437,000 previously unregistered substances were added during 1973.

Each *CAS* Registry Number designates only one substance so far as that substance has been elucidated and defined in terms of atoms, valence bonds, and stereochemistry. It is the hope that Registry Numbers, with or without associated names, may furnish an efficient means of structure identification in technical publications and in communication between chemists. Their use in research articles, patent specifications, etc., can aid in identifying unique, unambiguous chemical substances without resort to complex chemical nomenclature. The Registry Numbers are cited after the names of specific compounds in the Chemical Substance Index as well as in the Formula Index and the Index Guide.

One advantage of the Registry is, of course, that changes in the nomenclature rules or *CA* indexing practices will not affect the Registry Number. A disadvantage is that there is presently only an incomplete list of Registry Numbers. Five parts of the *CAS* Registry Handbook Number Section have been published. They cover the period 1965 through 1971 and include the Number, name, and molecular formula for Registry Numbers 35-66-5 through 26499–99–0.

In summary, *Chemical Abstracts* and the accompanying indexes provide the chemist with a powerful tool in searching the world's chemical literature for needed information.

Chapter Five

NATURAL PRODUCTS

Organic compounds derived from living or once-living plant and animal organisms are called natural products. An almost infinite variety of these compounds has been isolated and identified. They can be grouped into about a dozen classes on the basis of structural features and chemical reactions. Three classes to be discussed in this chapter are the three main foodstuffs: carbohydrates, proteins, and fats. The odorous components of many plants consist of carbocyclic hydrocarbons with the molecular formula $C_{10}H_{16}$. These and other closely related compounds are known as terpenes. Steroids are defined as compounds with a fused ring system largely parallelling the one in cholesterol. Sterols, bile acids, sex hormones, adrenal steroids, toad poisons, and the steroid sapogenins belong to this class.

The complex structures of natural products lead to long systematic names. The long names, coupled with the fact that many of them were isolated and given trivial names early in the game, have caused workers in the field to rely heavily on the trivial names for everyday use. Specialists in each field have developed shorthand notation to suit their needs.

Because of this specialization in nomenclature, the approach in this chapter differs from earlier ones. Except for the section on carbohydrates, which covers ideas of considerable importance in biochemistry, you'll find few questions or exercises. Instead, only fundamentals and a few examples of the rules will be cited — enough, hopefully, for you to recognize the easier names and structures and to understand the more complex when you see them. If you become a serious student of natural products, the references given in the appendix will provide additional information.

5.1 CARBOHYDRATES

Carbohydrates are compounds of carbon, hydrogen, and oxygen in which the hydrogen and oxygen are ordinarily present in the same ratio as in water, 2 to 1. The term carbohydrate stems from this fact. Many carbohydrates have the empirical formula $C_nH_{2n}O_n$. Plants synthesize them from CO_2 and H_2O in the process called photosynthesis. Sugar, starch, and cellulose are all carbohydrates. Sugars, besides being a foodstuff, are important parts of all living organisms. Starch is the principal constituent of seeds and grain, while cellulose is the main constituent of wood.

Structurally, carbohydrates can be considered as hydroxylated aldehydes and ketones. Those with three to nine carbon atoms are called monosaccharides. If more than one unit is condensed through an acetal linkage, the carbohydrate may be a disaccharide (two units), trisaccharide or, in general, a polysaccharide. Saccharides are further classified as aldoses or ketoses, depending upon whether an aldehyde or ketone group is effectively or potentially present. The monosaccharides are classified according to the number of carbon atoms present as tetroses (4), pentoses (5), hexoses (6), etc.

Which of the following Fischer projections represents a ketohexose?

A _____

$$
\begin{array}{c}
\text{CHO} \\
| \\
\text{H} - \text{C} - \text{OH} \\
| \\
\text{HO} - \text{C} - \text{H} \\
| \\
\text{H} - \text{C} - \text{OH} \\
| \\
\text{H} - \text{C} - \text{OH} \\
| \\
\text{CH}_2\text{OH}
\end{array}
$$

B _____

$$
\begin{array}{c}
\text{CH}_2\text{OH} \\
| \\
\text{HO} - \text{C} - \text{H} \\
| \\
\text{C} = \text{O} \\
| \\
\text{HO} - \text{C} - \text{H} \\
| \\
\text{H} - \text{C} - \text{OH} \\
| \\
\text{CH}_2\text{OH}
\end{array}
$$

C _____

$$
\begin{array}{c}
\text{CH}_2\text{OH} \\
| \\
\text{C} = \text{O} \\
| \\
\text{H} - \text{C} - \text{OH} \\
| \\
\text{H} - \text{C} - \text{OH} \\
| \\
\text{CH}_2\text{OH}
\end{array}
$$

• •

A _____

Incorrect. The Fischer projection that you chose does represent a hexose, but it's an aldohexose, not a ketohexose. It has a $- \overset{\text{O}}{\overset{||}{\text{C}}} - \text{H}$ group, not a $- \overset{\text{O}}{\overset{||}{\text{C}}} -$ group. Go back and choose another answer.

B _____

Right. You chose a Fischer projection with six carbon atoms and a ketone group, $- \overset{\text{O}}{\overset{||}{\text{C}}} -$. Go on to the next section.

C _____

Incorrect. Your choice is a ketose, all right, but not a ketohexose. It has only five carbon atoms and is a ketopentose. Go back and select another answer.

Here are two Fischer projections for the simple sugar, D-glucose. The first is the acyclic representation and the second is the cyclic pyranose (hemiacetal) form.

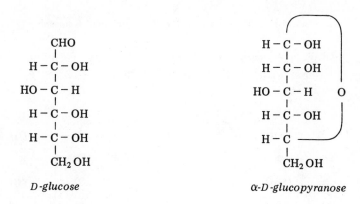

D-glucose   α-D-glucopyranose

The IUPAC rules for carbohydrate nomenclature are designed to name first a parent monosaccharide represented in the Fischer projection of the acyclic form and then to name its cyclic forms and derivatives. We'll study the acyclic forms first. Because all carbohydrates have at least one chiral center, proper use of the Fischer projections is absolutely essential.

Glucose is an aldohexose and, since it is not easily degraded to a simpler carbohydrate, a monosaccharide. Nearly all monosaccharides have three, four, five, or six carbon atoms.

The great organic chemist Emil Fischer laid down the rules for representing the various carbohydrate stereoisomers. The Fischer projection is named in his honor. First, write the chain of carbon atoms vertically with the $-CHO$ group, if present, at the top. If a $-\overset{\overset{O}{\|}}{C}-$ group is present, it should be as near the top as possible. Next, in your mind's eye, bring the $-CH_2OH$ group at the bottom up *behind* the chain until it touches the top end of the molecule. (If you have a set of molecular models with tetrahedral angles, the chain will automatically make the ring for you.)

Now you have a somewhat rigid ring. The H's and OH's on each of the chiral centers will be directed outward (toward your eyes) and to the left or right. They are indicated to the left or right of the C in the Fischer projection. Figure 5.1 attempts to illustrate the procedure.

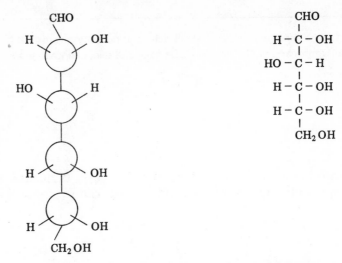

Figure 5.1. Ball-and-stick and Fischer projection of D-glucose.

Some authors use what are known as "cross formulas" instead of conventional Fischer projections. In them, intersecting lines replace the C to represent a carbon atom.

```
   CHO                        CHO
    |
H - C - OH              H ——————OH
    |
HO - C - H             HO ——————H
    |
H - C - OH              H ——————OH
    |
H - C - OH              H ——————OH
    |
  CH₂OH                     CH₂OH
```

Figure 5.2. Fischer projection and "cross formula" for D-glucose.

Recall that horizontal bonds come out; vertical bonds go back. The IUPAC rules use the Fischer projections to avoid any possible ambiguity. Since neither is a chiral center, − CHO and − CH₂ OH groups are usually written on a single line to avoid confusion with the chiral centers.

Let's see if you have the hang of it. Which of the following is the correct Fischer projection for this monosaccharide?

A _____

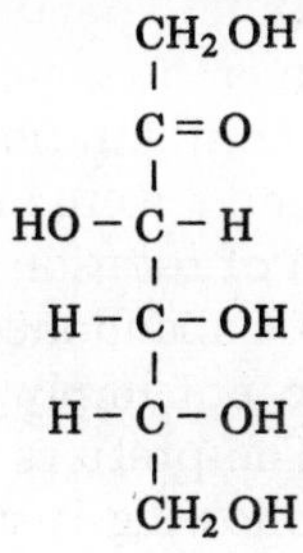

$$CH_2\,OH$$
$$|$$
$$C = O$$
$$|$$
$$HO - C - H$$
$$|$$
$$H - C - OH$$
$$|$$
$$H - C - OH$$
$$|$$
$$CH_2\,OH$$

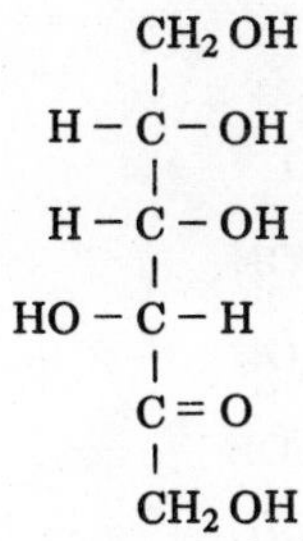

B _____

$$CH_2\,OH$$
$$|$$
$$H - C - OH$$
$$|$$
$$H - C - OH$$
$$|$$
$$HO - C - H$$
$$|$$
$$C = O$$
$$|$$
$$CH_2\,OH$$

C _____

$$CH_2\,OH$$
$$|$$
$$C = O$$
$$|$$
$$H - C - OH$$
$$|$$
$$HO - C - H$$
$$|$$
$$HO - C - H$$
$$|$$
$$CH_2\,OH$$

• •

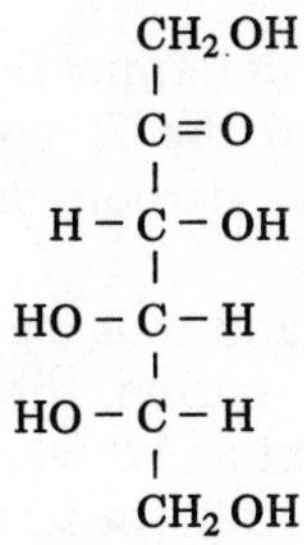

A _____

You are right. The projection you have chosen correctly represents the monosaccharide shown. The ketone group is nearer the top. Go on to the next section.

B _____

Incorrect. While the projection you have chosen correctly portrays the configuration of the chiral centers, it has the ketone group nearer the bottom than the top. Go back and choose another projection.

C _____

You are wrong. The projection you chose has the carbon chain correctly represented with the ketone group nearer the top, but the configurations of the chiral centers are backwards. Take another look at Figure 5.1 before you choose another answer.

While the (R)-(S) system is the only truly general method for describing configuration, the descriptors D and L are used in the nomenclature of carbohydrates to indicate the configuration of the chiral center nearest the bottom of the Fischer projection. The D configuration is assigned to the stereoisomer whose configuration is the same as the glyceraldehyde isomer with positive optical rotation. The L configuration represents the stereoisomer with the opposite rotation. These isomers were originally designated by the prefixes *d* and *l*, referring to the direction of rotation. Because of the confusion that has arisen, the use of *d* and *l* should be abandoned.

The designations D and L do not imply a direct relationship with the sign (direction) of optical rotation. The prefixes are based solely on configuration. For our present purposes, the following is sufficient: if the − OH on the chiral center nearest the bottom is written on the right, the prefix is D. If it's written on the left, the prefix is L. For example:

CHO	CHO	CHO	CHO
H − C − OH	H − C − OH	HO − C − H	HO − C − H
CH₂OH	H − C − OH	CH₂OH	HO − C − H
	CH₂OH		CH₂OH
D-glyceraldehyde	*D-erythrose*	*L-glyceraldehyde*	*L-erythrose*

As mentioned earlier (p. 91) determination of absolute configuration has shown D-glyceraldehyde to have the *R* configuration. It could properly be called (*R*)-glyceraldehyde, but carbohydrate chemists continue to use D-glyceraldehyde.

The relationship between *R* and D is a general one for monosaccharides. If a Fischer projection is correctly written, the configuration of a chiral center is *R* when the − OH lies to the right and *S* when the − OH lies to the left. It does not hold for all derivatives (see p. 292 for tartaric acid). For example:

$$CHO$$
$$H − C − OH$$
$$H − C − OH$$
$$CH_2OH$$

D-erythrose
(2R,3R)-2,3,4-trihydroxybutanal

The principle for numbering monosaccharides gives the carbonyl or potential carbonyl group the lower of the possible locants. If it's a − CHO group, there's no problem − it's number 1. If it's a $-\overset{\displaystyle O}{\overset{\|}{C}}-$ group, start numbering from the nearer end. There's a rule for breaking ties, but we'll skip it for now.

Now, what about this *potential* carbonyl group? Monosaccharides are frequently written in a cyclic form. The carbonyl group and one of the hydroxy groups may react to form a cyclic hemiacetal or hemiketal.

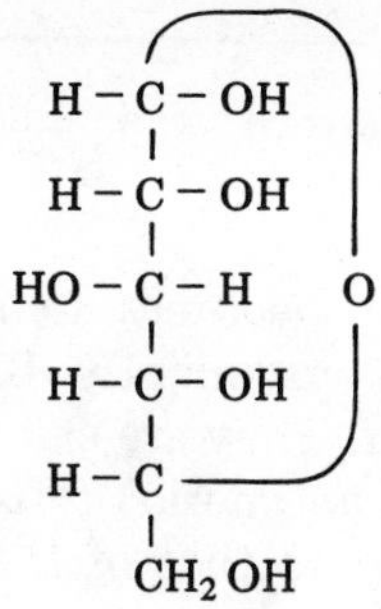

$$H - C - OH$$
$$H - C - OH$$
$$HO - C - H \qquad O$$
$$H - C - OH$$
$$H - C$$
$$CH_2 OH$$

hemiacetal

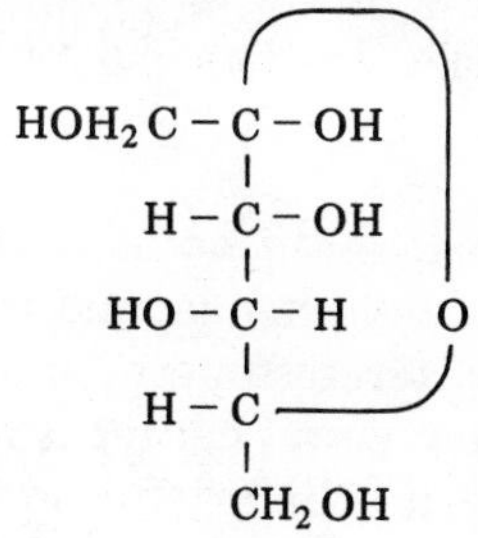

$$HOH_2 C - C - OH$$
$$H - C - OH$$
$$HO - C - H \qquad O$$
$$H - C$$
$$CH_2 OH$$

hemiketal

The potential carbonyl carbon is clearly next to the oxygen in the ring structure. But which one? It's the one with the − OH group on it.

Another, but not the last, term you need to know is the reference carbon atom. It is the highest numbered chiral center in the carbon chain.

Consider this Fischer projection. Which of the following statements is true?

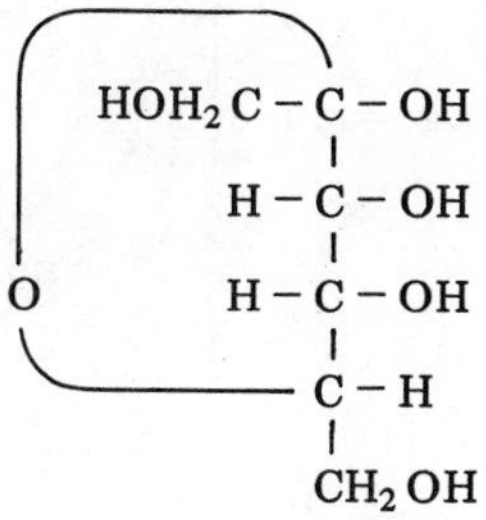

$$HOH_2 C - C - OH$$
$$H - C - OH$$
$$O \qquad H - C - OH$$
$$C - H$$
$$CH_2 OH$$

A _____

Carbonyl carbon is number 1; reference carbon is number 5.

B _____

Carbonyl carbon is number 2; reference carbon is number 6.

C _____

Carbonyl carbon is number 2; reference carbon is number 5.

• •

A _____

Incorrect. You correctly identified the reference carbon but missed the carbonyl carbon. Look carefully and you'll see that carbon number 1 is not part of the ring structure. It can't be the potential carbonyl group. Choose another answer.

B _____

You are wrong. You picked the number 2 carbon as the potential carbonyl group all right, but missed the reference carbon. Remember that the reference carbon is the highest numbered *chiral center.* Go back and choose another answer.

C _____

Right. The carbonyl carbon is number 2 and the reference carbon is number
5.

You may have noticed that the carbonyl carbon becomes a chiral center when the cyclic hemiacetal or hemiketal is formed. Consequently, there can be two cyclic stereoisomers for each acyclic stereoisomer. They are called *anomers* and the new chiral center is the *anomeric center*. An anomeric prefix, α or β, is used to relate the orientation of the anomeric center to that of the reference carbon.

The designation α is applied to the anomer in which the $-OH$ groups on the anomeric center and reference carbon atom are on the same side of the carbon chain in the Fischer projection. If they are on opposite sides, the anomer is designated β. The anomeric prefixes (α or β) are used only in conjunction with the configurational prefixes (D or L) and with the relationship defined here. Some examples to illustrate the point are shown in Figure 5.3.

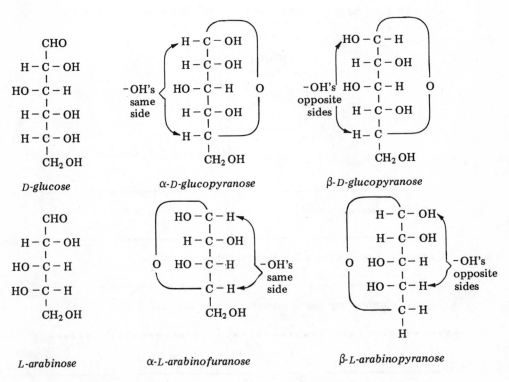

Figure 5.3. Relationship of D-glucose and L-arabinose to their α and β anomers.

The cyclic forms of monosaccharides nearly always form rings with five or six members (including the oxygen atom) and less often with seven. There are two ways to indicate the ring size. In the first, the terminal letters *se* of the acyclic form are replaced by *furanose* for the five-atom ring, *pyranose* for the six-atom ring, and *septanose* for the seven-atom ring. Furanose rings are nearly planar; a chair conformer of the pyranose ring is the most stable. Furanose and pyranose rings are shown in Figure 5.3.

The second method of indicating ring size uses locants. They denote the two carbon atoms to which the ring oxygen is attached. The potential carbonyl carbon is cited first. The numerals are separated by a comma, placed in parentheses, and joined to the end of the name of the acyclic compound by a hyphen. Here are three examples, one of each ring size:

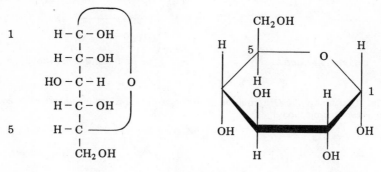

β-L-arabinofuranose	β-D-fructopyranose	α-D-glucoseptanose
or	or	or
β-L-arabinose-(1,4)	β-D-fructose-(2,6)	α-D-glucose-(1,6)

It is worth noting that the difference between the locants is 3 for furanoses, 4 for pyranoses, and 5 for septanoses.

As you can see in Figure 5.3, cyclic structures are represented in the Fischer projection by drawing the open-chain form and then constructing the ring. The ring is drawn on the same side of the carbon chain as the − OH group involved in ring formation.

While the Fischer projections are more than adequate for relating structures to names of monosaccharides and their derivatives, they don't give us much of an idea of the actual shape of the cyclic forms. For this reason the Haworth representation has become popular. In the Haworth projection a monosaccharide is depicted with the carbon chain horizontal and with the potential carbonyl group to the right. The carbon chain projects in front of the plane of the paper and the oxygen bridge is behind.

Here are the Fischer and Haworth projections for α-D-glucopyranose side by side:

α-D-glucopyranose

In the Haworth representation the ring is in a plane approximately perpendicular to the plane of the paper and the groups attached to the carbon atoms of the ring project away from the center of the ring and are either above or below the plane of the ring. The carbon atoms of the ring are shown merely by the intersections of lines representing the bonds in the ring.

Let's compare Fischer and Haworth representations. Remember the conventions that the carbonyl group is nearer the top of the Fischer projection and that the ring oxygen atom is to the right and behind the plane of the paper in the Haworth projection. Thus groups which appear to the right in the Fischer projection appear below the ring in the Haworth projection except for the highest numbered carbon linked to the ring oxygen (at carbon 5 in the example above). To preserve its configuration, the groups must be reversed: left-hand group below the ring.

There is a simple rule for assigning the anomeric designation α or β in Haworth projections. Check first to verify that the ring oxygen atom is in the proper place. If so, the anomeric center will be at the right end of the projection. For members of the D family the $-OH$ group on the anomeric center will be below the ring for the α anomer and above the ring for the β anomer. When members of the L family are encountered, the opposite holds true: the $-OH$ group on the anomeric center is above the ring for the α anomer and below it for the β anomer. The Haworth projections in Figure 5.4 illustrate this point.

Figure 5.4. Orientation of $-OH$ group on anomeric center for α and β anomers of D- and L-monosaccharides in Haworth projections.

You should recall the enantiomers are molecules that are mirror images of one another. Notice that the α-D and α-L forms are enantiomers, as are the β-D and β-L forms. This may seem odd until one remembers that the anomeric prefix defines a relationship between the configuration of two carbon atoms within a molecule. Reversing the configurations of all the chiral centers in changing from the D family to the L family (see p. 281) does not alter the anomeric relationship.

Galactose is an aldohexose. Which of the following pairs of projections might both correctly represent β-L-galactopyranose?

A _____

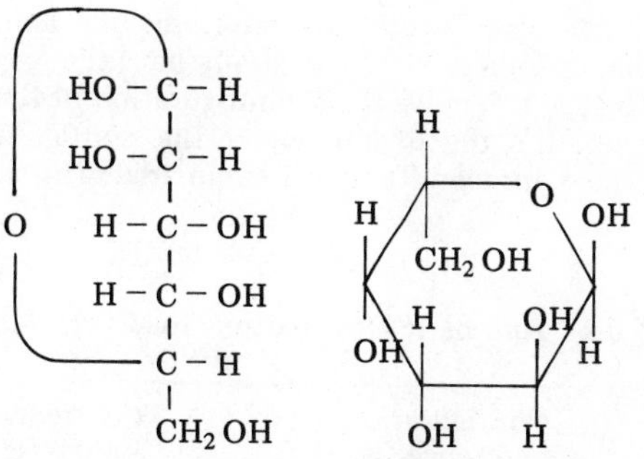

B _____

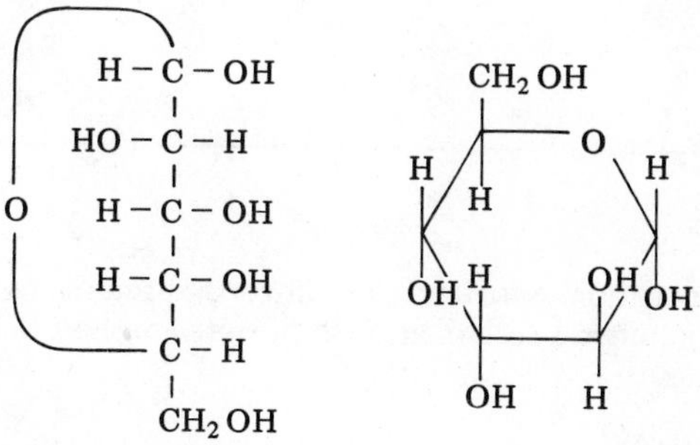

C _____

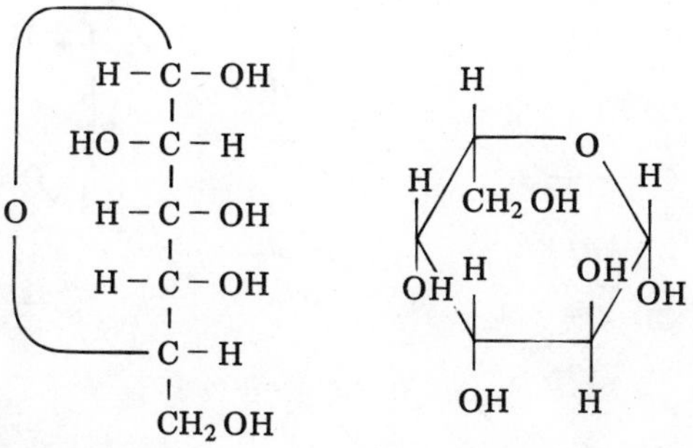

• •

A _____

You are incorrect. Both of these projections represent α-L-galactopyranose. Notice that in the Fischer projection the −OH groups on the anomeric and reference carbons are on the same side of the carbon chain. For the β isomer they must be on opposite sides. Go back and choose another answer.

B _____

Wrong. The Fischer projection is all right, but the Haworth is not. Look at the configuration of the reference carbon (5). It is D. The Haworth projection is α-D-galactopyranose. Pick another answer.

C _____

　　　Right. Both projections correctly depict β-L-galactopyranose. Go on to the next section.

　　　The sequence-rule can also be used to determine the anomeric prefix, but it's a little tedious. You know that members of the D family have the R configuration at the reference carbon while it is S for the L family. If the configuration of the anomeric center is the same as the reference carbon, it's the β anomer. If the configuration is the opposite of the reference carbon, it's the α anomer. Table 5.1 summarizes the situation.

TABLE 5.1　Anomeric Designations for Cyclic Monosaccharides

Configurational Prefix	Configuration of Reference Carbon	Configuration of Anomeric Center	Anomeric Prefix
D	R	R	β
D	R	S	α
L	S	R	α
L	S	S	β

　　　A minor complication arises in the Haworth projections for the furanose forms of hexoses when the number 4 carbon atom is D. Here are the Fischer and Haworth projections for α-D-glucofuranose:

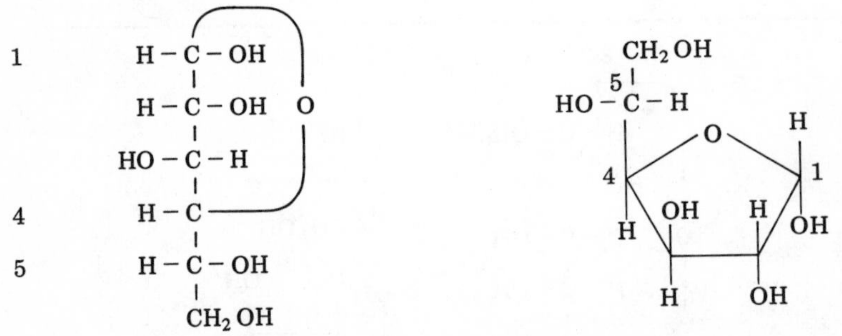

α-D-glucofuranose

　　　In order to preserve the D configuration of the reference carbon atom (number 5), the − H and − OH must be reversed in the Haworth projection.

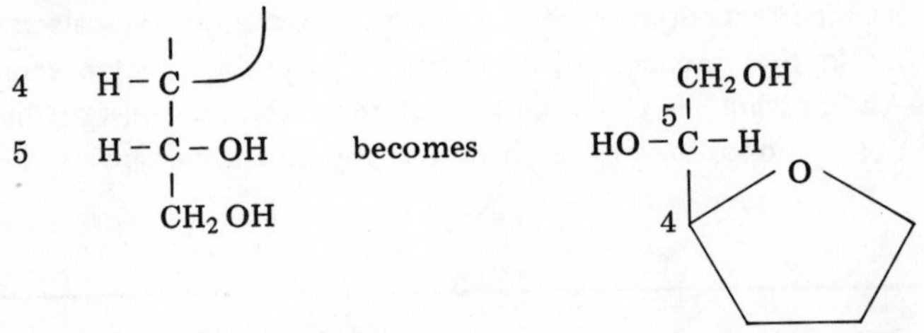

The previously described procedure works perfectly when the number 4 carbon is L.

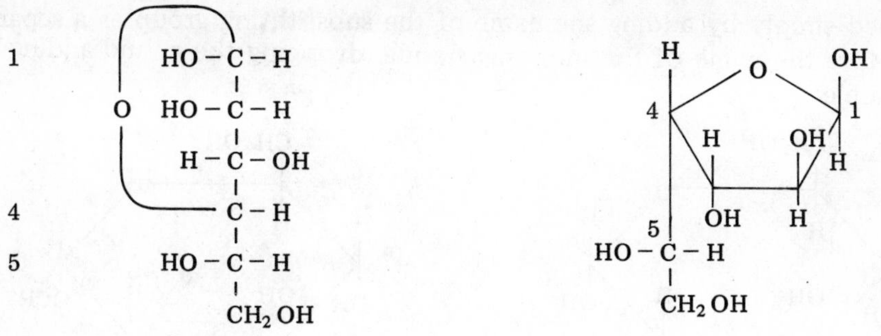

α-L-glucofuranose

Because of the importance of molecular shape to chemical and biological activity, many authors represent six-membered rings with perspective outlines like this:

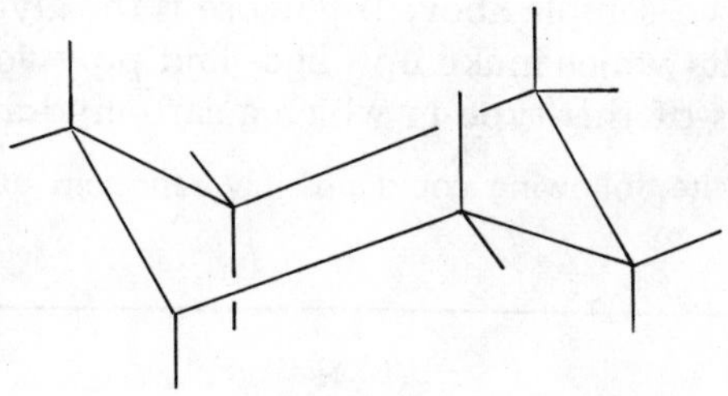

You can see that there is one bond oriented generally upward and one oriented generally downward from each ring member. Substituting O for the right rear carbon atom and following the up-down pattern of the Haworth projection gives a correct perspective formula. For example:

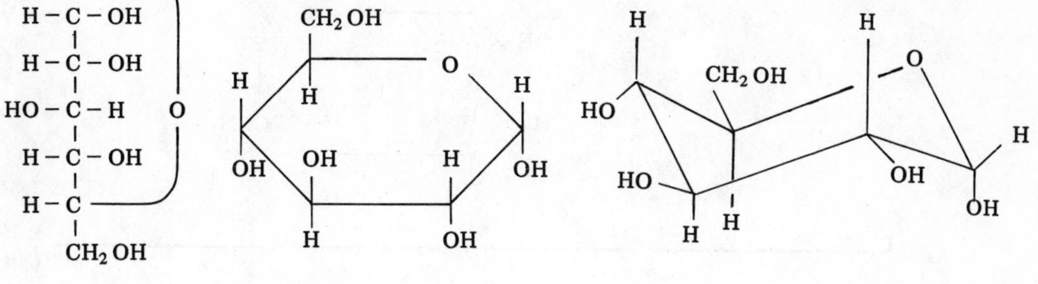

α-D-glucopyranose

The most common derivatives of monosaccharides are *glycosides*. They often result from the replacement of the hydrogen on the anomeric hydroxy group with a substituting group derived from an alcohol or phenol. Glycosides are named simply by adding the name of the substituting group as a separate word before the name of the monosaccharide, dropping the *e*, and adding *ide*. For example:

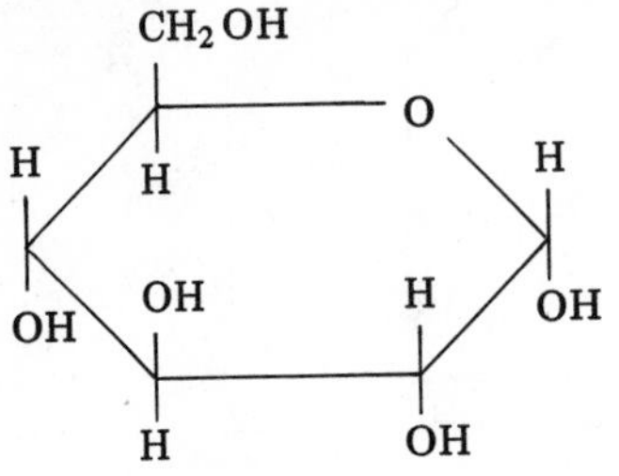

α-D-glucopyranose 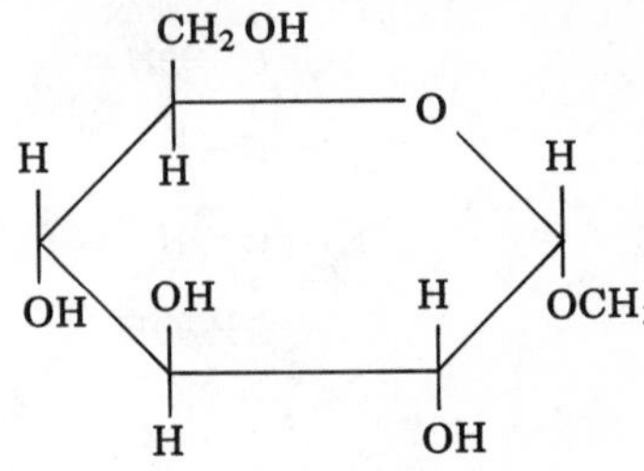   methyl α-D-glucopyranoside

Glycosides containing five-member rings are furanosides; those with six-member rings are pyranosides. The carbohydrate portion of a glycoside is termed the *glycone* and the alcohol or other non-carbohydrate portion is called the *aglycone*. In the sample above D-glucose is the glycone and methanol is the *aglycone*. The units which make up oligo- and polysaccharides are joined with glycosidic linkages of this type in which a carbohydrate serves as the aglycone.

Which of the following could possibly represent ethyl β-L-galactopyrano-side?

A _____

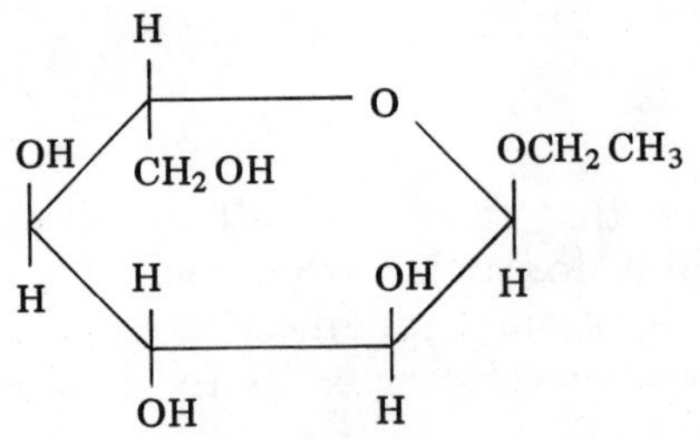

B _____

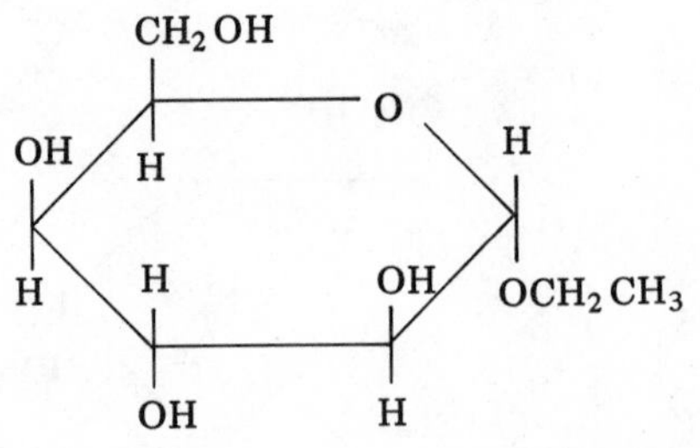

C _____

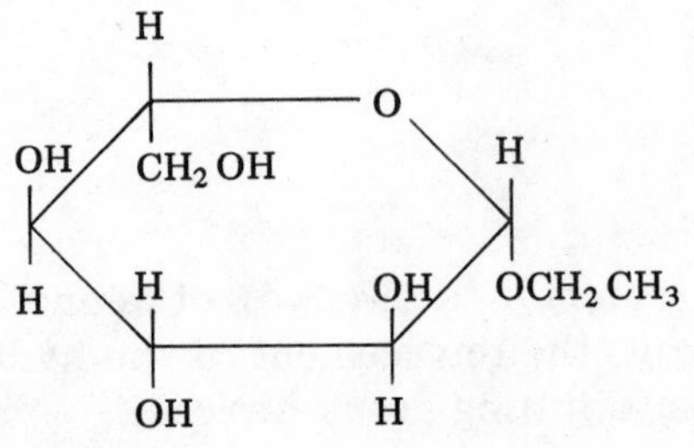

●●●

A _____

Incorrect. The configuration of the reference carbon is indeed L, but the anomeric center on your choice is α. Choose another answer.

B _____

You are wrong. There are two carbon atoms whose configurations are important. The reference carbon must be L and the anomeric center must be β. In this projection they are not. Look it over carefully and then choose another answer.

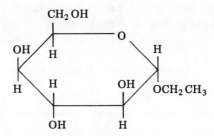

ethyl α-D-galactopyranoside

C _____

Right. You correctly recognized the L configuration of the reference carbon and the β configuration of the anomeric center. Go on to the next section.

So far we've covered the configurational prefix, anomeric prefix, and ring size. You should be wondering how to tell if a given projection represents glucopyranose, galactopyranose, or something else. That's the next step: learning to recognize the stem name for a monosaccharide. The next few paragraphs will discuss this point.

Both the IUPAC and *Chemical Abstracts* recognize a number of trivial names for monosaccharide stereoisomers. The two differ in some minor respects in the formation of systematic names. The differences will be pointed out.

As we begin to examine the naming of aldoses in more detail, you will recognize many of the trivial names that have been in use for a long time.

The terms aldose and ketose are used to denote monosaccharides in which the carbonyl group, or potential carbonyl group, is terminal or non-terminal. The carbonyl carbon is number 1 in aldoses and has the lower possible number in ketoses.

Under the IUPAC rules the trivial names of the acyclic aldoses with three, four, five, or six carbon atoms are preferred to the systematic names. *Chemical Abstracts* uses the trivial names for those with five or six, but indexes those with three or four as hydroxylated aldehydes. The systematic names for aldoses use a stem name as shown in Table 5.2.

TABLE 5.2 Systematic Names for Aldoses

Number of Carbons	Stem Name	Preferred Names
3	triose	IUPAC — glyceraldehyde; CA — 2,3-dihydroxypropanal
4	tetrose	IUPAC — erythrose, threose; CA — 2,3-dihydroxybutanal
5	pentose	arabinose, lyxose, ribose, xylose
6	hexose	allose, altrose, galactose, glucose, gulose, idose, mannose, talose
7	heptose	
8	octose	
	etc.	

The use of a specific trivial name depends on the configurations of the chiral centers. We'll get to that in a moment.

Which of the following might possibly be the Fischer projection for D-arabinose?

A ─────────────────────────────────

$$
\begin{array}{c}
\text{CHO} \\
|\\
\text{HO} - \text{C} - \text{H} \\
|\\
\text{H} - \text{C} - \text{OH} \\
|\\
\text{H} - \text{C} - \text{OH} \\
|\\
\text{CH}_2\text{OH}
\end{array}
$$

B ─────────────────────────────────

$$
\begin{array}{c}
\text{CHO} \\
|\\
\text{H} - \text{C} - \text{OH} \\
|\\
\text{HO} - \text{C} - \text{H} \\
|\\
\text{HO} - \text{C} - \text{H} \\
|\\
\text{CH}_2\text{OH}
\end{array}
$$

C ─────────────────────────────────

$$
\begin{array}{c}
\text{CH}_2\text{OH} \\
|\\
\text{C} = \text{O} \\
|\\
\text{H} - \text{C} - \text{OH} \\
|\\
\text{H} - \text{C} - \text{OH} \\
|\\
\text{CH}_2\text{OH}
\end{array}
$$

• •

A ─────────────────────────────────

Right. The Fischer projection represents a D aldose with five carbon atoms and is D-arabinose. Go on to the next section.

B _____

Incorrect. The Fischer projection you have selected does represent an aldopentose, but it is an L-aldopentose. Go back and select another answer.

C _____

You are incorrect. D-Arabinose is an aldopentose. The projection you have chosen represents a ketopentose. Go back and carefully select another answer.

The configuration of one to four contiguous $>$CHOH groups, or derivative groups such as $>$CHOCH$_3$ or $>$CHNH$_2$, is designated by one of the following configurational prefixes which, except for *glycero*, are derived from the trivial names listed in Table 5.2. When used in systematic names the prefixes are uncapitalized, italicized, and affixed with a hyphen to the proper stem name. As we'll see later, there may be more than one such prefix in a name.

The Fischer projections of the D-prefixes are shown in Table 5.3; X is the group containing the lowest numbered carbon atom.

TABLE 5.3   Configurations of Aldoses

One $>$CHOH Group	Two $>$CHOH Groups	
H–C–OH (X above, Y below) *D-glycero-*	X / H–C–OH / H–C–OH / Y *D-erythro-*	X / HO–C–H / H–C–OH / Y *D-threo-*

Three $>$CHOH Groups			
X / HO–C–H / H–C–OH / H–C–OH / Y *D-arabino-*	X / HO–C–H / HO–C–H / H–C–OH / Y *D-lyxo-*	X / H–C–OH / H–C–OH / H–C–OH / Y *D-ribo-*	X / H–C–OH / HO–C–H / H–C–OH / Y *D-xylo-*

Four $\rangle$ CHOH *Groups*

X	X	X	X
H – C – OH	HO – C – H	H – C – OH	H – C – OH
H – C – OH	H – C – OH	HO – C – H	HO – C – H
H – C – OH	H – C – OH	HO – C – H	H – C – OH
H – C – OH	H – C – OH	H – C – OH	H – C – OH
Y	Y	Y	Y
D-allo-	*D-altro-*	*D-galacto-*	*D-gluco-*

X	X	X	X
H – C – OH	HO – C – H	HO – C – H	HO – C – H
H – C – OH	H – C – OH	HO – C – H	HO – C – H
HO – C – H	HO – C – H	H – C – OH	HO – C – H
H – C – OH	H – C – OH	H – C – OH	H – C – OH
Y	Y	Y	Y
D-gulo-	*D-ido-*	*D-manno-*	*D-talo-*

The systematic name for an aldose is formed by using the configurational prefix, the appropriate stem, and the ending -*ose*. (For convenience and clarity the acyclic forms will be used. The endings for cyclic forms are furanose, pyranose, septanose, etc.) One tetrose, one pentose, and one hexose will serve as examples:

		CHO
		H – C – OH
	CHO	HO – C – H
CHO	H – C – OH	H – C – OH
HO – C – H	H – C – OH	HO – C – H
H – C – OH	H – C – OH	H – C – OH
CH$_2$OH	CH$_2$OH	CH$_2$OH

Trivial name (preferred):	*D-threose*	*D-ribose*	*D-galactose*
Systematic name:	*D-threo-tetrose*	*D-ribo-pentose*	*D-galacto-hexose*

Members of the L family have the opposite configuration at *every one* of the chiral centers ($\rangle$CHOH groups). Compare the two mannoses:

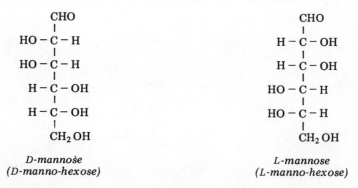

	CHO			CHO
	HO – C – H			H – C – OH
	HO – C – H			H – C – OH
	H – C – OH			HO – C – H
	H – C – OH			HO – C – H
	CH_2 OH			CH_2 OH

D-mannose   *L-mannose*
(D-manno-hexose)   *(L-manno-hexose)*

Can you put it all together? What is the trivial name for this aldose?

CHO
H – C – OH
H – C – OH
HO – C – H
CH_2 OH

A _____

L-lyxose

B _____

D-lyxose

C _____

L-gulose

• •

A _____

Right. The configurations of all three chiral centers are opposite from those shown in Table 5.3 for D-lyxose. The projection correctly represents L-lyxose. Go on to the next section.

B _____

You are wrong. Look at the projection again. You'll see that the configuration of the reference carbon (number 4) is L. Comparison with Table 5.3 will show that it has the opposite configuration from D-lyxose at every chiral center. It is L-lyxose. When you are sure you understand, go on to the next section.

C _____

Incorrect. The projection represents an aldose with three $\rangle$ CHOH groups. According to Table 5.3, L-gulose should have four of them. Choose another answer.

Stereoisomers that differ only in the configuration of one chiral center are called *epimers*. For example, D-ribose and D-xylose are epimers since they differ only in the configuration of the number 3 carbon atom.

```
        CHO                              CHO
         |                                |
    H – C – OH                       H – C – OH
         |                                |
    H – C – OH                      HO – C – H
         |                                |
    H – C – OH                       H – C – OH
         |                                |
       CH₂OH                           CH₂OH

      D-ribose                        D-xylose
```

It follows, of course, that L-ribose and L-xylose are epimers also.

Usage since 1952 in the United States and the United Kingdom has attached a different significance to the stem names. They have been related to a series of consecutive, but not necessarily contiguous, $\angle$CHOH groups. All three of these, for example, are D-*arabino:*

```
                                                         X
         X                      X                        |
         |                      |                    HO – C – H
     HO – C – H            HO – C – H                     |
         |                      |                         N
         N                  H – C – OH                    |
         |                      |                    H – C – OH
     H – C – OH                 N                         |
         |                      |                         M
     H – C – OH             H – C – OH                    |
         |                      |                    H – C – OH
         Y                      Y                         |
                                                          Y
```

N and M may represent a single non-chiral carbon atom or a sequence of non-chiral carbon atoms. N or M might be a methylene group, $-CH_2-$, in a deoxy compound or a carbonyl group, $-\overset{\overset{\textstyle O}{\|}}{C}-$, in a ketose. Names formed according to this usage will appear in the next few paragraphs. Now let's move on to discuss ketoses.

It shouldn't surprise you to learn that monosaccharides with the $-\overset{\overset{\textstyle O}{\|}}{C}-$ group are called *ketoses*. They are further classified as 2-ketoses, 3-ketoses, etc., according to the position of the carbonyl or potential carbonyl group. For example:

```
    CH₂OH        HOH₂C – C – OH                CH₂OH          CH₂OH
     |                  |                       |          HO – C – H
    C = O          H – C – OH              HO – C – H       HO – C
     |                  |                       |               |
  H – C – OH       H – C – OH   O            C = O         H – C – OH   O
     |                  |                       |               |
  H – C – OH       H – C                    H – C – OH      H – C – OH
     |                  |                       |               |
    CH₂OH              H                     H – C – OH      H – C
                                                 |               |
                                               CH₂OH            H

    a 2-ketose                                      a 3-ketose
```

Remember that the potential carbonyl group is the carbon atom that has the hydroxy group and is adjacent to the ring oxygen. When no ambiguity can arise, the locant 2 may be omitted from the name of a 2-ketose, but it's usually a good idea to retain it.

The stem names for the acyclic ketoses are given in this list:

No. of C Atoms	Stem Name
4	tetrulose
5	pentulose
6	hexulose
7	heptulose
8	octulose
9	nonulose
10	deculose
etc.	etc.

There are a number of trivial names so well established by use that they are retained. Four are shown here:

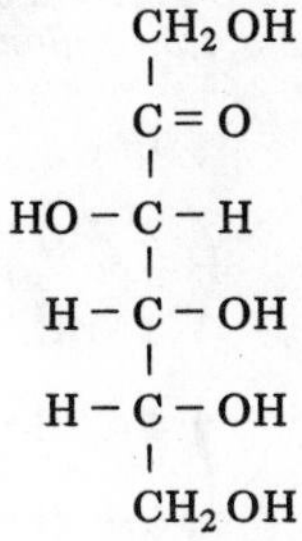

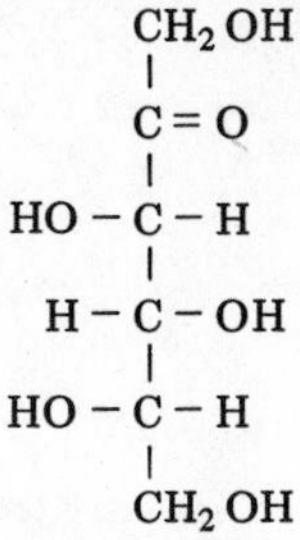

```
        CH2 OH                  CH2 OH
          |                       |
        C = O                   C = O
          |                       |
   HO - C - H              HO - C - H
          |                       |
    H - C - OH              H - C - OH
          |                       |
    H - C - OH             HO - C - H
          |                       |
        CH2 OH                  CH2 OH
```

Trivial name: *D-fructose*   *L-sorbose*
Systematic name: *D-arabino-2-hexulose*   *L-xylo-2-hexulose*

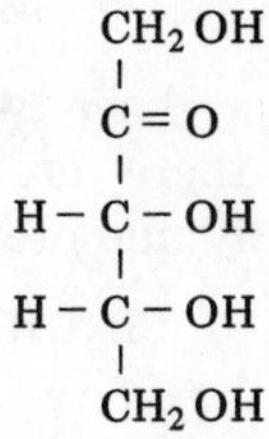

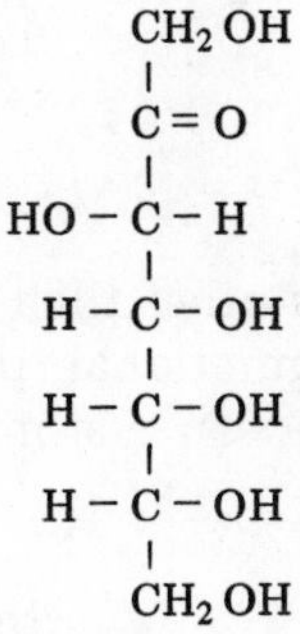

```
                                CH2 OH
                                  |
                                C = O
                                  |
        CH2 OH             HO - C - H
          |                       |
        C = O               H - C - OH
          |                       |
    H - C - OH              H - C - OH
          |                       |
    H - C - OH              H - C - OH
          |                       |
        CH2 OH                  CH2 OH
```

Trivial name: *D-ribulose*   *D-sedoheptulose*
Systematic name: *D-erythro-2-pentulose*   *D-altro-2-heptulose*

D-Fructose is the most abundant ketose. It is found in honey and fruit juices. Although it does not occur naturally, L-sorbose is an important intermediate for the commerical synthesis of vitamin C. D-Ribulose and D-sedoheptulose play a part in the conversion of carbon dioxide and water to carbohydrates by plants.

Occasionally two configurational prefixes are required for a compound that may be an aldose, a ketose that is not a 2-ketose, or a diketose. It may or may not have more than four chiral centers. The same configurational prefixes are utilized.

Let's look first at an aldose with more than four chiral centers. Beginning with the chiral center next to the $-$CHO group, the chain is divided into as many groups of four as possible, perhaps with some left over. Each group is assigned a prefix from the list. The order of citation in the name is from the end farthest from the $-$CHO group. Locants may be inserted if desired. If they are, all are given and they immediately precede the configurational prefix. Here are two examples taken directly from the IUPAC rules:

$$
\begin{array}{rl}
1 & \mathrm{CHO} \\
2 & \mathrm{H-C-OH} \\
3 & \mathrm{HO-C-H} \\
4 & \mathrm{H-C-OH} \\
5 & \mathrm{H-C-OH} \\
6 & \mathrm{H-C-OH} \\
7 & \mathrm{CH_2OH}
\end{array}
$$

D-gluco { 2, 3, 4, 5 }
D-glycero { 6 }

D-glycero-D-gluco-heptose
6-D-glycero-2,3,4,5,-D-gluco-heptose

$$
\begin{array}{rl}
1 & \mathrm{CHO} \\
2 & \mathrm{HO-C-H} \\
3 & \mathrm{HO-C-H} \\
4 & \mathrm{H-C-OH} \\
5 & \mathrm{H-C-OH} \\
6 & \mathrm{HO-C-H} \\
7 & \mathrm{HO-C-H} \\
8 & \mathrm{HO-C-H} \\
9 & \mathrm{CH_2OH}
\end{array}
$$

D-manno { 2, 3, 4, 5 }
L-ribo { 6, 7, 8 }

L-ribo-D-manno-nonose
6,7,8-L-ribo-2,3,4,5,-D-manno-nonose

Ketoses other than 2-ketoses may have two acceptable names. One will use two configurational prefixes. If there are four or fewer chiral centers, Anglo-American custom is to use only one prefix. An example is:

$$
\begin{array}{rl}
& \mathrm{CH_2OH} \\
& \mathrm{HO-C-H} \\
& \mathrm{C=O} \\
& \mathrm{H-C-OH} \\
& \mathrm{H-C-OH} \\
& \mathrm{CH_2OH}
\end{array}
$$

L-glycero { HO-C-H }
D-erythro { H-C-OH, H-C-OH }
D-arabino

Systematic name: *D-erythro-L-glycero-3-hexulose*
Anglo-American name: *D-arabino-3-hexulose*

Diketoses or their derivatives containing two ketonic carbonyl groups, actual or potential, are named by replacing *ulose* with *odiulose*. The smallest possible locants are used and inserted together with a hyphen between the configurational prefix and the stem name. The configurational prefixes used depend upon whether the carbonyl groups are adjacent, and possibly upon your nationality. Here are some examples:

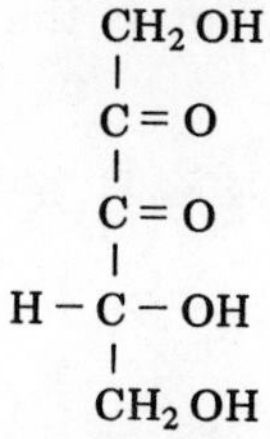

$$
\begin{array}{l}
CH_2\,OH \\
| \\
C = O \\
| \\
C = O \\
| \\
H - C - OH \\
| \\
CH_2\,OH
\end{array}
\qquad \text{Easy.}
$$

D-glycero-2,3-pentodiulose

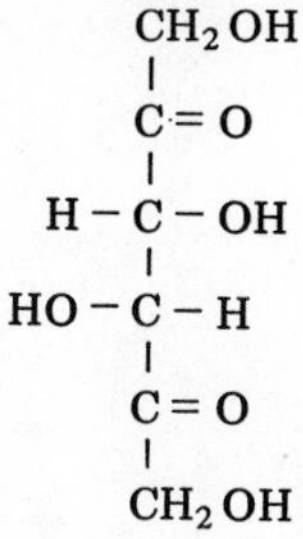

$$
\begin{array}{l}
CH_2\,OH \\
| \\
C = O \\
| \\
H - C - OH \\
| \\
HO - C - H \\
| \\
C = O \\
| \\
CH_2\,OH
\end{array}
\qquad \text{Number the chain from either end.}
$$

L-threo-2,5-hexodiulose

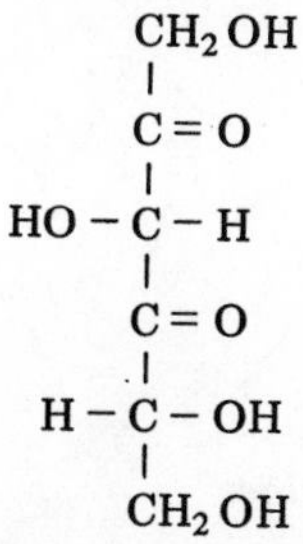

$$
\begin{array}{l}
CH_2\,OH \\
| \\
C = O \\
| \\
HO - C - H \\
| \\
C = O \\
| \\
H - C - OH \\
| \\
CH_2\,OH
\end{array}
\qquad
\begin{array}{l}
\text{Remember to cite the configuration} \\
\text{prefixes from the end opposite the} \\
\text{number 1 carbon.}
\end{array}
$$

Systematic name: *D-glycero-L-glycero-2,4-hexodiulose*
Anglo-American name: *D-threo-2,4-hexodiulose*

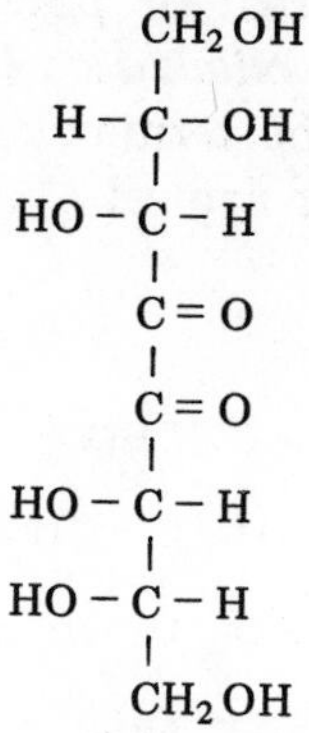

$$
\begin{array}{l}
CH_2\,OH \\
| \\
H - C - OH \\
| \\
HO - C - H \\
| \\
C = O \\
| \\
C = O \\
| \\
HO - C - H \\
| \\
HO - C - H \\
| \\
CH_2\,OH
\end{array}
$$

Systematic name: *L-erythro-L-threo-4,5-octodiulose*
Anglo-American name: *L-altro-4,5-octodiulose*

L-*Threo*-D-*erythro*-4,5-octodiulose might seem a proper name for the last example. It has the same set of locants. The correct name, however, lists the prefixes in alphabetical order.

Aldoketoses and their derivatives have a $-CHO$ group and a $-\overset{\overset{O}{\parallel}}{C}-$ group, actual or potential. They are named in the same manner as diketoses using the termination *osulose*. The $-CHO$ group is always number 1; the locant is not cited in the name. If the locant of the $-\overset{\overset{O}{\parallel}}{C}-$ group is 2, it may be omitted. Examples are:

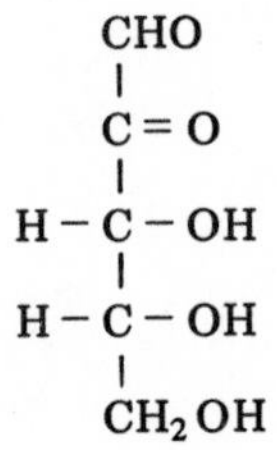

$$
\begin{array}{c}
CHO \\
| \\
C = O \\
| \\
H-C-OH \\
| \\
H-C-OH \\
| \\
CH_2OH
\end{array}
$$

D-erythro-2-pentosulose

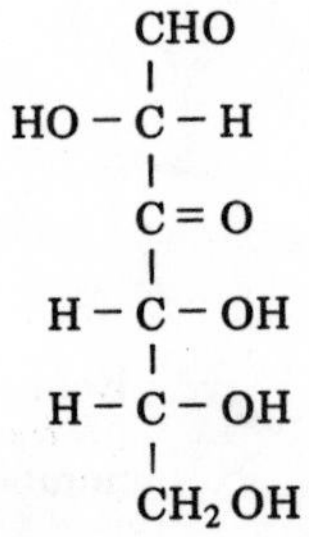

$$
\begin{array}{c}
CHO \\
| \\
HO-C-H \\
| \\
C = O \\
| \\
H-C-OH \\
| \\
H-C-OH \\
| \\
CH_2OH
\end{array}
$$

D-erythro-L-glycero-3-hexosulose
Anglo-American: *D-arabino-3-hexosulose*

Finally, the dialdoses. These are monosaccharides or their derivatives containing two $-CHO$ groups, real or potential. Names are formed as for diketoses with the termination *odialdose*. Locants are not needed for the $-CHO$ groups since they must be terminal. Here are three examples, two of which have special aspects to their names:

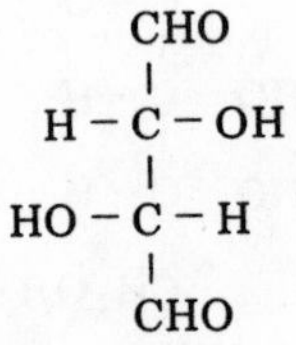

$$
\begin{array}{c}
CHO \\
| \\
H-C-OH \\
| \\
HO-C-H \\
| \\
CHO
\end{array}
$$

L-threo-tetrodialdose

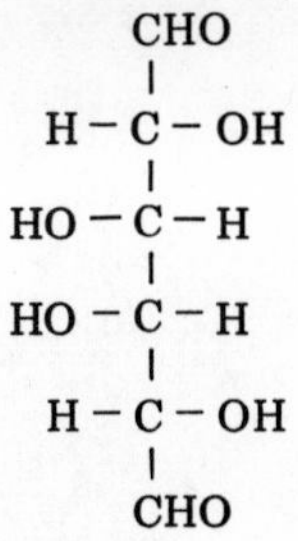

```
            CHO
             |
       H  -  C  -  OH
             |
   HO  -  C  -  H
             |
   HO  -  C  -  H
             |
       H  -  C  -  OH
             |
            CHO
```

meso-galacto-hexodialdose

Configuration is *D-galacto* from one end and *L-galacto* from the other. The compound has a plane of symmetry and *meso-* is used to make this clear.

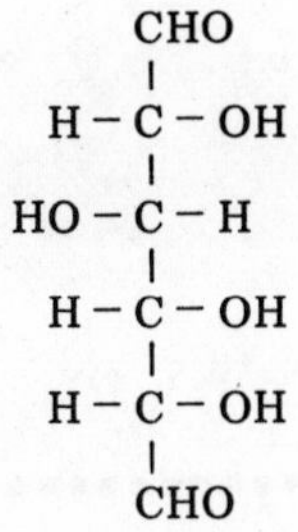

```
            CHO
             |
       H  -  C  -  OH
             |
   HO  -  C  -  H
             |
       H  -  C  -  OH
             |
       H  -  C  -  OH
             |
            CHO
```

D-gluco-hexodialdose

Numbered from the top, this would be *L-gulo*-hexodialdose. The other name is preferred because *gluco-* precedes *gulo-* alphabetically.

Now it's time for you to exercise your brain. Here are three Fischer projections and three names. Work out systematic names for the former and Fischer projections for the latter. Check your answers on the next page.

1.
```
            CHO
             |
       H  -  C  -  OH
             |
             C  =  O
             |
   HO  -  C  -  H
             |
   HO  -  C  -  H
             |
            CH₂OH
```

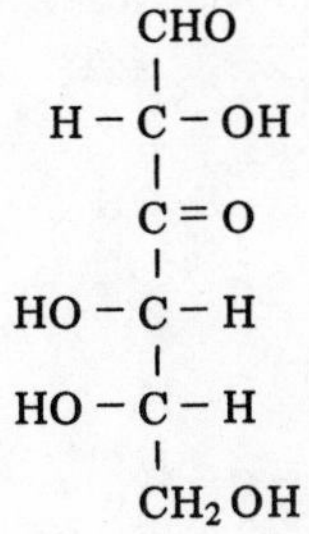

2.
```
            CHO
             |
   HO  -  C  -  H
             |
       H  -  C  -  OH
             |
   HO  -  C  -  H
             |
       H  -  C  -  OH
             |
   HO  -  C  -  H
             |
       H  -  C  -  OH
             |
            CH₂OH
```

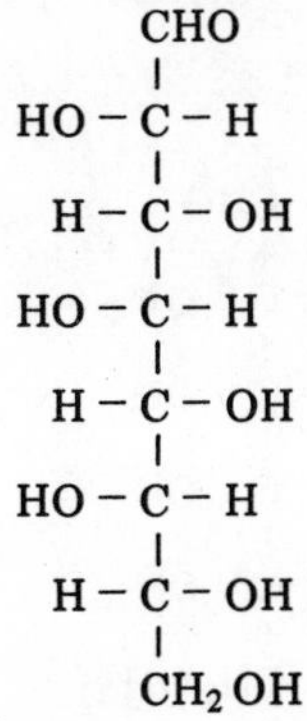

3.

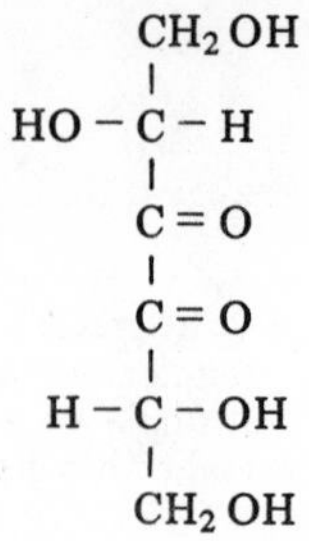

```
            CH₂OH
             |
    HO ─ C ─ H
             |
            C = O
             |
            C = O
             |
    H ─ C ─ OH
             |
            CH₂OH
```

4. L-*ribo*-3-hexulose (Anglo-American)

or

L-*erythro*-L-*glycero*-3-hexulose

5. D-*allo*-hexose

or

D-allose (trivial name)

6. L-*glycero*-D-*gulo*-dialdose

••

1. L-*erythro*-D-*glycero*-3-hexosulose

or

L-*arabino*-3-hexosulose (Anglo-American)

2. D-*threo*-D-*ido*-octose

or

6,7-D-*threo*-2,3,4,5-D-*ido*-octose

3. D-*glycero*-L-*glycero*-3,4-hexodiulose

or

D-*threo*-3,4-hexodiulose (Anglo-American)

4.

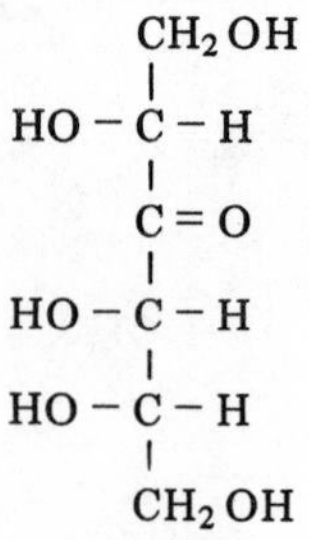

```
            CH₂OH
             |
    HO ─ C ─ H
             |
            C = O
             |
    HO ─ C ─ H
             |
    HO ─ C ─ H
             |
            CH₂OH
```

5.

```
            CHO
             |
    H ─ C ─ OH
             |
    H ─ C ─ OH
             |
    H ─ C ─ OH
             |
    H ─ C ─ OH
             |
            CH₂OH
```

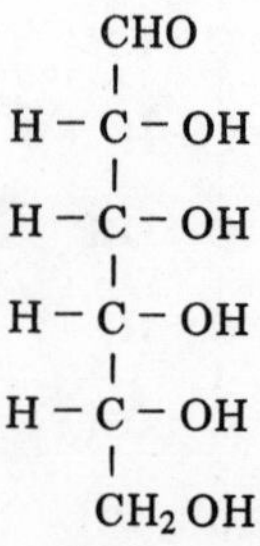

6.

$$
\begin{array}{c}
CHO \\
| \\
H-C-OH \\
| \\
H-C-OH \\
| \\
HO-C-H \\
| \\
H-C-OH \\
| \\
HO-C-H \\
| \\
CHO
\end{array}
$$

Derivatives of carbohydrates are important in biochemistry. The next few paragraphs will discuss some of them. The treatment is by no means exhaustive, but should serve to alert you to some of the possibilities.

The replacement of an alcoholic hydroxy group of a monosaccharide or derivative by a hydrogen atom is denoted by the prefix *deoxy*. It is preceded by a numerical locant and followed by a hyphen and a systematic or trivial name. The systematic name consists of a stem and appropriate configurational prefixes. A trivial name may be used only if the transformation to the deoxy compound does not alter the configuration of any chiral center. For instance:

$$
\begin{array}{c}
CHO \\
| \\
HO-C-H \\
| \\
H-C-OH \\
| \\
H-C-OH \\
| \\
HO-C-H \\
| \\
CH_3
\end{array}
\qquad\qquad
\begin{array}{c}
CHO \\
| \\
H-C-OH \\
| \\
H-C-OH \\
| \\
HO-C-H \\
| \\
HO-C-H \\
| \\
CH_3
\end{array}
$$

$\qquad\qquad$ *L-fucose* $\qquad\qquad\qquad\qquad$ *L-rhamnose*
$\qquad$ *(6-deoxy-L-galactose)* $\qquad\qquad\quad$ *(6-deoxy-L-mannose)*

L-Rhamnose is the most common naturally occurring deoxy sugar. L-Fucose is one of the products of the hydrolysis of the cell walls of marine algae and certain carbohydrates of animal origin.

A special trivial name has been established for 2-deoxy-D-*erythro*-pentose. It is *deoxyribose*. Deoxyribose is a hydrolytic product of deoxyribonucleic acids, *DNA*, that are present in chromosomes. It may also be called 2-deoxy-D-ribose.

$$
\begin{array}{c}
CHO \\
| \\
CH_2 \\
| \\
H-C-OH \\
| \\
H-C-OH \\
| \\
CH_2OH
\end{array}
$$

2-deoxy-D-erythro-pentose
(deoxyribose)

Amino-monosaccharides or amino sugars are formed by the replacement of an alcoholic hydroxy group by an amino group. For naming purposes the replacement is viewed as the substitution of an amino group for the appropriate hydrogen of the deoxy compound. The carbon atom remains a chiral center and is included in the configurational prefix. Several have accepted trivial names. D-Glucosamine and D-galactosamine are constituents of polysaccharides, brain glycosides, and glycoproteins. Other amino sugars have been isolated from the hydrolysis products of fungal and bacterial metabolites that often have antibiotic properties. N-methyl-L-glucosamine is obtained from streptomycin and 3-amino-3-deoxy-D-ribose from puromycin.

```
        CHO                            H   CHO
         |                             |    |
   H – C – NH2                  CH3 – N – C – H
         |                                  |
  HO – C – H                         H – C – OH
         |                                  |
   H – C – OH                       HO – C – H
         |                                  |
   H – C – OH                       HO – C – H
         |                                  |
      CH2 OH                            CH2 OH
```

D-glucosamine   *N-methyl-L-glucosamine*
(2-amino-2-deoxy-D-glucose)   *(2-deoxy-2-(methylamino)-L-glucose)*

```
        CHO
         |
   H – C – NH2                        CHO
         |                             |
  HO – C – H                     H – C – OH
         |                             |
  HO – C – H                     H – C – NH2
         |                             |
   H – C – OH                    H – C – OH
         |                             |
      CH2 OH                        CH2 OH
```

D-galactosamine   *3-amino-3-deoxy-D-ribose*
(2-amino-2-deoxy-D-galactose)

     Carbohydrates are frequently subject to oxidation and reduction in biochemical processes. Some of the resulting compounds will be treated in the next few paragraphs.

     Reduction of the carbonyl group in an aldose or ketose leads to a polyhydric alcohol called an *alditol*. Names for the alditols are derived by changing the ending *ose* to *itol*. If the same alditol can be derived from two different aldoses, the preferred name is the one which comes first in alphabetical order.

     D-Glucitol, which has the non-preferred trivial name D-sorbitol, is found in many plants from algae to the higher orders. D-Mannitol occurs in many plants and, unlike D-glucitol, is present in plant exudates known as mannas. Galactitol is present also in many plants and plant exudates.

```
        CH₂OH              CH₂OH              CH₂OH
         |                  |                  |
    H – C – OH        HO – C – H         H – C – OH
         |                  |                  |
   HO – C – H         HO – C – H         HO – C – H
         |                  |                  |
    H – C – OH         H – C – OH        HO – C – H
         |                  |                  |
    H – C – OH         H – C – OH         H – C – OH
         |                  |                  |
        CH₂OH              CH₂OH              CH₂OH

     D-glucitol          D-mannitol         galactitol
    (D-sorbitol)
```

For the sake of clarity, galactitol, which has a plane of symmetry, may be named *meso*-galactitol.

Oxidation of the − CHO group of an aldose to − COOH yields an *aldonic acid*. They have also been known as *glyconic acids*. They may be classified as aldotrionic, etc., depending upon the number of carbon atoms in the chain. Names of individual compounds are formed by replacing -*ose* with -*onic acid*.

D-Gluconic acid is used to inhibit foaming in alkaline cleaning compounds. The common biochemical name for 2-amino-2-deoxy-D-gluconic acid is D-glucosaminic acid.

```
            COOH                         COOH
             |                            |
        H – C – OH                   H – C – NH₂
             |                            |
       HO – C – H                   HO – C – H
             |                            |
        H – C – OH                    H – C – OH
             |                            |
        H – C – OH                    H – C – OH
             |                            |
           CH₂OH                        CH₂OH

        D-gluconic acid       2-amino-2-deoxy-D-gluconic acid
                                   (D-glucosaminic acid)
```

Oxidation of the terminal − CH_2OH group of aldoses leads to *uronic acids*, also called *glycouronic acids*. Names are formed by changing the ending -*ose* to -*uronic acid*. The carbon atom of the carbonyl group is numbered 1, not that of the carboxy group. D-Galactouronic acid is found in the pectins of fruits and berries.

```
            CHO
             |
        H – C – OH
             |
       HO – C – H
             |
       HO – C – H
             |
        H – C – OH
             |
           COOH

     D-galactouronic acid
```

The dicarboxylic acids resulting from the oxidation of both terminal groups of an aldose are called *aldaric acids*. In the past they were known as glycaric acids. Names are formed by changing *-ose* to *-aric acid*. Tartaric acid is an aldaric acid; the trivial names of its three forms are preferred to the systematic names.

$$COOH$$
$$H-C-OH$$
$$HO-C-H$$
$$COOH$$

$$COOH$$
$$HO-C-H$$
$$H-C-OH$$
$$COOH$$

$$COOH$$
$$H-C-OH$$
$$H-C-OH$$
$$COOH$$

L(+)-tartaric acid
(L-threaric acid)
(RR-tartaric acid)

D(−)-tartaric acid
(D-threaric acid)
(SS-tartaric acid)

meso-tartaric acid
(erythraric acid)
(RS-tartaric acid)

Oligosaccharides and polysaccharides are polymers of monosaccharide units joined by glycosidic linkages, most often in the 1,4 or 1,6 positions. Oligosaccharides yield a few monosaccharide molecules on hydrolysis while polysaccharides yield many. The most common monomer among the naturally occurring polymers is D-glucose. Most of the others found in nature are in the D family. While it is possible to assign systematic names to the smaller oligosaccharides, trivial names are generally used.

Chemically, oligosaccharides can be classified into two groups: non-reducing and reducing. As we shall see, the non-reducing sugars do not have an − OH group on any anomeric center while the reducing sugars do have an − OH group on an anomeric center. As the number of monomer units in the oligosaccharide increases, the overall effect of the free − OH group decreases.

Non-reducing sugars are named as glycosyl glycosides and the names have two words. Reducing sugars are named as glycosylglycoses and the names are a single word.

Sucrose, our ordinary table sugar, is a non-reducing disaccharide. This means that it has no free − OH group on an anomeric center and that two monosaccharide molecules are produced by the hydrolysis of sucrose. The Haworth projection for sucrose is

sucrose

In this and succeeding projections the anomeric centers will be shown as C in the ring.

Splitting the glycosidic linkage between the two rings yields these two structures:

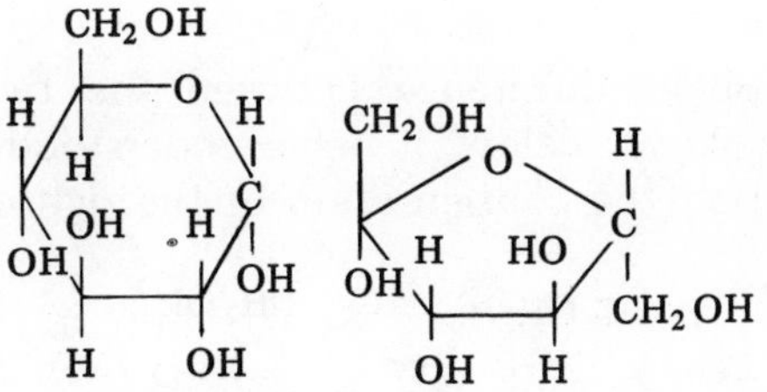

CH$_2$OH

CH$_2$OH

The one on the left can be identified as α-D-glucopyranose. The other must be reoriented to place its anomeric center on the right if we are to examine it easily.

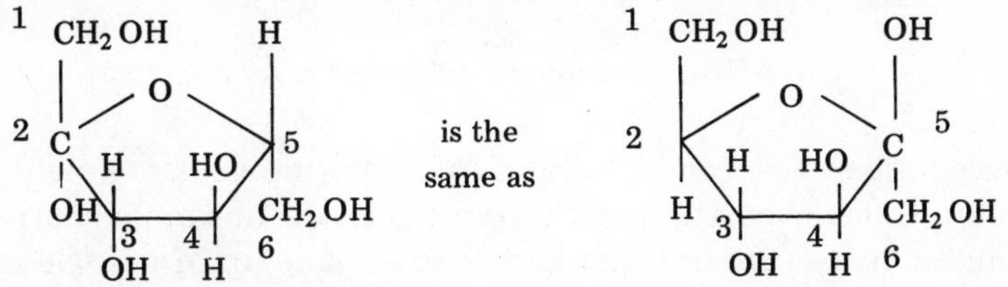

is the same as

It can now be identified as β-D-fructofuranoside. Since *fructo* precedes *gluco* alphabetically, the correct systematic name for sucrose is β-D-fructofuranosyl α-D-glucopyranoside.

Maltose is a reducing disaccharide formed by the enzyme-catalyzed hydrolysis of starch. Hydrolysis of maltose shows it to be composed of two D-glucose units. Other studies have shown that the two glucose units are linked through the 1 position of the non-reducing half and the 4 position of the reducing half of the molecule. This evidence leads to these formulas for α-maltose and β-maltose. They differ only in the configuration of the anomeric center of the reducing half of the molecules.

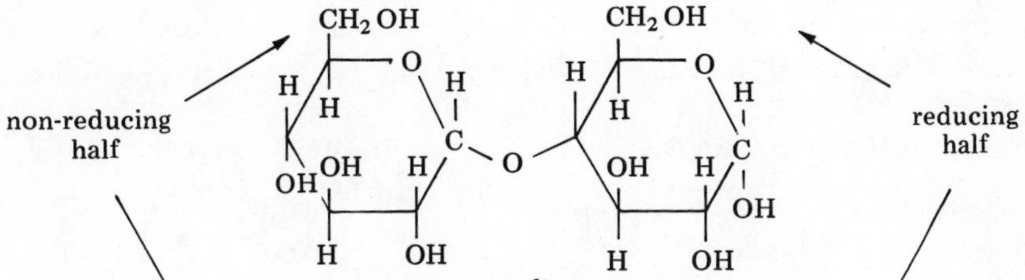

non-reducing half

reducing half

α-maltose
4-O-α-D-glucopyranosyl-α-D-glucopyranose

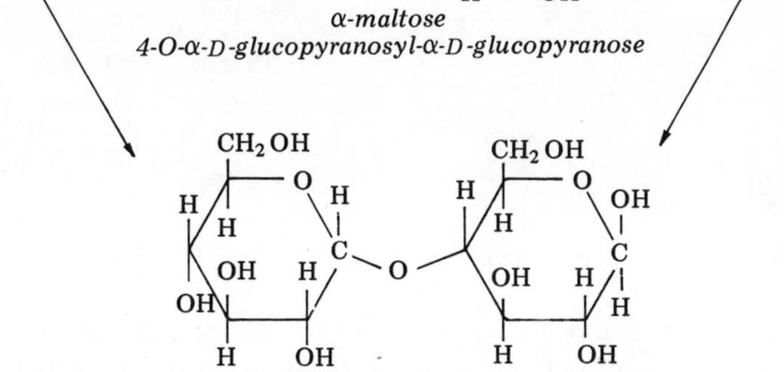

β-maltose
4-O-α-D-glucopyranosyl-β-D-glucopyranose

The systematic names are clearly more descriptive inasmuch as they indicate the point of attachment as well as the ring size, optical family, and configuration of each half. *Chemical Abstracts* uses the systematic names for indexing.

Because of the slight difference between the two isomers, maltose is frequently written as shown below, it being understood that there are α and β isomers depending upon the configuration of the right-hand anomeric center.

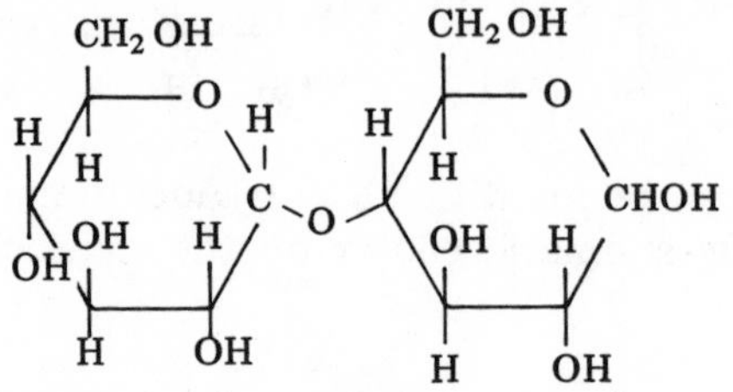

maltose
4-O-α-D-glucopyranosyl-D-glucopyranose

Polysaccharides are carbohydrates of high molecular weight, 30,000 to 400,000,000. Since even the purified products are not molecularly homogeneous, systematic names are not assigned. Those made up of a single monomer are called homopolysaccharides while those that yield two or more monomers on hydrolysis are heteropolysaccharides. Many polysaccharides have well established trivial names. The others are given names ending in *an* and have the generic name *glycans*.

Homopolysaccharides of D-glucose are of considerable biological interest. Starch is the main chemical energy reserve of plants while glycogen serves the same purpose for animals. Cellulose is the major constituent of the supporting structures of plants. Both starch and glycogen are composed of D-glucose units linked by α-1$\rightarrow$4 linkages as in maltose and branches formed by α-1$\rightarrow$6 linkages.

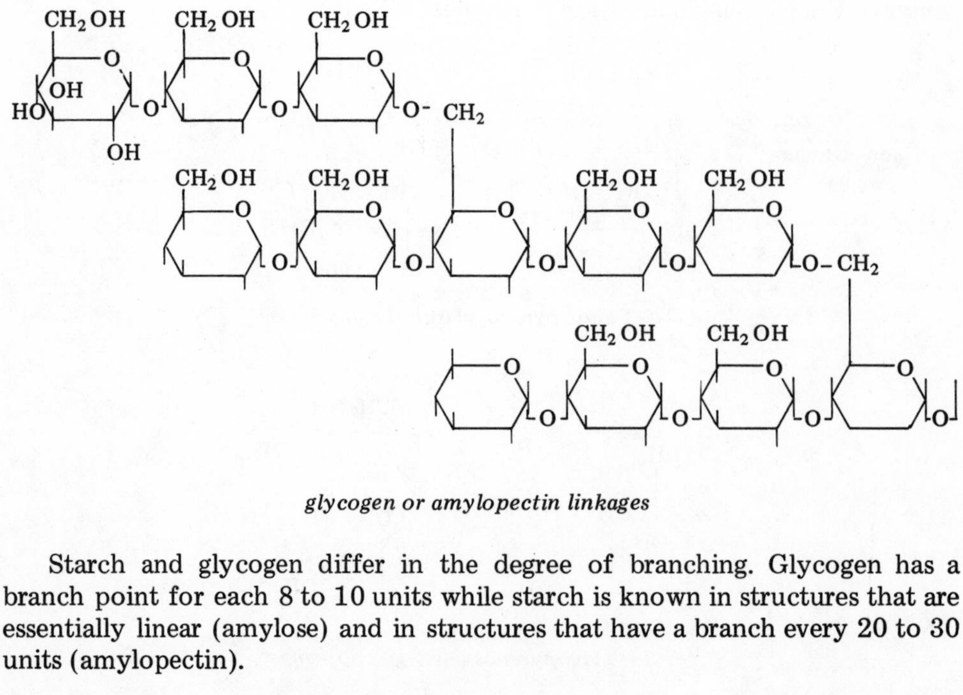

glycogen or amylopectin linkages

Starch and glycogen differ in the degree of branching. Glycogen has a branch point for each 8 to 10 units while starch is known in structures that are essentially linear (amylose) and in structures that have a branch every 20 to 30 units (amylopectin).

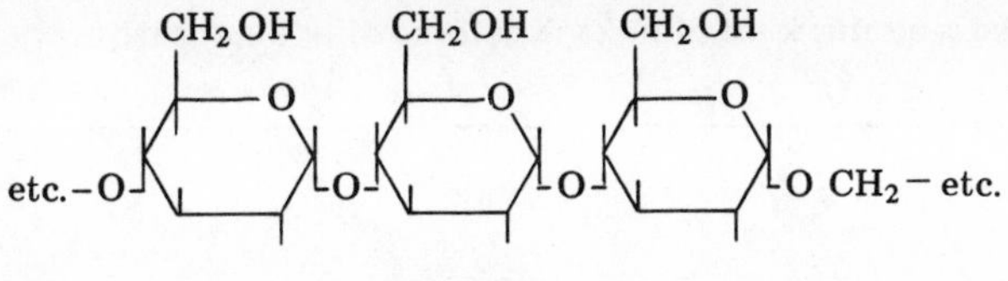

amylose

Cellulose is a linear polymer of D-glucose units joined by α-1→4 linkages.

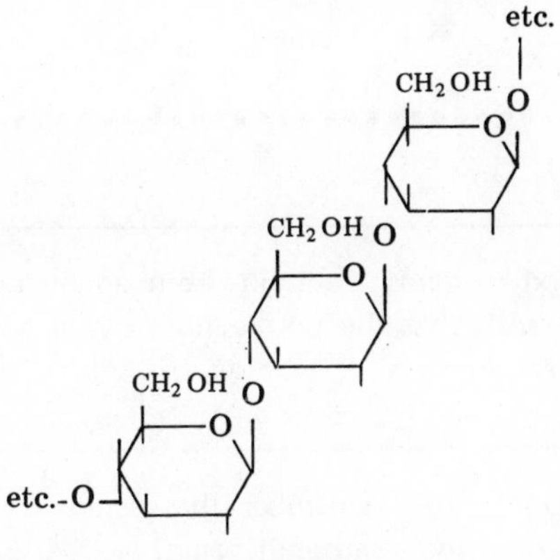

cellulose

The difference in configuration between amylose and cellulose, α-1→4 as opposed to β-1→4 linkage, is responsible for a pronounced difference in physical properties and an even more striking difference in chemical properties. Starch is almost universally acceptable as a foodstuff, but cellulose is digested only by some microorganisms. Even the wood termite depends upon microorganisms in its digestive tract to supply the enzymes necessary to cleave the β-1→4 linkage.

5.2 PEPTIDES AND PROTEINS

Proteins may well be the most important constituents of living systems. All cells and viruses contain proteins, so that no matter how one defines living systems, proteins are present. Muscle fiber, skin, nerves, and blood are largely composed of proteins. Certain hormones, enzymes, and antibodies are proteins.

The proteins are polymers of amino acids (sec. 3.9) joined by peptide bonds. For example:

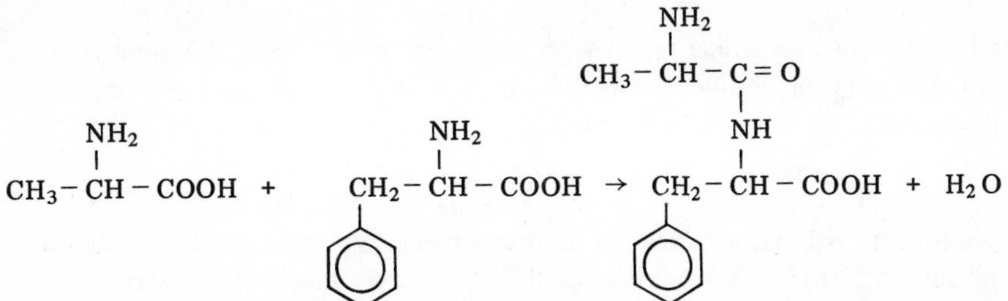

amino acid   +   *amino acid*   →   *peptide*   +   *water*

You have seen this kind of linkage before. What kind is it?

A _____

amide

B _____

ester

C _____

ether

•••

A _____

Right. Peptides and proteins are built from amino acids through amide linkages. You should recall that the textile fiber nylon has the same linkage. Continue to the next section.

B _____

You are wrong. Don't you remember that esters are formed from the reaction between an alcohol and a carboxylic acid? Neither of them has nitrogen in its functional group.

$$R-OH \;+\; HO-\overset{\overset{\displaystyle O}{\|}}{C}-R' \;\rightarrow\; R-O-\overset{\overset{\displaystyle O}{\|}}{C}-R' \;+\; H_2O$$

alcohol   *acid*   *ester*   *water*

Perhaps you need to go back and review Chapter 3. If this was only a careless mistake, go back and choose another answer.

C _____

Don't you remember that the functional group of the ether family is $- O -$? Here is the peptide and protein linkage again.

$$R-\overset{\overset{\displaystyle O}{\|}}{C}-\overset{\overset{\displaystyle H}{|}}{N}-R'$$

The functional group of the linkage is

$$-\overset{\overset{\displaystyle O}{\|}}{C}-\overset{\overset{\displaystyle H}{|}}{N}-$$

It is formed when an amine reacts with a carboxylic acid. Does this spur your memory? If not, you should review Chapter 3. If it does, turn back and choose another answer.

The members of this class of compounds are also called *peptides* or *polypeptides*, and there is no universally accepted distinction between these terms and *protein*. Protein usually designates naturally occurring polypeptides with molecular weights of a few thousand or more. Proteins occasionally have non-amino acid components.

A considerable amount of specialized terminology is used by workers in the life sciences, but there are some conventions that are used by all. Although systematic names for polypeptides are used by *Chemical Abstracts* for indexing, sequence names are in general use. For example:

$$\underset{\text{glycylphenylalanylglycine}}{\overset{\displaystyle \overset{NH_2}{|} \quad \overset{O}{\|} \quad \overset{H}{|} \qquad \overset{O}{\|} \quad \overset{H}{|}}{CH_2 - C - N - CH - C - N - CH_2 - COOH}}$$

Note that the sequence name is formed by adding the prefix names of the amino acid units in sequence from the *N*-terminus (unit with free $- NH_2$ group) to the *C*-terminus (unit with free $-$ COOH group).

Symbols are commonly used to represent not only the names of the amino acids but also their structural formulas. There are two sets of symbols. The first, commonly used for sequence formulas, assigns a three-letter symbol to each acid. The symbol is ordinarily the first three letters of the trivial name. They are printed or typed as one upper case letter followed by two lower case letters, and the L configuration is assumed unless otherwise indicated. The second set of symbols assigns a single capital letter to represent each acid. As you can see from Table 5.4, the single letters are less suggestive of the names.

TABLE 5.4 Common Amino Acids

Alanine	Ala	A	Leucine	Leu	L
Arginine	Arg	R	Lysine	Lys	K
Asparagine	Asn	N	Methionine	Met	M
Aspartic acid	Asp	D	Phenylalanine	Phe	F
Cysteine	Cys	C	Proline	Pro	P
Glutamic acid	Glu	E	Serine	Ser	S
Glutamine	Gln	Q	Threonine	Thr	T
Glycine	Gly	G	Tryptophan	Trp	W
Histidine	His	H	Tyrosine	Tyr	Y
Isoleucine	Ile	I	Valine	Val	V

For consistency and ease of typing as well as economy in printing, the hyphen is the standard connecting symbol and represents the peptide bond. Using glycine as an example, there is only one distinct form for the free amino acid but there are three corresponding residues:

Gly	=	$NH_2 - CH_2 - COOH$	the free amino acid
Gly-	=	$NH_2 - CH_2 - CO -$	the left-hand unit (*N*-terminal)
-Gly-	=	$- NH - CH_2 - CO -$	the non-terminal unit
-Gly	=	$- NH - CH_2 - COOH$	the right-hand unit (*C*-terminal)

Which of the following structural formulas might be represented by Ala-?

A _____

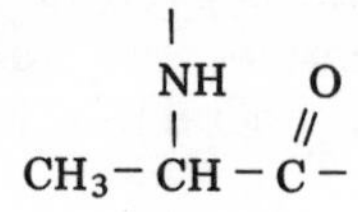

$$CH_3 - \underset{\underset{NH}{|}}{CH} - \overset{\overset{O}{\|}}{C} -$$

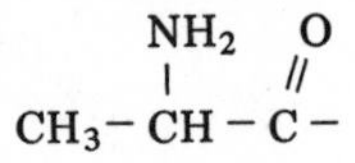

B _____

$$CH_3 - \underset{\underset{NH_2}{|}}{CH} - \overset{\overset{O}{\|}}{C} -$$

C _____

$$CH_3 - \underset{\underset{NH}{|}}{CH} - COOH$$

• •

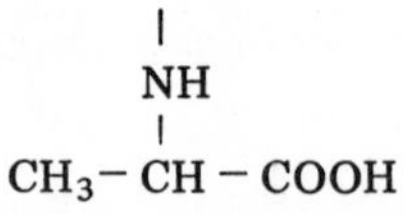

A _____

Incorrect. As you can see from the symbol, Ala- has only one open bond and represents the left-hand unit. The structural formula you have chosen is the middle unit with two open bonds. Go back and choose another answer.

B _____

You are correct. Ala- represents the left-hand unit with the $- NH_2$ group intact and the open bond on the $- \overset{\overset{O}{\|}}{C} -$ group. Go on to the next section.

C _____

You are wrong. The symbol Ala- represents the left-hand unit. It should have the $- NH_2$ group intact. The formula you chose has $- \overset{\overset{H}{|}}{N} -$ and is the right-hand unit. Go back and choose another answer.

Although the sequence formula Gly-Phe-Gly is adequate for the tripeptide shown above, it may be written as H-Gly-Phe-Gly-OH as a means of emphasizing the left-hand and right-hand units.

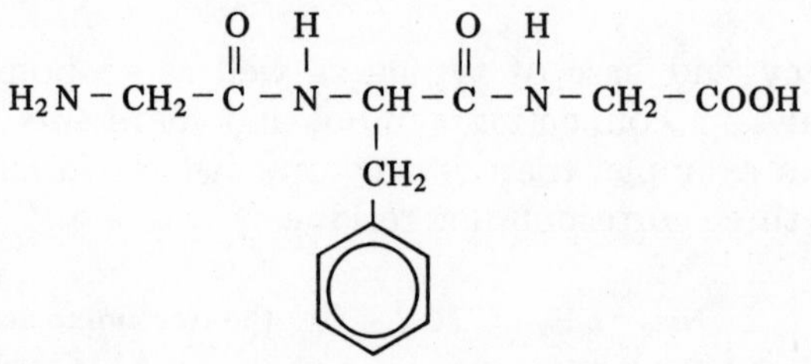

$$H_2N - CH_2 - \overset{\overset{O}{\|}}{C} - \underset{\underset{}{}}{\overset{\overset{H}{|}}{N}} - \underset{\underset{CH_2}{|}}{CH} - \overset{\overset{O}{\|}}{C} - \overset{\overset{H}{|}}{N} - CH_2 - COOH$$

The practice of *Chemical Abstracts* is to index linear peptides of 2 to 12 units at "peptide systematic names." The heading parent is the right-hand unit.

Configurational prefixes (D and L) are included. You should note that none is needed for glycine because it has no chiral center. For example

H-Gly-L-Phe-L-Pro-L-Phe-OH

is named

N-[1-(*N*-glycyl-L-phenylalanyl)-L-prolyl]-L-phenylalanine

and indexed as

L-Phenylalanine, *N*-[1-(*N*-glycyl-L-phenylalanyl)-L-prolyl]-

Look at the name carefully. Notice that the three units enclosed by the square brackets are joined to N on the terminal phenylalanine unit. The two units enclosed by parentheses are joined to position number 1 on the proline ring. The glycine unit is attached to N on the non-terminal phenylalanine unit. Here is the structural formula for your study and comparison:

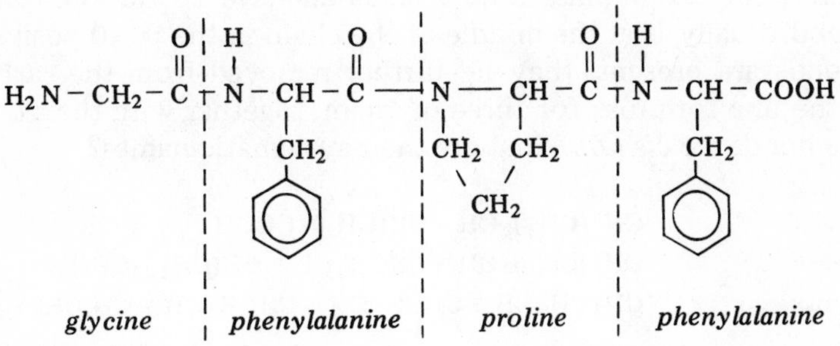

$$H_2N-CH_2-\overset{\overset{\displaystyle O}{\|}}{C} \vdots \overset{\overset{\displaystyle H}{|}}{N}-CH-\overset{\overset{\displaystyle O}{\|}}{C} \longrightarrow N-CH-\overset{\overset{\displaystyle O}{\|}}{C} \vdots \overset{\overset{\displaystyle H}{|}}{N}-CH-COOH$$

glycine · phenylalanine · proline · phenylalanine

Linear peptides with 13 units or more are indexed at "amino-acid-sequence names," similar to the peptide systematic names but with omission of enclosing marks and locants.

Naturally occurring biologically active peptides with five or fewer amino acid units are indexed at peptide systematic names. Those with 6 to 50 units are assigned the trivial names commonly used in the literature, with a cross-reference to the peptide sequence name. Here is an example taken from the *Chemical Abstracts* Index Guide:

H-L-Arg-L-Pro-L-Pro-Gly-L-Phe-L-Ser-L-Pro-L-Phe-L-Arg-OH
 Bradykinin (cross-reference from L-
  Arginine, N^2-[*N*-[1-[*N*-[*N*-
  [*N*-[1-(1-L-arginyl-L-prolyl)-
  L-prolyl]glycyl]-L-
  phenylalanyl]-L-seryl]-L-
  prolyl]-L-phenylalanyl]-)

It's easy to see why those who work in the field prefer the trivial names.

Chemical Abstracts arbitrarily defines proteins as natural peptides containing more than 50 amino acid units. A protein for which the complete amino acid sequence has been reported is indexed as a chemical substance at the trivial name with auxiliary information such as the species name cited in parentheses after the name.

5.3 FATS AND OILS

Fats and oils constitute the third large class of human foodstuffs. They are esters of higher fat (fatty) acids and 1,2,3-propanetriol. The trivial name glycerol is accepted by the IUPAC, but *Chemical Abstracts* uses the systematic name. The difference between fats and oils is that the former are solid or semisolid at ordinary temperatures in their country of origin while oils are liquid at the same temperatures.

Since all fats and oils are esters of glycerol, their differences must be caused by the nature of the acids with which glycerol is esterified. Both saturated and unsaturated acids occur, with the latter predominant in oils. The most important saturated acids are:

dodecanoic acid	lauric acid	$CH_3(CH_2)_{10}COOH$
tetradecanoic acid	myristic acid	$CH_3(CH_2)_{12}COOH$
hexadecanoic acid	palmitic acid	$CH_3(CH_2)_{14}COOH$
octadecanoic acid	stearic acid	$CH_3(CH_2)_{16}COOH$

You will notice that these acids contain an even number of carbon atoms.

The most important unsaturated acids all have 18 carbon atoms and one double bond usually is at the middle of the chain in the 9–10 position. If other double bonds are present, they lie further removed from the carboxy group. Here are the line formulas for three of them, together with their trivial names. All double bonds are *cis (Z)*. What are their systematic names?

oleic acid	$CH_3(CH_2)_7 CH = CH(CH_2)_7 COOH$
linoleic acid	$CH_3(CH_2)_4 CH = CHCH_2 CH = CH(CH_2)_7 COOH$
linolenic acid	$CH_3 CH_2 CH = CHCH_2 CH = CHCH_2 CH = CH(CH_2)_7 COOH$

Their systematic names are:

oleic acid	(Z)-9-octadecenoic acid
linoleic acid	(Z,Z)-9,12-octadecadienoic acid
linolenic acid	(Z,Z,Z)-9,12,15-octadecatrienoic acid

Esters of glycerol (1,2,3-propanetriol) are commonly called glycerides. Their names are formed in the same manner as other esters. For example:

$$CH_3(CH_2)_{14}COOCH_2$$
$$CH_3(CH_2)_{14}COOCH$$
$$CH_3(CH_2)_{14}COOCH_2$$

1,2,3-propanetriyl trihexadecanoate

$$CH_3 CH_2 CH_2 COOCH_2$$
$$CH_3(CH_2)_{15}COOCH$$
$$CH_3 COOCH_2$$

2-(butanoyloxy)-1-[(ethanoyloxy)methyl]-ethyl heptadecanoate

For indexing purposes, *Chemical Abstracts* assumes triesters and lists them under the heading of the acid (e.g., **Acid, hexadecanoic,** esters). Mixed glycerides are listed under the heading of each acid.

Since fats and oils are mixed glycerides and generally do not have exact compositions, they are commonly called by trivial names based on their origin. Palmitic acid is the most abundant fat acid in nature while oleic acid is the most widely distributed.

Some natural products other than fats and oils yield fat acids on hydrolysis. These include waxes, phosphatides, and cholesterol esters (see p. 306). Biochemists group the fat acids and substances that yield them together with other fat-soluble compounds under the general term lipid.

5.4 TERPENES

In the broadest sense, the term terpene includes all compounds isolated from the essential oils of plants whose skeletons can be divided evenly into isopentane (iso-C_5) units. In a more limited sense terpene refers only to compounds containing two of these iso-C_5 units. The broad class, then, includes hemiterpenes, C_5, terpenes, C_{10}, sesquiterpenes, C_{15}, diterpenes, C_{20}, etc. The isopentane units are generally joined in a head-to-tail fashion.

The formulas used to represent terpenes and their derivatives utilize an extension of the common skeleton formulas. Not only is a carbon atom present at the intersection of two or more lines, but also at the end of each line. Symbols for carbon and hydrogen are not used unless particular attention is to be called to them. Unsaturation and atoms other than carbon and hydrogen are always shown. Enough hydrogens are always attached to each carbon atom to satisfy its normal valence. These formulas are called ultimate skeleton formulas. Both common and ultimate skeleton formulas will be shown for the next few examples to let you familiarize yourself with them.

Acyclic terpene hydrocarbons are named in the same manner as other unsaturated acyclic hydrocarbons. For example:

7-methyl-3-methylene-1,6-octadiene

The dashed lines show that the compound contains two iso-C_5 units and is a terpene. You should also notice that each of the double bonds is of the type

$$\underset{a}{\overset{a}{\searrow}} C = C \underset{c}{\overset{b}{\nearrow}}$$. There are no *E-Z* (*cis-trans*) isomers.

Phytol, $C_{20}H_{39}OH$, is an acyclic diterpene that constitutes about one third of the chlorophyll molecule. This alcohol has been isolated from the chlorophyll of more than 200 species of plants. The configuration of both chiral centers has been shown to be R; the double bond is *trans*.

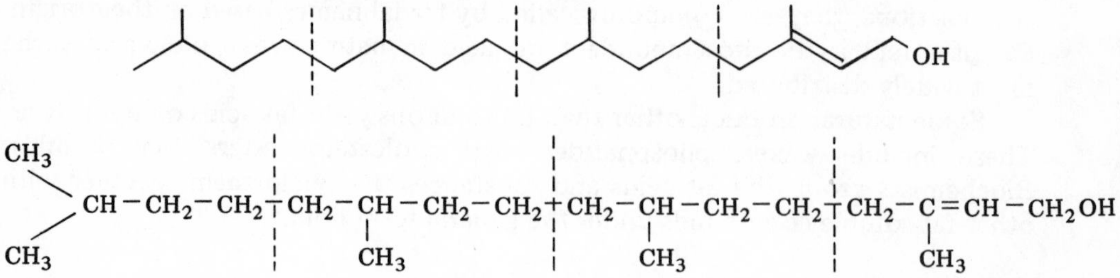

phytol
CA: [R-[R*,R*-(E)]]-3,7,11,15-tetramethyl-2-hexadecen-1-ol

Terpenes are given exceptional treatment in the IUPAC rules because of long-established custom and the fact that the systematic names are frequently unmanageably long and complicated, difficult to say, and conceal the terpene nature of the compounds. Five fundamental structural types and special systems of numbering are used as the basis for naming monocyclic and bicyclic terpene hydrocarbons. They are shown here with both their IUPAC-accepted trivial names and the systematic names used for indexing by *Chemical Abstracts* (cross-referenced to the IUPAC name in the Index Guide).

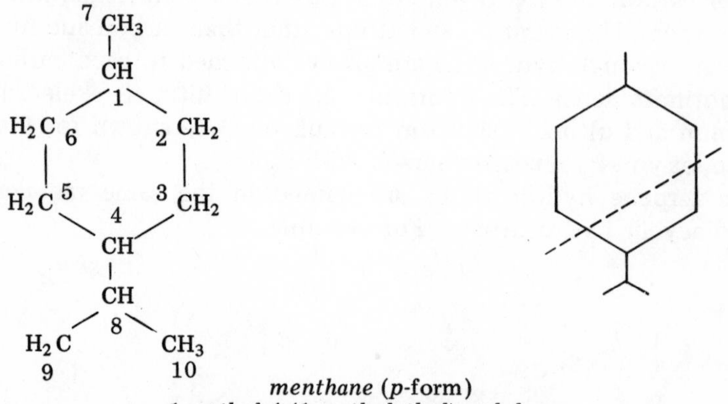

menthane (p-form)
1-methyl-4-(1-methylethyl)cyclohexane

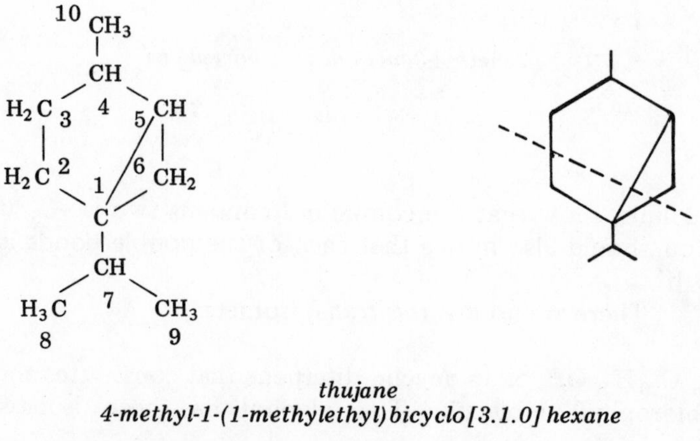

thujane
4-methyl-1-(1-methylethyl)bicyclo[3.1.0]hexane

carane
3,7,7-trimethylbicyclo[4.1.0] heptane

pinane
2,6,6-trimethylbicyclo[3.1.1] heptane

bornane
1,7,7-trimethylbicyclo[2.2.1] heptane

In terpene nomenclature the prefix *nor* signifies the complete replacement by hydrogen of all methyl groups on the ring structure. Such replacement leads to these three additional structural types:

norcarane
bicyclo[4.1.0] heptane

norpinane
bicyclo[3.1.1]heptane

norbornane
bicyclo[2.2.1]heptane

Most of the monocyclic terpenes are derivatives of *p*-menthane. Limonene is the main terpene component of lemon, orange, caraway, dill, and many other oils. It contains a chiral center; both stereoisomers are found in nature.

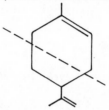

limonene
1-methyl-4-(1-methylethenyl)cyclohexene

Menthol is an oxygenated derivative of limonene. It has three chiral centers and all eight optical isomers are known. Three possible ways to represent one of the isomers are shown here:

I

II

III

(+)-menthol
[1S-(1α,2β,5α)]-5-methyl-2-(1-methylethyl)cyclohexanol

In **I** the ring is in the plane of the page. Bonds to groups in front of the page are shown by the heavy wedge and those to the back by a dashed line. In **II** the ring is perpendicular to the page, as in the Haworth projections for carbohydrates. The perspective formula (**III**) emphasizes the stable chair conformation.

α and β are used as relative descriptors for ring positions of cyclic compounds in the indexing system of *Chemical Abstracts.* This diagram illustrates their use:

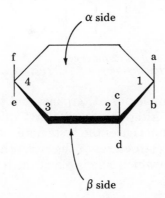

The preferred (or reference) groups at the three chiral centers as determined by the sequence-rule are a, c, and e. The functional group $(-OH)$ determines the number 1 position on the ring. The α side of the reference plane is the side on which the preferred substituent lies at the lowest-numbered chiral center. In this instance, a determines the α side; c lies on the same side and is designated α; e lies on the opposite side and is designated β. The complete description is $1\alpha,2\alpha,4\beta$.

Applying this principle to (+)-menthol, we find the following to be true: $CH_3 - \overset{|}{CH} - CH_3 > -H, -OH > -H, CH_3 - > -H$. The α side is determined by $-OH$, the preferred group on the number 1 carbon atom. The designations 2β and 5α follow.

Vitamin A_1, the fat-soluble vitamin that plays a part in resistance to infection and is required for the production of visual purple (rhodopsin), is a monocyclic diterpene.

all-trans vitamin A$_1$
CA index name: *all-trans retinol*
Systematic name: *(all-E)-3,7-dimethyl-9-(2,6,6-trimethyl-1-cyclohexen-1-yl)-2,4,6,8-nonatetraen-1-ol*

Vitamin A is a physiological activity, not a specific chemical compound. Literature on vitamin A will be found in *Chemical Abstracts* indexed chemically under carotenes, retinol, didehydroretinol, etc. The species isolated as vitamin A_1 has been identified as all-*trans* retinol.

An important bicyclic terpene is α-pinene, the principal component of oil of turpentine, which is obtained from pine trees. Its name according to the IUPAC rules is 2-pinene, but is indexed under its systematic name by *Chemical Abstracts*.

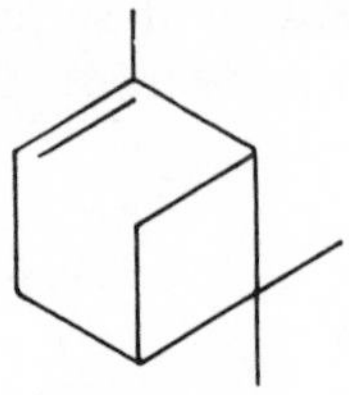

2-pinene
2,6,6-trimethylbicyclo[3.1.1]hept-2-ene

As you can see, the complexity of terpenes and their derivatives leads to complicated nomenclature. Although the IUPAC rules accept a number of trivial names, *Chemical Abstracts* uses the systematic names for indexing unless there are more than four ring structures or the stereochemistry is complex. The only substance in this section which is indexed at the trivial name is retinol.

5.5 STEROIDS

Steroids are compounds that contain a ring system largely similar to that present in cholesterol. The group includes the sterols, the bile acids, the cardiac aglycones, the sex hormones, the adrenal steroids, the toad poisons, and the steroid sapogenins.

From the viewpoint of systematic nomenclature, steroids are derivatives of cyclopenta[a]phenanthrene. The [a] as used here specifies the location of the fusion of the cyclopentane and phenanthrene ring systems. Because of the large number of steroids and the complexity of their stereochemistry, a detailed body of specialized terminology has been devised. The fundamentals, together with examples of specific substances, will be given in this section.

Steroids are numbered and the rings are lettered as shown in this structural formula:

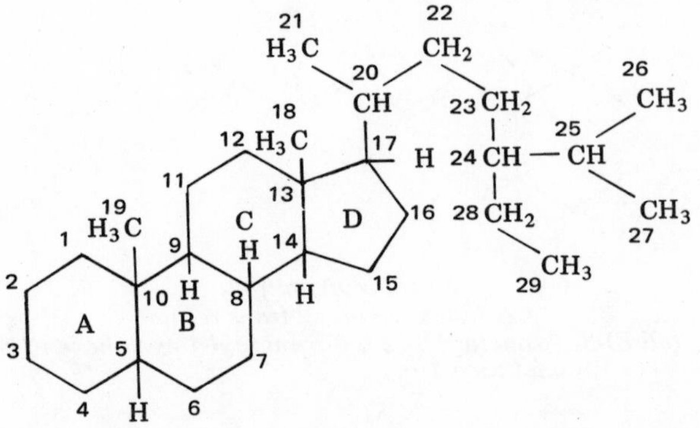

If one or more of the carbon atoms shown in the formula is not present and a steroid name is used, the number of the remaining carbon atoms is unchanged.

Unless stated to the contrary, the use of a steroid name implies that atoms or groups attached to the ring-junction positions 8, 9, 10, 13, and 14 are oriented as shown here. A carbon chain attached to 17 is assumed to project above the ring. When formulas are drawn in this way, bonds pointing upwards are drawn bold and those pointing downwards are drawn broken. Those pointing upward are designated β and those downward as α. Hence the configurations, unless stated to the contrary, are 8β, 9α, 10β, 13β, 14α, and 17β.

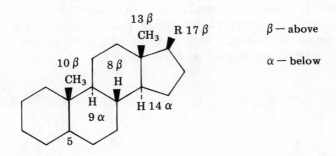

If known, the configuration of hydrogen or a substituent group at ring-junction position 5 is designated as α or β after the numeral 5, and both are placed before the stem name. While steroids may have many stereoisomers, the configuration at position 5 determines the geometry of the fusion of rings A and B and has important consequences in the biological and chemical reactivity of the substances. Perspective representation of two of the stereoisomers of the steroid shown above are presented here. They differ only in the configuration at position 5; all other chiral centers remain the same.

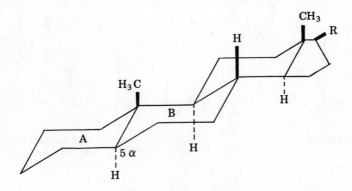

a 5α-steroid

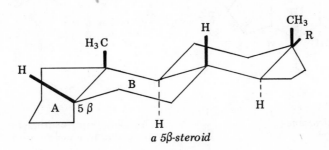

a 5β-steroid

When trivial names are assigned to steroid derivatives, the systematic suffixes are used in their correct sense. For example, estradiol has two − OH groups, testosterone has a $-\overset{\overset{\displaystyle O}{\displaystyle \|}}{C}-$ group, etc. Even though they may have widely-used trivial names, derivatives of steroids are now indexed by *Chemical Abstracts* at their systematic names. These are based on the parent steroid compound and there is a cross-reference to the trivial name in the Index Guide. The Index Guide also contains a diagram showing the structure of each parent.

The following pairs of formulas are five examples of steroids and derivatives. The first in each pair is the steroid parent with its trivial name. The second is the derivative, together with its systematic name as used by *Chemical Abstracts* and the trivial name used by most authors.

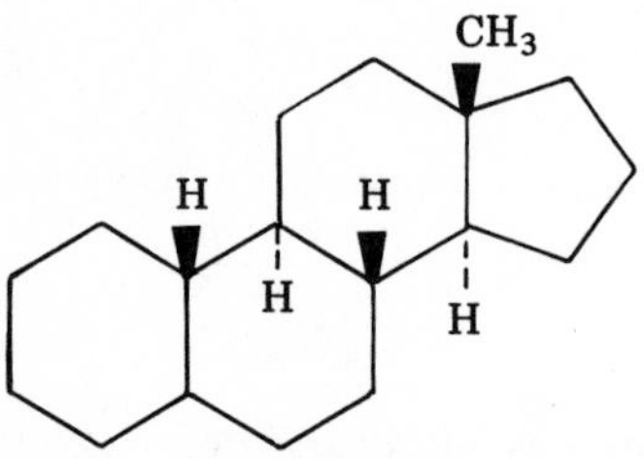

estrane

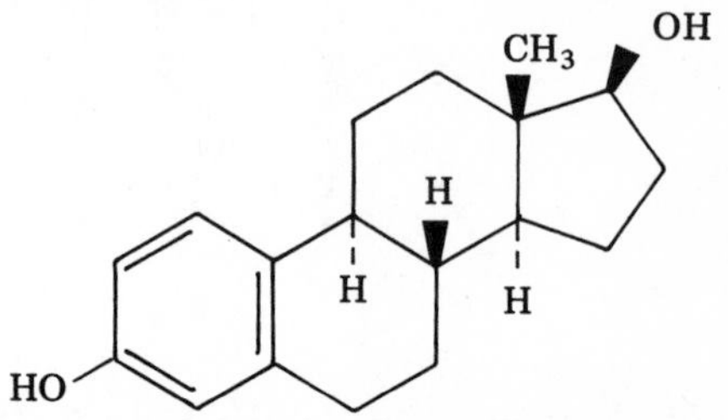

estra-1,3,5(10)-trien-3,17-diol
estradiol, a female sex hormone

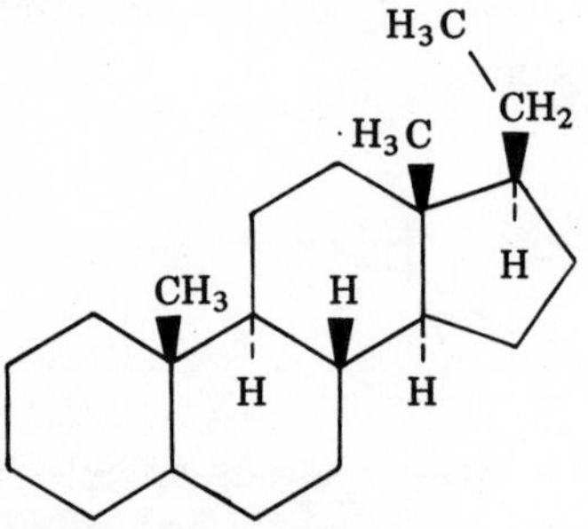

pregnane

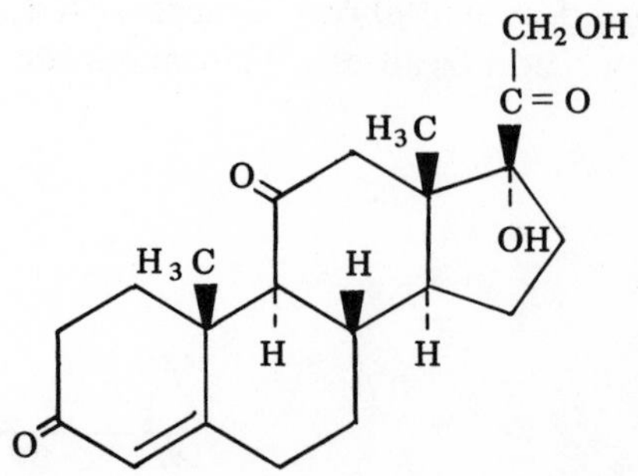

17,21-dihydroxy-pregn-4-en-3,11,20-trione
cortisone, an adrenal hormone

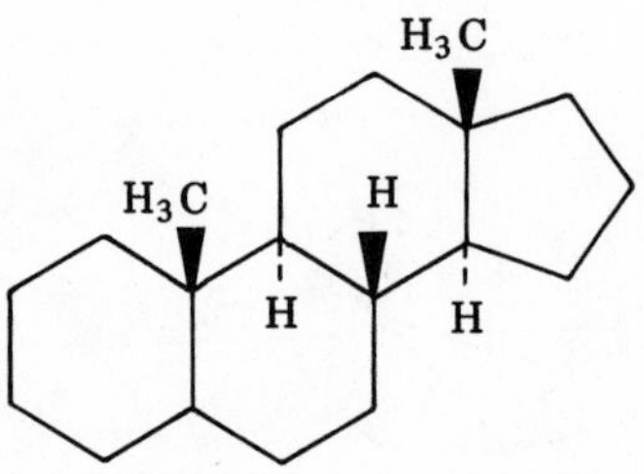

androstane

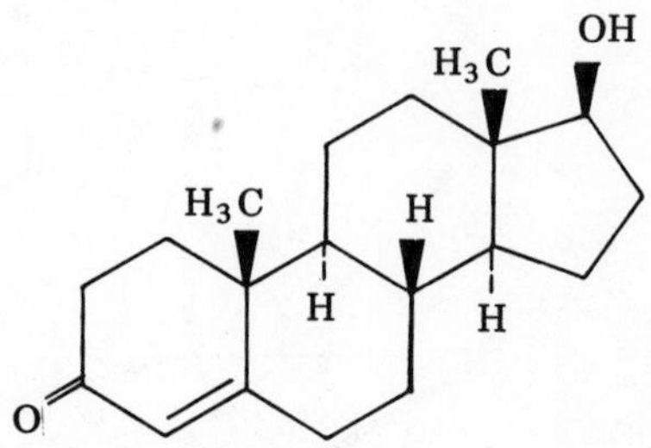

(17β)-17-hydroxy-androst-4-en-3-one
testosterone, a male sex hormone

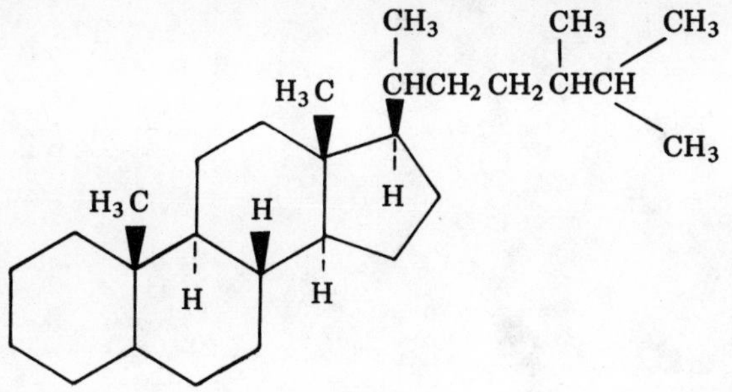

ergostane

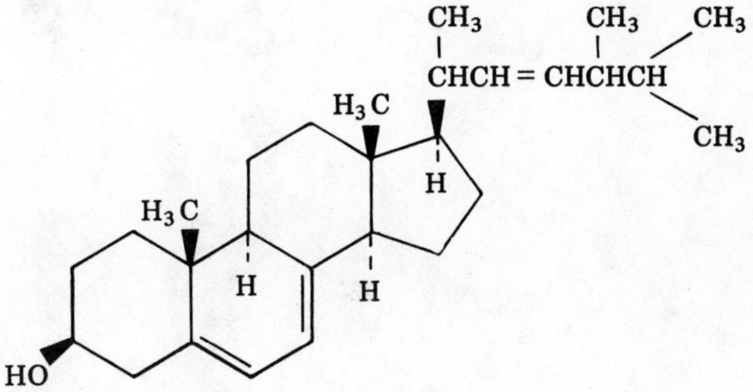

(3β)-ergosta-5,7,22-trien-3-ol
ergosterol, a plant sterol

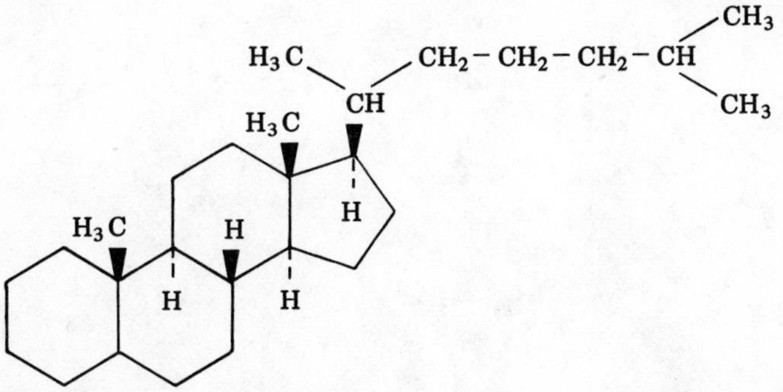

cholestane

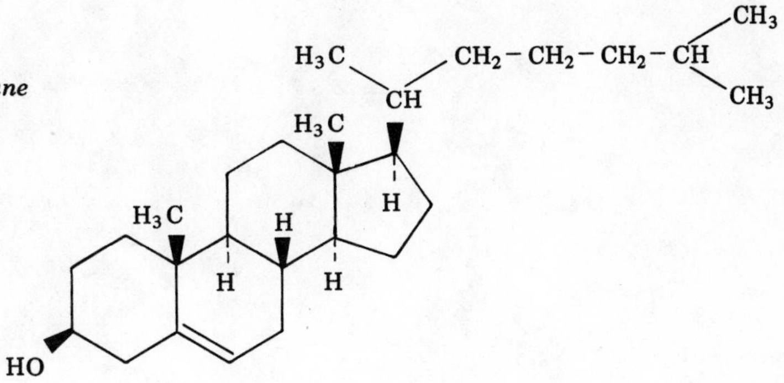

(3β)-cholest-5-en-3-ol
cholesterol, an animal sterol

SOME COMMON GROUPS IN ORDER OF SEQUENCE-RULE PREFERENCE*

APPENDIX 1* Some Common Groups in Order of Sequence-Rule Preference†

A. Alphabetical Order (Higher Number Denotes Greater Preference)

64	Acetoxy	38	Carboxyl	9	Isobutyl	55	Nitroso
36	Acetyl	74	Chloro	8	Isopentyl	6	n-Pentyl
48	Acetylamino	17	Cyclohexyl	20	Isopropenyl	61	Phenoxy
21	Acetylenyl	52	Diethylamino	14	Isopropyl	22	Phenyl
10	Allyl	51	Dimethylamino	69	Mercapto	47	Phenylamino
43	Amino	34	2,4-Dinitrophenyl	58	Methoxy	54	Phenylazo
44	Ammonio +H₃N −	28	3,5-Dinitrophenyl	39	Methoxycarbonyl	18	Propenyl
37	Benzoyl	59	Ethoxy	2	Methyl	4	n-Propyl
49	Benzoylamino	40	Ethoxycarbonyl	45	Methylamino	29	1-Propynyl
65	Benzoyloxy	3	Ethyl	71	Methylsulfinyl	12	2-Propynyl
50	Benzyloxycarbonylamino	46	Ethylamino	66	Methylsulfinyloxy	73	Sulfo
13	Benzyl	68	Fluoro	72	Methylsulfonyl	25	m-Tolyl
60	Benzyloxy	35	Formyl	67	Methylsulfonyloxy	30	o-Tolyl
41	Benzyloxycarbonyl	63	Formyloxy	70	Methylthio	23	p-Tolyl
75	Bromo	62	Glycosyloxy	11	Neopentyl	53	Trimethylammonio
42	tert-Butoxycarbonyl	7	n-Hexyl	56	Nitro	32	Trityl
5	n-Butyl	1	Hydrogen	27	m-Nitrophenyl	15	Vinyl
16	sec-Butyl	57	Hydroxy	33	o-Nitrophenyl	31	2,6-Xylyl
19	tert-Butyl	76	Iodo	24	p-Nitrophenyl	26	3,5-Xylyl

B. Increasing Order of Sequence-Rule Preference

1	Hydrogen	20	Isopropenyl	39	Methoxycarbonyl**	58	Methoxy
2	Methyl	21	Acetylenyl	40	Ethoxycarbonyl**	59	Ethoxy
3	Ethyl	22	Phenyl	41	Benzyloxycarbonyl**	60	Benzyloxy
4	n-Propyl	23	p-Tolyl	42	tert-Butoxycarbonyl**	61	Phenoxy
5	n-Butyl	24	p-Nitrophenyl	43	Amino	62	Glycosyloxy
6	n-Pentyl	25	m-Tolyl	44	Ammonio +H₃N −	63	Formyloxy
7	n-Hexyl	26	3,5-Xylyl	45	Methylamino	64	Acetoxy
8	Isopentyl	27	m-Nitrophenyl	46	Ethylamino	65	Benzoyloxy
9	Isobutyl	28	3,5-Dinitrophenyl	47	Phenylamino	66	Methylsulfinyloxy
10	Allyl	29	1-Propynyl	48	Acetylamino	67	Methylsulfonyloxy
11	Neopentyl	30	o-Tolyl	49	Benzoylamino	68	Fluoro
12	2-Propynyl	31	2,6-Xylyl	50	Benzyloxycarbonylamino	69	Mercapto HS −
13	Benzyl	32	Trityl	51	Dimethylamino	70	Methylthio CH₃S −
14	Isopropyl	33	o-Nitrophenyl	52	Diethylamino	71	Methylsulfinyl
15	Vinyl	34	2,4-Dinitrophenyl	53	Trimethylammonio	72	Methylsulfonyl
16	sec-Butyl	35	Formyl	54	Phenylazo	73	Sulfo HO₃S −
17	Cyclohexyl	36	Acetyl	55	Nitroso	74	Chloro
18	1-Propenyl	37	Benzoyl	56	Nitro	75	Bromo
19	tert-Butyl	38	Carboxyl	57	Hydroxy	76	Iodo

*From IUPAC 1968 Tentative Rules, Section E, "Fundamental Stereochemistry," *Journal of Organic Chemistry*, 35:2866, 1970. By permission of the American Chemical Society.

†ANY alteration to structure, or substitution, etc., may alter the order or preference.

**These groups are ROC (= O) − .

Appendix 2

SUBSTITUTIVE NAMES OF FUNCTIONAL GROUPS

The following groups are cited only as prefixes. The order in this list has no significance insofar as their use in naming compounds is concerned.

Formula	Prefix Name	Formula	Prefix Name
$-Br$	bromo	$-OOH$	hydroperoxy
$-Cl$	chloro	$-OR*$	R-oxy
$-F$	fluoro	$-OOR$	R-dioxy
$-I$	iodo	$-SR$	R-thio
$=N_2$	diazo	$-S(O)R$	R-sulfinyl
$-N_3$	azido	$-SO_2 R$	R-sulfonyl
$-NO$	nitroso	$-SSR$	R-dithio
$-NO_2$	nitro		

*R indicates an alkyl or aryl group. For example, $-OCH_3$ is methoxy.

The following groups may be cited as either prefixes or suffixes. They are listed in descending order of precedence for citation as suffixes. An italic C in a formula indicates that the carbon atom so marked is not included in the corresponding name of the prefix or suffix.

Formula	Suffix Name	Prefix Name
Cations	-onium	-onio
Anions	-ate, -ide	-ato, -ido
$-C$OOH	-oic acid	
$-$COOH	-carboxylic acid	carboxy
$-$SO$_2$OH	-sulfonic acid	sulfo
$-C$OX	-oyl (-yl) halide	
$-$COX	-carbonyl halide	haloformyl (*CA:* halocarbonyl)
$-C$ONH$_2$	-amide	
$-$CONH$_2$	-carboxamide	carbamoyl (*CA:* aminocarbonyl)
$-C$ONHCO$-$	-imide	
$-$CONHCO$-$	-dicarboximide	iminodicarbonyl
$-C\equiv$N	-nitrile	
$-$C$\equiv$N	-carbonitrile	cyano
$-C$HO	-al	oxo
$-$CHO	-carbaldehyde	formyl
$>$C$=$O	-one	oxo
$>$C$=$S	-thione	thioxo
$-$OH	-ol	hydroxy
$-$SH	-thiol	mercapto
$-$NH$_2$	-amine	amino
$=$NH	-imine	imino

Appendix 3

ORDER OF PRECEDENCE OF COMPOUND CLASSES

The following is an abbreviated list, in descending order, of the order of precedence for compound classes used by *Chemical Abstracts* during the ninth collective period (1972–1976).

Free radicals
Cationic compounds: onium (aminium) cations
Anionic compounds: carbanions
Acids
Acid halides
Amides
Nitriles
Aldehydes
Ketones
Alcohols and phenols (of equal rank), thiols
Hydroperoxides
Amines
Imines
Nitrogen compounds
Oxygen compounds
Sulfur compounds
Carbon compounds, including carbocyclic and acyclic hydrocarbons

WISWESSER LINE-FORMULA NOTATION

The function of both names and formulas of various types is to communicate information about chemical substances. As the number of substances has grown, so has the complexity of the nomenclature systems used to give each a unique name. This book has been devoted to an explanation of the systematic names of organic compounds.

Concurrent with this growth in chemical knowledge, there have been profound changes in other areas, from elaborate descriptions to greatly compressed identifications, as society attempts to cope with the information explosion. Witness the change of familiar telephone numbers from, for example, PEnnsylvania 6-5000 to 736-5000 and the alphabet soup of acronyms for the names of organizations such as ACS, AMA, EPA, DOT, FDA, HUD, NASA, TVA, etc.

From the early days chemists have recognized the economy and convenience of being able to write a line formula such as $CH_3 CH_2 OH$ to represent a chemical substance. The great Jons Jacob Berzelius, often called the "organizer of chemistry," insisted that "letters should be used for chemical symbols because they could be written more easily than other signs and did not 'disfigure' the printed book."* The advent of modern automatic data processing equipment has placed even more stringent limitations on notation because of the limited number of characters available and the lack of subscripts and superscripts.

In 1950 William J. Wiswesser proposed a line-formula notation that is compatible with data processing equipment. After Wiswesser had worked out the fundamentals, Elbert G. Smith and others began to develop large lists of substances encoded in Wiswesser Line-Formula Notation (WLN). Today control of WLN rules rests in the hands of the Chemical Notation Association, an international group with more than a hundred members.

WLN has been adopted for information-management use by a large number of academic, industrial, and governmental organizations. Most chemists now in training can be expected to use WLN to some extent during their careers.

Wiswesser Line-Formula Notation seeks to assign a unique and unambiguous notation to any chemical structure using a simple 42-symbol character set consisting of the 26 capital letters A to Z, the ten numerals Ø to 9, the five punctuation marks &, –, /, *, ·, and the blank space.

---

*From A. J. Ihde, *The Development of Modern Chemistry*. (New York: Harper & Row, 1964), p. 114. By permission of Harper & Row Publishers, Inc.

A complete explanation of the WLN system is beyond the scope of this book. A discussion of the symbols and a few rules will show the value of this notation in reducing the complexity and amount of space required for lists of compounds. Just as it is easier to read a foreign language than to write it, so it is easier to decode a WLN to the corresponding structural formula than to encode such a formula to its WLN.

Knowing only the following equivalences, one can decode or encode unbranched hydrocarbons, alcohols, aldehydes, carboxylic acids, and ketones.

Q	represents	OH	H represents H

$$V \quad \text{represents} \quad -\overset{\overset{\displaystyle O}{\|}}{C}- \qquad 1,2,\ldots \text{ represent number of carbon}$$
atoms in an unbranched alkyl chain

We immediately recognize these familiar substances:

1H	CH_4	methane
Q1	$CH_3 OH$	methanol
Q2	$CH_3 CH_2 OH$	ethanol
1V1	$CH_3 \overset{\overset{\displaystyle O}{\|}}{C} CH_3$	acetone (*CA*: 2-propanone)
VH1	$CH_3 CHO$	acetaldehyde
QV1	$CH_3 COOH$	acetic acid
QH	$H_2 O$	water

You should be able to encode $CH_3 CH_2 CH_2 OH$ as Q3, $CH_3 CH_2 \overset{\overset{\displaystyle O}{\|}}{C} CH_3$ as 2V1, etc. If you are alert, you'll wonder why these two aren't 3Q and 1V2. The order of citation is specified by the WLN system rules. Before considering the rules, let's look at all of the special symbols of WLN and their meanings.

With seven exceptions, international atomic symbols are used. The letters K, U, V, W, and Y are used for special purposes in WLN so KA, UR, VA, WO and YT are used to represent the five elements ordinarily symbolized by the single letters. Because of their frequent occurrence, Cl and Br are represented by G and E, respectively. When used in WLN, all two-letter atomic symbols are set off by hyphens, e.g., -CR-.

Here is a summary of the special meanings assigned to the single letter symbols:

C   Carbon atom attached to no more than two other atoms and either multiply bonded to at least one atom other than carbon, or doubly bonded to each of two other carbon atoms

E   Br

F   F

G   Cl

H   H (when expressed)

I   I

J   generic halogen

K $-\overset{|}{\underset{|}{N}}^{\pm}$ (hydrogen-free and positively charged)

M NH

N $-\overset{|}{N}-$ or $-N=$ or $N\equiv$ (hydrogen-free and not positively charged)

O O

Q OH

R

U $=$ (UU means $\equiv$)

V $-\overset{\overset{O}{\|}}{C}-$

W O_2

X $-\overset{|}{\underset{|}{C}}-$ (attached to four atoms other than hydrogen)

Y $-\overset{|}{C}H$ or $=\overset{|}{\underset{|}{C}}$ (attached to three atoms other than hydrogen)

Z NH_2

Arabic numerals denote alkyl carbon chain lengths.*

The first step in encoding any structural formula is to make a literal transcription of the structure into WLN symbols as shown here:

$$BrCH=CHC\overset{\overset{O}{\|}}{}CH_2NH_2 \quad \text{becomes} \quad E\ 1\ U\ 1\ V\ 1\ Z$$

All chains of structural units are cited symbol by symbol as connected, beginning at the end that occurs *latest* in this alphanumeric list:

$\cdot$ & - / Ø 1 2 3 . . . 9 1Ø 11 . . . (etc.)   A B C . . . (etc) . . . X Y Z *

The correct WLN for the structure cited above is Z1V1U1E. Just as we can often recognize the meaning of misspelled words, so this notation would be recognizable if written E1U1V1Z. Written in this way, however, it would appear out of place in an ordered list or directory of substances.

---

*Prior to 1975, K, X, and Y symbols were considered methyl-branched unless otherwise specified. Some examples of this former practice may still be seen. Here are two illustrations:

Correct WLN		Obsolete WLN
QY1&1	$HO-\overset{\overset{H}{\|}}{\underset{\underset{CH_3}{\|}}{C}}-CH_3$	QY
QX1&1&1	$HO-\overset{\overset{CH_3}{\|}}{\underset{\underset{CH_3}{\|}}{C}}-CH_3$	QX

The symbols H and R are accorded exceptional treatment in the citation order because they occur so often. R is subordinated to all other symbols, including &. For example,

CH₃C $-$ ⬡   1VR   acetophenone
(*CA*: 1-phenylethanone)

When a hydrogen symbol, H, is not implied as part of another symbol, it is cited after the symbol to which it is attached and does not affect the citing order for the chain of symbols. Here are some examples of the subordination of R and H:

Line Formula	WLN	Name
HBr	EH	hydrogen bromide
$CH_3 CH_3$	2H	ethane
$CH_3 SH$	SH1	methanethiol
$CH_3 CHO$	VH1	acetaldehyde
$C_6 H_5 OH$	QR	phenol
$C_6 H_5 CH_3$	1R	methylbenzene
$C_6 H_5 Cl$	GR	chlorobenzene

To avoid confusion in the notation of branched chain and substituted ring compounds, symbols are needed to indicate the end of a chain and for ring locants. The ampersand, &, is used to terminate a branched chain that does not end in one of the strictly terminal symbols E, F, G, H, I, Q, W, or Z.

Line Formula	Tran-scription	WLN	Name
$CHCl_3$	G G Y G	GYGG	chloroform (*CA*: trichloromethane)
CCl_4	G G X G G	GXGGG	carbon tetrachloride
$CH_3 CH_2 C(CH_3)OHCH_2 CH_3$	2 1 X Q 2	QX2&2&1	3-methyl-3-pentanol
$Cl_2 CHCH_2 OH$	G 1Q Y G	Q1YGG	2,2-dichloroethanol
$C_6 H_5 CHBrCOOH$	E VQ Y R	QVYER	2-bromo-2-phenylacetic acid (*CA*: α-bromo-benzeneacetic acid)

A.8

$C_6H_5C(CH_3)NH_2COOH$	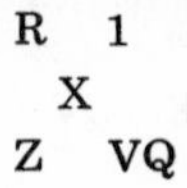 R 1 X Z VQ	QVXZ1&R	2-amino-2-methyl-2-phenylacetic acid (*CA*: α-amino-α-methylbenzeneacetic acid)

$$CH_3CH_2CH_2C(CHBrF)OHCH_2CH_3$$

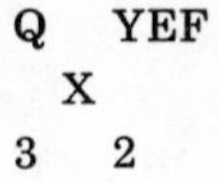

	Q YEF X 3 2	QX3&2&YFE	3-(bromofluoromethyl)-3-hexanol

$$\begin{array}{c} CH_3\ CH_3 \\ |\ \ \ \ | \\ CH_3-C-CH-NH_2 \\ | \\ CH_3 \end{array}$$

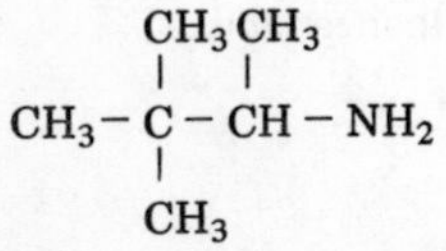	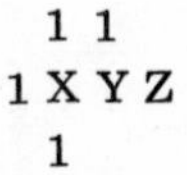 1 1 1 X Y Z 1	ZY1&X1&1&1	2,3,3-trimethylpropanamine (*CA*: 3,3-dimethyl-2-butanamine)

$$\begin{array}{c} O\ \ \ \ \ \ \ O \\ \|\ \ \ \ \ \ \ \| \\ CH_3-C-CH-C-CHCH_3 \\ |\ \ \ \ \ \ \ | \\ CH_3\ \ \ \ \ CH_3 \end{array}$$

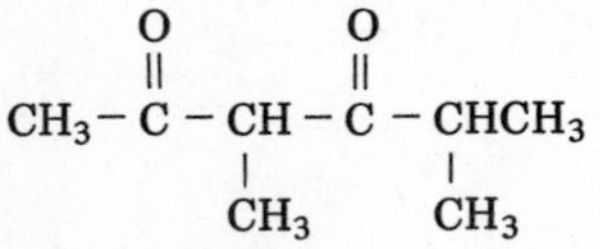	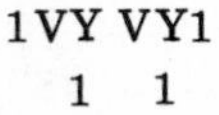 1VY VY1 1 1	1Y1&VY1&V1	3,5-dimethyl-2,4-hexanedione

$$\begin{array}{c} CH_3\ CH_3 \\ |\ \ \ \ | \\ CH_3-C-C-CH_3 \\ |\ \ \ \ | \\ CH_3\ CH_3 \end{array}$$

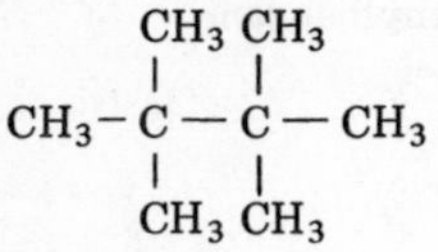	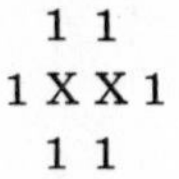 1 1 1 X X 1 1 1	1X1&1&X1&1&1	2,2,3,3-tetramethylbutane

$$\begin{array}{c} CH_3\ \ \ \ \ \ \ \ \ CH_3 \\ |\ \ \ \ \ \ \ \ \ \ \ \ \ \ \ | \\ CH_3-CH-CH_2-CH-CH_3 \end{array}$$

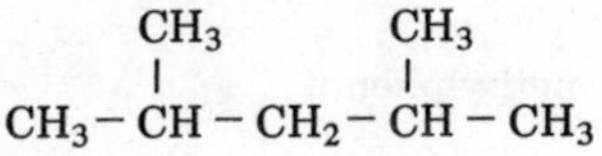	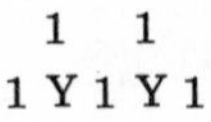 1 1 1 Y 1 Y 1	1Y1&1Y1&1	2,4-dimethylpentane

Multiple substituents on benzene rings are assigned "lower case" letters (designated as a capital letter preceded by a blank space) as locants according to the same scheme used for systematic names, with the "a" position assigned to the group whose symbol occurs latest in the WLN character set. For instance:

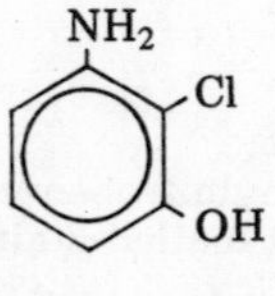 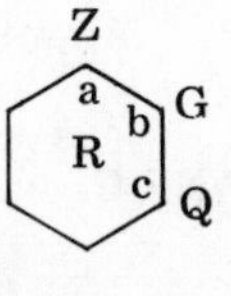 ZR CQ BG	3-amino-2-chlorophenol

In the notation shown above, the symbol Z is assigned the "a" locant because it occurs later in the character set than G or Q. The other locants are assigned in the usual manner to give the "lowest set" of locants. In the notation, however, Q precedes G even though Q has a "later" locant. Here are some more examples:

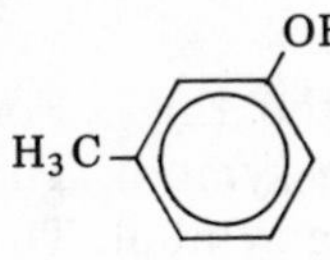 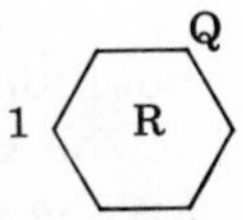 QR C1	3-methylphenol

Structure	WLN ring	WLN	Name
$CH_3\overset{\displaystyle O}{\overset{\displaystyle \|}{C}}$—⟨ring with Br⟩ 1V ⟨R⟩ E	ER CV1	1-(3-bromophenyl)ethanone
$CH_3\overset{\displaystyle H}{\overset{\displaystyle \|}{N}}$—⟨ring with CH_2CH_3⟩ 1M ⟨R⟩ 2	2R CM1	N-(3-ethylphenyl)methan-amine (CA: 3-ethyl-N-methylbenzenamine)
Br—⟨ring with Cl, Cl, CH_2Br⟩ E⟨R⟩1E with G, G	GR CG EE B1E	5-bromo-2-(bromomethyl)-1,3-dichlorobenzene
H_3C—⟨ring with CH_3⟩ 1⟨R⟩1	1R C1	1,3-dimethylbenzene
HO—⟨ring with CH_3⟩ Q⟨R⟩1	QR B1	2-methylphenol
⟨ring with CH_3, OCH_3⟩ 1⟨R⟩O1	1OR B1	2-methylmethoxybenzene (CA: 1-methoxy-2-methyl-benzene)
⟨ring with CH_3, NO_2⟩ 1⟨R⟩NW	WNR C1	3-methylnitrobenzene (CA: 1-methyl-3-nitrobenzene)

Aliphatic and benzenoid compounds can be described in WLN using the symbols and principles outlined to this point. Somewhat longer notations are needed to describe even the simplest of other types of cyclic compounds. The reason is that these other types have several independent variables: carbocyclic or heterocyclic, aromatic or having a lone saturated carbon, and saturated or locally unsaturated. All of these situations can be handled, but more symbols are required.

The first need is to set boundaries for the notation of the rings. For carbocyclic rings L is the opening symbol and J is the closing symbol. The opener for heterocyclic rings is T and the same J is the closing symbol. The opening symbol is followed by a numeral indicating the number of atoms in the ring. Maximum unsaturation is assumed.

Hetero atoms, V and sometimes X or Y, or both, are cited in ascending alphabetic order of their ring locants. In monocyclic notation the "a" locant is understood and the symbol for the hetero atom immediately follows the ring size numeral.

T5OJ   oxole (*CA:* furan)

Notice that the ring has the maximum unsaturation possible without cumulative double bonds.

Locants for other ring segments may be omitted if they are directly connected to the one preceding; otherwise, the locant is cited.

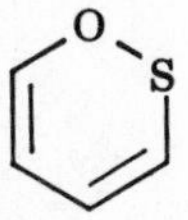

T6OSJ   1,2-oxathiin

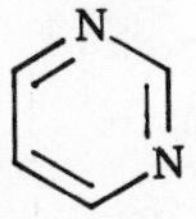

T6N CNJ   1,3-diazine (*CA:* pyrimidine)

Unlike substituted benzenes, the "a" locant is assigned to the symbol *earliest* in the character sequence when there are two otherwise equivalent "lowest sets." In T6OSJ the lowest set is "a,b" whether O or S is cited first. O is assigned the "a" because it precedes S in the character set. Rings that are dehydrogenated to the aromatic limit and have just one saturated or "extra-H" carbon atom are shown by the added symbol H, always with a locant.

L5 AHJ   1,3-cyclopentadiene

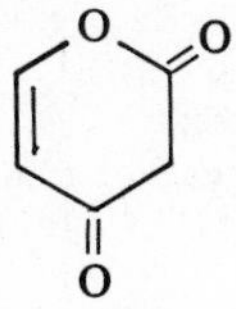

T6OV DV CHJ   2*H*-pyran-2,4(3*H*)-dione

Fully saturated rings are designated with the symbol T just before the closing symbol J. All such notations describe rings with more than one saturated carbon atom and no C − C unsaturations.

L3TJ   cyclopropane

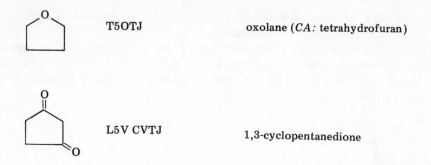

T5OTJ   oxolane (*CA:* tetrahydrofuran)

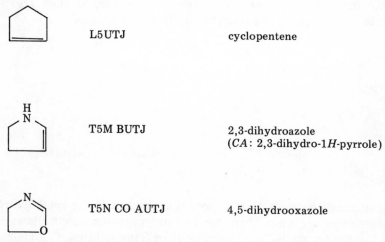

L5V CVTJ   1,3-cyclopentanedione

The customary symbol U is used to indicate localized unsaturation in ring compounds. Except for simple cycloalkenes and cycloalkynes, a locant is required.

L5UTJ   cyclopentene

T5M BUTJ   2,3-dihydroazole
                               (*CA:* 2,3-dihydro-1*H*-pyrrole)

T5N CO AUTJ   4,5-dihydrooxazole

Thio-, imino- or other methylene-type substituents provide another kind of ring unsaturation. A cyclic Y symbol replaces the V and the required U appears outside the ring boundaries.

L5YTJ AUM   cyclopentanimine

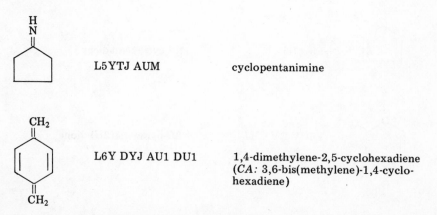

L6Y DYJ AU1 DU1   1,4-dimethylene-2,5-cyclohexadiene
                               (*CA:* 3,6-bis(methylene)-1,4-cyclo-
                               hexadiene)

Many proposed nomenclature and notation systems eventually founder on polycyclic substances where seemingly endless complications are found. The entire Ring Index has been successfully encoded by Elbert G. Smith and Tommy Ebe. This single feat attests to the power of WLN. Readers who wish to pursue this notation further are directed to Smith's book listed in the references at the end of this book.

A.12

The two great strengths of WLN are the compactness achieved and the ease with which structures can be retrieved from a file. In most large files the average length of notations is 20 to 25 characters and more than 98 per cent have less than 50 characters.

With the aid of computers and appropriate programming, searches for particular structures and substructures represented by a specific combination of symbols can be carried out readily. Such searches are proving especially valuable in the elucidation of relationships between structure and biological activity.

REFERENCES

1. Barker, R., *Organic Chemistry of Biological Compounds*, Prentice-Hall, Englewood Cliffs, New Jersey, 1971.
   Concise discussion of the chemistry and nomenclature of biologically active organic compounds.

2. Cahn, R. S., *An Introduction to Chemical Nomenclature*, 3rd edition. Plenum Press, New York, 1968.
   Brief treatment of both inorganic and organic nomenclature. Useful for comparison of American and British usage.

3. Chemical Abstracts Service, "Naming and Indexing of Chemical Substances for *Chemical Abstracts* during the Ninth Collective Period (1972–1976)." Marketing Department, Chemical Abstracts Service, Columbus, Ohio, 1973.
   Introduction, with key and discussion, to the naming of chemical compounds for indexing in *Chemical Abstracts*. Includes a comprehensive bibliography on chemical nomenclature.

4. Dean, J. A., ed., *Lange's Handbook of Chemistry*, 11th edition. McGraw-Hill, New York, 1973.
   A reference volume containing physical constants of organic compounds and cross-references to trivial and systematic names. Also a list of organic ring systems.

5. Fletcher, J. H., et al., *Nomenclature of Organic Compounds, Principles and Practice*. American Chemical Society, Washington, D.C., 1974.
   The principles of organic chemical nomenclature with discussion of the differences between the practice of the IUPAC and *Chemical Abstracts*. Many examples.

6. IUPAC Commission on the Nomenclature of Organic Chemistry, *Nomenclature of Organic Chemistry*. Sections A, B, and C. Butterworth's, London, 1971.
   The complete IUPAC rules for naming hydrocarbons, fundamental heterocyclic systems, and characteristic groups containing carbon, hydrogen, oxygen, nitrogen, halogen, sulfur, selenium, and/or tellurium.

7. IUPAC Commission on the Nomenclature of Organic Chemistry, "IUPAC 1968 Tentative Rules, Section E, Fundamental Stereochemistry," *Journal of Organic Chemistry*, 35:2849, 1970.
   Definitive discussion of the types of isomerism, chirality and the sequence rule, conformation, and stereoformulas. Reprints available from Chemical Abstracts Service, Columbus, Ohio.

8. *The Merck Index of Chemicals and Drugs*, 8th edition. Merck and Company, Rahway, New Jersey, 1968.
   Extensive list of organic and inorganic compounds used in medicine and pharmacy. Particularly good listing of trademarked names and generic names.

9. Smith, E. G., *The Wiswesser Line-Formula Chemical Notation*. McGraw-Hill, New York, 1968.
   Complete discussion with many examples.

10. Ternay, A. L., *Contemporary Organic Chemistry*, W. B. Saunders Co., Philadelphia, 1976.
    Textbook treatment of organic chemistry with emphasis on compounds of biochemical and medical interest.

INDEX

Note: Page references in *italics* denote illustrations or structural formulas. The symbol (t) following a reference indicates information contained in a table.

A note concerning inverted names of compounds:
Ordering in the index is based on the index heading parent, followed by a comma (the *comma of inversion*), substituents in alphabetical order, and then stereochemical information if needed. For example,

Butane, 2,3-dichloro-, (*Z*)-
becomes
(*Z*)-2,3-Dichlorobutane when uninverted.

Hyphens at the end of the set of substituents in the inverted part of an index name signify that no space is intended when the name is uninverted. Conversely, absence of a hyphen after substituents indicates that a space appears at that point in the uninverted name. For example,

(1) Acetic acid, chloro-
becomes
Chloroacetic acid when uninverted.
(2) Disulfide, bis(2-chloroethyl)
becomes
Bis(2-chloroethyl) disulfide when uninverted.

When names are divided between lines, a special punctuation mark, ◦, is used to indicate that there is no space or hyphen in the name when undivided. If a hyphen appears, it is used in the undivided name. For example,

(1) Benzoic acid, 3-(3-chloro-4-iodo◦
benzoyl)-
3-(3-Chloro-4-iodobenzoyl)benzoic acid

(2) Heptanedioic acid, 4-[2-(acetyloxy)-
2-oxoethyl]-
4-[2-(Acetyloxy)-2-oxoethyl]heptanedioic acid

Teflon. *See* Ethene, tetrafluoro-, polymers.
Terephthalic acid. *See* 1,4-Benzenedicarbox◦
ylic acid.
Terpenes, 301–306
Tertiary carbon atom, 35
  in bridged systems, 111
Testosterone. *See* Androst-4-en-3-one, 17-
  hydroxy-, (17β)-.
Tetradecanoic acid, 300
Tetrahedron, model of, *5*
Thia, prefix, 235, 238
Thio, prefix, 235
Thiol, suffix, 235
Thiole. *See* Thiophene.
Thiols, 235
Thione, suffix, 235
Thiophene, 235, 249
Thiophenol. *See* Benzenethiol.
Threo enantiomers, 93, 132
Threonine, 226
Thujane. *See* Bicyclo[3.1.0]hexane,
  4-methyl-1-(1-methylethyl)-.
Thymine. *See* 2,4(1*H*,3*H*)-Pyrimidinedione,
  5-methyl-.
TNT. *See* Benzene, 2-methyl-1,3,5-trinitro-.
Toluene. *See* Benzene, methyl-.
o-Toluidine. *See* Benzenamine, 2-methyl-.
Tolyl group. *See* Methylphenyl group.
2,4,6-Trimethylphenyl group, *122*
Triphenylmethyl group, *123*
Trisaccharide, definition, 263
Trityl group. *See* Triphenylmethyl group.
Tryptophan, *226*
Twist conformation. *See* Cyclohexane,
  conformers of.
2,4-D. *See* Acetic acid,
  (2,4-dichlorophenoxy)-.

Unsaturation, 55
  cumulative, definition, 248
  in heterocyclic compounds, 252
  in Wiswesser Line-Formula Notation, A.11
  non-cumulative, definition, 248
Uracil. *See* 2,4(1*H*,3*H*)-Pyrimidinedione.
Uronic acids, 291

Valeryl group. *See* 1-Oxopentyl group.
Valine, 226
Vanillin. *See* Benzaldehyde, 4-hydroxy-3-
  methoxy-.
Veratrole. *See* Benzene, 1,2-dimethoxy-.
Vinyl chloride. *See* Ethene, chloro-.
Vinyl group. *See* Ethenyl group.
Vitamin A, 305

Wiswesser Line-Formula Notation, A.5–A.13
WLN. *See* Wiswesser-Line-Formula Notation.

Xylene. *See* Benzene, dimethyl-.
2,3-Xylidine. *See* Benzenamine,
  2,3-dimethyl-.
Xylyl Group. *See* Dimethylphenyl group.

Z, as stereodescriptor, definition, 74